FREEDOM AND RATIONALITY

BOSTON STUDIES IN THE PHILOSOPHY OF SCIENCE

VOLUME 117

John Watkins

FREEDOM AND RATIONALITY

Essays in Honor of John Watkins

From his Colleagues and Friends

Edited by

FRED D'AGOSTINO

University of New England, Australia

and

I.C. JARVIE

York University, Toronto, Canada

KLUWER ACADEMIC PUBLISHERS

DORDRECHT / BOSTON / LONDON

Library of Congress Cataloging in Publication Data

Freedom and rationality : essays in honor of John Watkins / edited by
Fred D'Agostino and I.C. Jarvie.
p. cm. -- (Boston studies in the philosophy of science ; v.
117)
Bibliography: p.
Includes index.
ISBN 0-7923-0264-8
1. Philosophy. 2. Science--Philosophy. 3. Political science.
4. Watkins, John W. N. I. Watkins, John W. N. II. D'Agostino,
Fred. III. Jarvie, I. C. (Ian Charles), 1937- . IV. Series.
Q174.B67 vol. 117
[B29]
001'.01 s--dc20
[100] 89-8024
CIP

ISBN 0-7923-0264-8

Published by Kluwer Academic Publishers,
P.O. Box 17, 3300 AA Dordrecht, The Netherlands.

Kluwer Academic Publishers incorporates
the publishing programmes of
Martinus Nijhoff, Dr W. Junk, D. Reidel, and MTP Press.

Sold and distributed in the U.S.A. and Canada
by Kluwer Academic Publishers,
101 Philip Drive, Norwell, MA 02061, U.S.A.

In all other countries, sold and distributed
by Kluwer Academic Publishers Group,
P.O. Box 322, 3300 AH Dordrecht, The Netherlands.

printed on acid free paper

Printed in the Netherlands

TABLE OF CONTENTS

PART II. STUDIES IN THE HISTORY AND METHODOLOGY OF THE NATURAL SCIENCES

INTRODUCTION

The editors of this volume – Jarvie and D'Agostino – encountered John Watkins at such different times in his career that they have never encountered each other. The common purpose of wishing to present to a teacher, colleague and friend this *Festschrift* for his retirement has brought them together. Then known as J.W.N., Watkins was the third member of the London School of Economics Department of Logic and Scientific Method when Jarvie first encountered him in 1957. He was running the Department of Philosophy, Logic and Scientific Method in 1973 when D'Agostino first encountered him. In Jarvie's day the department was a service department teaching logic and scientific method papers to students taking honours degrees in other fields. By D'Agostino's day it was a full-fledged department of philosophy, teaching that subject all the way from undergraduate honours to the Ph.D. level. Yet both at the time when it was a small band, and latterly, when it was much grown, we each found that the contribution of Watkins to our education and to our careers was of the utmost value. Thus it was a great pleasure to make contact with the many others from our generations and those between and subsequent to us to invite them to participate in this celebration of Watkins's wide-ranging contributions to philosophy, and in the course of doing so to non-inductively corroborate in what affection and respect he is held by such a broad cross-section of professionals. What they have written signals their attitudes.

John Watkins was twenty-five years old and a decorated veteran of the Second World War (DSC) when in 1949 he won, for his essay 'Laski on Conscience and Counter-Revolution' (1949), the Century Essay Prize, and began his career as a politically- (and later scientifically-) minded philosopher. At the LSE he had studied with Sir Karl Popper, before taking an M.A. at Yale University.

Watkins was, successively, Assistant Lecturer and Lecturer in Government at the LSE, then Reader in the History of Philosophy, and, from 1966, Professor of Philosophy in the University of London. Upon gaining his personal chair he served as coordinator of the department formerly headed by Sir Karl.

In 1952 Watkins made his first waves in the world of professional

Fred D'Agostino & I. C. Jarvie, Freedom and Rationality: Essays in Honor of John Watkins, ix–xi.

philosophy when he inaugurated a debate about the principle of methodological individualism, a debate which continues to this day, and which has inspired a literature as great as any in contemporary philosophy. Few collections of material in the general area of philosophy of social science would be considered complete unless they contained at least one of Watkins's many contributions to the discussion of this issue.

In 1957 Watkins published the first of a series of three papers (1957b, 1958d and 1960a) in which he tried to codify and rehabilitate metaphysics within the Popperian philosophy, placing it somewhere between the analytic and the empirical. He thus signalled the emergence of an important implication of Popper's thought that had not to that point been stressed by Sir Karl himself, and which marked off his followers from the antimetaphysical ideas of the regnant logical positivists.

In 1965 years of work in political philosophy and in the history of philosophy in the seventeenth century were brought to fruition in Watkins's widely cited and admired *Hobbes's System of Ideas* (1965a, second edition 1973d). This book is an important contribution not just to our understanding of Hobbes's political thinking, but, perhaps more importantly, to our understanding of the way in which a system of ideas is constituted and applied. Watkins built on earlier work in developing an account of Hobbes's ideas in which was revealed and clarified the unity of Hobbes's metaphysical, epistemological and political ideas.

During this whole period Watkins made notable contributions as well to decision theory and to discussions of human freedom. He developed in some detail Popper's doctrine of negative utilitarianism (still perhaps insufficiently explored as an alternative to more familiar versions of this now very influential moral doctrine), and was a frequent broadcaster, some of whose talks appeared in *The Listener*.

After being awarded a personal chair in philosophy at the London School of Economics in 1966, Watkins, under the growing influence of his colleague and friend Imre Lakatos, increasingly turned his attention to the philosophy of natural science, although it was also during these years that he participated in two important debates with which his name is still prominently associated – his critique of the proposal to replace critical rationalism with comprehensively critical rationalism (largely prosecuted against W.W. Bartley, III) and his defence of Popper's view that freedom excludes full predictability (largely in relation to the views of D.M. MacKay).

After the sudden death of Imre Lakatos in 1973 Watkins assumed for a brief period the editorship of *The British Journal for the Philosophy of*

Science, and, as head of the LSE Department, he participated in the attempts to articulate the full scope of Lakatos's Methodology of Scientific Research Programmes (1978f/g). He was joined in the Department by the noted philosopher and social scientist Ernest Gellner. Although he has been preoccupied in recent years with a project to rescue Popper's philosophy of science from the criticisms of Kuhn and Feyerabend and against the accusation that, despite protestations, it is inductivist (*Science and Scepticism*, 1984), he has also continued to explore many of the problems with which his name has long been associated: rationality, metaphysics, decision theory, freedom, and the philosophy of Thomas Hobbes.

Readers of Watkins's (1978f/g) (and especially of its precursor, a collective discussion paper entitled 'The LSE Position'), of his (1984), and of some of the critical comments on that (including those in this volume) may be as tickled as we are by the question of whether Watkins is still a follower of Popper's philosophy of science (given the great differences between his answer to Grünbaum's question, 'what is so good about a corroborated theory?' and the 'orthodox' answer Agassi published in the *Australasian Journal of Philosophy* in 1961), or whether he has shifted to Lakatos's (assuming they are different; and Lakatos did deny that Popper had solved the problem of induction – see Worrall's paper). While the question is not explicitly addressed in any paper in this volume, some of them provide ample material for discussion.

The question also has its institutional side. In the year of Watkins's retirement (1989) the School has created an established chair, to be known as the Sir Karl Popper Chair of Philosophy. This event may mark the entrenchment of a tradition of inquiry that derives from Sir Karl, or it may mark the point at which that tradition takes on a purely pro forma relation to his ideas. Whichever way it turns out, John Watkins's departure alas means that no further generations of students will benefit from his agreeable manner and sharp mind in supervision, enjoy his digressive and witty lecturing style, or gain an object lesson in how an attractive and good humoured chairman can control a meeting and summarise issues in a way that elegantly prepares the ground for questions and discussion. His friends and admirers whose work went into this volume join together to wish him a fruitful retirement with the anticipation of many scholarly contributions to come.

FD'A
ICJ
December 1988

PART I

STUDIES IN THE HISTORY AND METHODOLOGY OF SOCIAL AND POLITICAL THEORY

JOSEPH AGASSI

THE LOGIC OF CONSENSUS AND OF EXTREMES

PREFACE AND ABSTRACT

In this essay I wish to examine an individual citizen's set of possible attitudes to the body politic in which that individual is politically active. Since I discuss a diversity of individuals and of political systems, I shall have to present not explanations but explanatory schemata. I will argue that these are easy to handle both as (universalizable) empirical information and as testable hypotheses. My framework will consist of two claims. Generally but not always an individual citizen's sense of belonging to a political system is a central ingredient in that relationship; otherwise there is little point in discussing the variety of attitudes open to the citizen, much less any schema of varieties of attitudes, especially extreme and moderate ones. And prima facie leaderships see themselves as compromisers, yet at times they may find that an extremist part of the system has seceded, and has thus dropped out of the set of attitudes between which to compromise. These two claims invite models of the forces which regulate extremism, forces that present the extremist with an incentive to advocate as extreme a position as possible without thereby dropping out of the system. Staying within the system in national politics is, obviously, staying within the national consensus: one can drop out of the consensus and yet stay within the nation, and even take over and determine a new consensus. So much for my framework. My intent is to achieve the following conclusion. In a pluralist society both extremism and the branding of a group as extremist, must be considered most undesirable, as measures of a last resort.

0. PREAMBLE: THE CONTRIBUTION OF WATKINS TO POLITICAL THEORY

The present essay is dedicated to John Watkins in profound gratitude for the help, support and encouragement which he gave me so generously, in my early days, when this was most needed. I owe my success, such as it is, to many people, some of whom, Watkins included, gave me unique and indispensable help. Though the help he gave me was on many levels, only

Fred D'Agostino & I. C. Jarvie, Freedom and Rationality: Essays in Honor of John Watkins, 3–21.

some of his contributions to political theory will be discussed here briefly, those without which the present essay could not be attempted.

The significance of Watkins' works in general, or even of his work in social and political philosophy in general, is a topic too broad to discuss here. Even his position concerning the classical issue of methodological individualism is too intricate to handle here, and I will have to make do with one observation concerning it which pertains to the present essay. The classic debate in social and political philosophy concerns the individual and society and was held between two parties, both reductionist: the traditional rationalists were individualists who declared society and social institutions (in the broadest sense of the word) reducible to individual situations, attitudes and actions: sociological facts were explicable by psychological assumptions. (This is psychologism.) The irrationalists or romantic traditionalists were collectivists who declared individual situations, attitudes and actions to be reducible to social factors, including the different destinies of different societies or sub-societies: psychological facts were explicable by sociological assumptions. (This is sociologism.) I assume that Karl Popper's theory of the autonomy of sociology, as formulated in his *The Open Society and Its Enemies*, coupled with his expressed commitment to methodological individualism amounts to the rejection of both forms of reductionism (which he labels psychologism and sociologism). I do not know if this assumption is true. I also assume that Watkins, a leading disciple of Popper, shares this position. I do not know if this assumption is true either. Watkins's classic study of the topic is a special contribution in its noticing the important fact that, somehow or other, even if individual attitudes do not comprise societies, they are essential ingredients of their belonging to these societies and thus to the very existence of societies. (This important fact is taken for granted by all parties to the dispute; Watkins made it explicit and discusses its implications for the dispute.) This important fact is the topic of the next section.

The most significant specific contribution of Watkins to political theory is his contribution to *the theory of the national consensus*, to which the present essay belongs. Since the phenomenon of the national consensus is so important, socially and more so politically, the neglect of this phenomenon, the paucity of studies of this phenomenon, especially the almost total neglect of it in political philosophy, is truly amazing. This neglect is explicable by the observation that the phenomenon in question seems trivial to all reductionists. The very claim that society is reducible to individuals renders the consensus the sum total of individual characteristics: the general will, for example, which is the political motive force, should be the same as the common will, which is the sum total of citizens' wills, and when Rousseau said otherwise he sounded incomprehensible: how can the general will differ from the common will? The answer is that the general will is the collective will, the will of the collective as an individual. This is, of course, collectivist, so that the very claim that individuals are reducible to society renders the consensus into both the collective will and the collective destiny: being collective, then, is the same as having our consensus in the form of the general will. Only by refusing to accept either the individualist or the collectivist solution does the national consensus become problematic: when and how do the common will and the general will coincide? (In simpler words, how is democracy possible?)

Watkins has shown in his trail blazing study of epistemology and politics that democracy is rooted in the idea of cooperative criticism, one idea he acquired from Popper: science, according to Popper, is friendly–hostile cooperation; democracy, also according to Popper, is the (quasi-) scientific approach to social problems; yet Popper did not present democratic politics as a critical debate between contending parties: Watkins did that. And then the problem of the national consensus looms large: under what conditions does critical debate between contending political parties lead to consensus and thus to action, and when not? The reason Popper did not achieve this striking result of Watkins is rooted in an error in his classic *Logik der Forschung*: he did not allow for political pluralism in the physical sciences and even demanded scientific consensus despite his advocacy of the critical attitude without science. Thus the presence of political parties looked like a deviation from the scientific attitude. In my view we should permit parties even within science, and, in addition, we should distinguish clearly between science and scientific technology, thus permitting both conflicting opinions and conflicting interest within the national consensus. The problem of the national consensus then becomes more urgent and subsequently it avails itself of Watkins' solution – with interesting and useful results that will be shown below to be rather challenging.

1. AWARENESS AS ESSENTIAL TO SOCIAL GROUPINGS

One difficulty on the way to discussing any general point in the social sciences lies in the fact that social units are very diverse and are almost always defined *ad hoc*. Some social entities are utterly invisible. At times certain markets in special commodities, for example, are not noticed even by those participating in them, and at times, for another example, pressure groups have significant impact because they act as groups and yet each of their members or each of their subgroups may be utterly ignorant of the very existence of a group action. Nevertheless, these cases look *prima facie* extravagant and exceptional and quite marginal. It is therefore quite generally assumed that group membership includes, and even as a significant item, awareness of that membership. In the exeptional cases it is customary to speak of the group or its conduct as a mere corollary to the presence and conduct of other groups whose members are presumably fully cognizant of their conduct as members or affiliates of the groups within which they participate. This is the famous theory of unintended consequences put in a modern way.

This situation is so well-known, and well-known as significant, that it has led some social thinkers to the fashionable theory known as the social construction of reality. The reference in this phrase is not to physical reality, but to social reality, of course; yet the advocates of the social construction of reality do not say so – in order to emphasize the fact, which many philosophers overlook, that much of our perception of nature, including our perception of physical or biological nature, is socially conditioned: the way one perceives depends in part on one's social background. Yet the words "in part" are crucial: no amount of social background may make us ignore the given part of reality which is independent of society, and without which there is no society, nor even life itself. Now we need not complain about the exaggeration included in the social construction of reality; let us simply notice that it implies that social and political institutions exist merely because we take cognizance of them – perhaps even as identical with this cognizance. This is an exaggeration too: not only can there be unperceived social realities; there are perceptions of social affairs which are mere mirages: society often includes significant social 'constructs' which are not real – often known by the blanket title of fools' paradise. Politicians with no sense of responsibility and false prophets of all sorts encourage such misperceptions; and responsible politicians and spiritual leaders then rightly protest. (The popularity of the social construct of reality is thus explicable: it is a ticket to any fool's paradise in its very denial of the existence of fools' paradise: it excludes all social misperception, since it presents social perception as social reality.)

All this said, it becomes clear enough that the social construct of reality theory includes an important kernel of truth: the usual case is that social reality is perceived, and that its perception is usually a very important characteristic. Thus, it is hard to conceive a nation without citizens cognizant of their nationality or national affiliation, their sense of national identity, and even their perception of the importance – for them as individuals and for the nation – of their own sense of national identity. Thus, to conclude this point, it seems that usually the nation would not exist if it were not the case that most of its citizens consider themselves significantly citizens of this particular nation. Nevertheless, there can be exceptions to this, such as the Israeli nation. (The absence of national awareness in Israel is due to the confusion between the Israeli nation and the Jewish people, a confusion supported by Israeli national ideology and by Israeli law.) Moreover, awareness by itself does not suffice to produce a nation, or any other social entity one is aware of, much less the special characteristics one is aware of. Otherwise one could be sure of the existence of whatever the public is aware of in a clear and strong manner, and even reliably so. This is refuted by the frustration of national aspirations and of any other strong group aspirations. It is more obviously refuted by any observed case of conflicting awareness – even clear and strong ones. And these abound in history and in contemporary social and political life.

The point is not merely theoretical: advocates of the social construction of reality are mistaken not only on matters of principle but also on a very significant projection of this error in one very common aspect of everyday social and political life. They must deny the fact that quite generally social entities have members overseeing the cohesion of the entities in question by supervising the awareness of their members – both of the existence of these entities and of their own membership in them. These people are often called the leadership of the entities in question, especially national leaderships, and for a simple reason. One necessary condition, we remember, for the very survival of a normal social or political entity is usually awareness of its existence and of one's membership in it – awareness, for short. Awareness is central even for seemingly natural entities, such as families, lineages, and ethnic or linguistic groups, members of which at times have no awareness of the entity or of their membership: usually unless there is some awareness of family ties or of linguistic kinship, then the family does not function as a social entity, and the linguistic group is then an important sociological abstraction, but not a real grouping. Thus, where and when awareness of the extended family disappeared, so did the extended family. Now, when awareness is the only controllable factor, controlling awareness may be the same as controlling the very survival of the entity in question. The controller of awareness, then, is seen as the leader, since survival is usually the prime interest of every social and political entity which perseveres and which is active. The active part in charge of awareness, then, is the leadership. (This explains the known fact, that all extended family systems known are bound by systems of leaderships, usually of elders, so-called.)

Just as social entities are diverse and definable *ad hoc*, so are leaderships and so are their means of maintaining awareness. The naive newcomer to social studies is regularly surprised by the need for reaffirming awareness found in highly coherent, cohesive and compact social entities, such as preliterate tribes and extended families and small religious sects and the national psychoanalytic society and the national communist party and the communist international. The personages who happen to occupy positions of leadership and the rites by which they happen to enhance awareness are extremely varied. The naive newcomer's (often unarticulated) question is, why do members of coherent and cohesive and compact groups need their leadership to perform boring elaborate rituals affirming and reaffirming their already strong awareness? The answer to this question emerges from the logic of consensus and of extremes. For the most primitive and most widespread technique of maintaining unity is by maintaining awareness, especially the clear awareness of the alleged unacceptability of any deviance whatsoever.

2. THE LOGIC OF COMPROMISE

Leaderships vary immensely, and so do their poles, the extremes which define memberships in their social bodies, including the two extremes of uniformity and diversity. Yet in this latter case there is one pole which they all share: they are in charge of unity and its maintenance, chiefly by the means of maintaining the awareness of some (alleged or real) uniformity of some (alleged or real) importance. They therefore often undertake new tasks as the situation invites. For example, wherever national unity is threatened and the nation is on the brink of cessation, the national political leadership has to declare an emergency – if not an actual civil war – and declare to the public what steps it takes in an effort to maintain cohesion and unity – especially, what aspects of uniformity to emphasize. Alternatively, a leadership may recognize or declare cessation of a schism – political, religious or any other – and, again, they may do it by declaring what aspect of uniformity maintains unity or is expected to do so. The solution may be declared by the leadership of the old group or by the leadership of a seceding faction. All this indicates, again, the difficulty of generalizing due to the enormous variety of circumstances and the *ad hoc* nature of group identities, of cohesion, of the means of maintaining them, and of the measure of success of the employment of these means. A seceding faction, for example, may return, disband, unite with the mother group on a compromise or in full control, and the mother group may disband or regain control, etc. The leadership may act and react in similar fashions.

Nevertheless, the varieties of attitudes may be seen as products of two simple extreme strategies, one very primitive and extreme, the other fairly primitive and moderate – the strict maintenance of purity and the willingness to compromise. Of course, it will soon transpire that neither strategy can be applied in its pure form and that almost always one uses a mixture of both. This makes both strategies – extremity and moderation – a myth in the sense of Levi-Strauss' definition of myths as polar strategies which complement each other. The question of what purity the leadership is supposed to maintain does not avail itself of a general answer: for the Jewish people, for example, purity altered in time: racial purity, not in the modern racist sense, of course, was first important, and later played down to the point of disappearance, to reappear with a vengeance in the State of Israel. Hence, purity is an abstract entity. This should indicate that Levi-Strauss' claim that preliterate thinking is systematically and exclusively concrete is in error. One of the most central aspects of the most prevalent modes of primitive thinking is purity, as Mary Douglas has argued, and purity is definitely abstract, even though it is always presented as attached to concrete cases: most purity is cleanliness, and cleanliness is abstract enough; also, any pure concrete entity is concrete, but purity is abstract even in preliterate society – since it is not attached to one or another pure concrescent. (As Bertrand Russell has noted in his 'Zahatopolk', purity is made to seem concrete by creating strong feelings of approval and of revulsion towards concrete examples of purity and of contamination.)

The same, of course, holds for compromise, except that only in modern society has compromise gained public recognition as a quality comparable to purity in its achievement and maintenance of unity. Here, again, we can see the power of Levi-Strauss's scheme yet also its need for modification. According to Levi-Strauss, preliterate thinking is regularly that of compromise between extreme poles, yet without having a clear conception of the compromise, and, at times but not always, appealing *ad hoc* to a specified mediating pole to mask the act of compromise. In other words, purity is a pole whose presence looms large in the intellectual scheme of the preliterate, yet compromise is the tacit mode of thinking they employ within it. In later intellectual schemes purity and compromise themselves act as second-level poles. There can be no mediating pole between purity and compromise, since compromising compromise is compromise. Yet constantly the membership is supposed to know and the leadership is supposed constantly to reassert what must be kept pure at all cost and what may be compromised. Nevertheless, the empirical fact remains that the constancy referred to in the last sentence is not that constant. In particular, when survival is threatened, and even when only unity is threatened by a disastrous schism, a leadership may declare a compromise regarding an item hitherto uncompromisable – and in the interest of survival or of maintaining unity. Jewish history, again, offers many instances of both kinds.

Yet such instances, one must emphasize, are always the exceptions and must remain rare enough not to disrupt the awareness of the successful maintenance of purity.

Rare as these extreme cases must be, they are of daily significance. For, daily, the social or political entity has to decide which potential deviants and which possible deviances to view as tolerable rather than expel and exclude them at all cost. Unless we recognize the role of the leadership in this extreme aspect, we will not learn the logic of consensus and of extremes. For what we judge extremist, and what not, touches equally on matters of purity and compromise. The point at hand is thus significant not only in extreme situations but also in doctrinal matters every day. The extreme aspect invoked – that the leadership may alter the doctrine as to the allegedly unalterable principles of their flock as a means of preserving survival and unity – only illustrates a general fact about social entities: they are characterized both by their principles and also by other given aspects and also, above all, by their ability to survive. Thus, the extreme aspect here invoked is the situation in which the two clash, in which a social entity has the choice between what would be deemed extinction and what would be accepted, however reluctantly, as a drastic modification of a kind hitherto deemed out of the question. To conclude this point, let me comment on it. The point is obvious yet thus far it has not been stated in the theoretical literature. And it is clear why not. The modern theory of the social construction of reality precludes any possible clash between the defining characteristics of a social entity and its survival, since it ascribes the very existence of a social reality to awareness of its defining characteristics. The older alternatives to this philosophy are reductionist, whether of society to individual or *vice versa*, and, of course, reductionists could not raise the question of what a leadership can do in order to maintain a social entity: for the individualist saw social entities traditionally as derivative and the collectivist saw them as given; neither allows action specifically directed at the survival of a social entity proper.

3. THE LOGIC OF EXTREMISM

A leadership has to maintain both purity and unity. And both as understood by public awareness or by authorities endorsed by public awareness. Now, the leadership normally has a lot of leeway. Purity is never achieved, but the leadership repeatedly exhorts membership and at times institutes drives – revival drives – for the raising of the level of purity and of awareness concerning it. These drives usually are also meant to enhance unity. The more often they are used, the more cohesive the social entity, and *vice versa*. This, then, solves the problem mentioned earlier, which naive new students of the social sciences have, even though they cannot as yet articulate it: why do highly cohesive entities stress unity so much? They stress unity as purity and thus they heighten purity in an effort to strengthen and maintain inner stability. Yet they thereby also show poor adaptability. The more adaptable social entities sharply separate principles, to be kept pure, from other aspects of social and/or political life – other aspects on which one is at liberty to compromise or even to decide freely. On these other aspects leaderships often situate themselves comfortably in the middle between extremes. This is why, contrary to appearances, the leadership which has no support from any part of its membership, may be very stable. Indeed, when favoured by one group and rejected by another, the leadership may do something to court the other group even at the cost of losing some of the support of the one. In this balancing act of aspects other than purity, a leadership may act as Levi-Strauss's *bricoleur*, while taking as the gauge the guarding of unity through the maintenance of purity. (In Plato's *Gorgias* rhetoric is declared a kind of *bricolage*; if Levi-Strauss has consciously borrowed the term from Plato, then clearly he was more aware of the political aspect of his theory than he cared to show.)

This situation is the first indication of the preferability of developing an extremist attitude. Roughly, if the leadership is in the middle, then changing an extreme end is also changing the middle. The extremists, then, see this and find it useful to appear as more distant from the middle than they appear to themselves to be, in the hope of forcing the leadership to come closer to the position they see as the correct one. These days the ecological and the peace movements exploit this logic systematically and forcefully – as do the Israeli Jewish militants; they use positions as strong bargaining tactics, not as positions sincerely held.

Of course, this cannot possibly be the whole story. Were it the whole story then every social entity, in particular every nation, could develop two extremes, or rather pairs of extremes, and thus threaten unity as a matter of course. In other words, moving to the extreme has, as a major possible cost, the cost of being excluded from the body social, especially the body politic. Thus, the American Nazi party never influenced national politics simply because it was out, as were the Empire Loyalists in Britain in the post-World War II period. The movements just mentioned are examples of political parties using extremist positions as bargaining tactics: they feel safe from any step to eject or ignore them, and they know that their bargaining power is minute unless they become extreme embarrassments.

When, then, is extremism schismatic? When is it tolerated but ignored? When is its tactical polarization successful? These are general questions. For the reasons specified at the end of the previous section they do not avail themselves of general answers. But general schemes for specific answers are readily available, and they may be worded in testable versions.

The most important extremism which leads to schism is the one which is seen by the leadership as unnecessary adventurism, which is generally deemed a risk for survival, and which, when unchecked, risks survival one way or another. The anathematization of Spinoza is a simple example, but the more dramatic examples were created by the Spanish Inquisition, which deemed unity and purity to be one and the same. Now, it is clear that when conflict between unity and purity arises, then there may be those who side with purity and will not mind risking schism. They may be expelled and they may be tolerated. If tolerated, they may remain within the body politic and they may split. The split may damage the whole body or the splinter body or neither. It is also clear that the extremists may bank on the leadership's fear of split and they will then make demands in exchange for staying. It is also clear that extreme external conditions invite adventurism, i.e., extreme risk-taking, and that at times disaster is so inevitable one can say the body politic is doomed to extinction with or without the increase of extremism or unification. (This was the case in the last days of the schism-ridden ancient Judea.) Yet when schism is the greatest risk for survival, purists may yield to compromisers, as Rabbi Joshua illustrated when he yielded to Raban Gamaliel the President who had instituted the Julian Calendar within Judaism and abolished earlier practices which were thereby replaced.

If one looks at the harmony between purity and unity as the stable norm, then one looks at the conflict between purity and unity as the symptom of instability and risk, and then one tends to ascribe all causes of instability to changing external conditions. For example, the intrusion of liberal philosophy into the Jewish ethos was regulated by a threat to purity. Also the rise of the classic nationalist ideology in Europe in the nineteenth century shook the East European Jewish community.

One may say that societies are usually harmonious and thus stable, so that all threats to harmony are due to external factors. This view is known as functionalism. It may be advisable to approach all intrusion of instability this way, yet, as Ernest Gellner has argued ('Time and Theory in Social Anthropology'), it is dogmatic to deny *a priori* the possibility of instability generated from the inside. For there are instances to the contrary; at least *prima facie* a whole class of instances comes to mind: any case in which instability is caused by the evolution of new ideas which alter the set of options will do; any case which Max Weber has characterized as change introduced by the charismatic personality; any case in which some inconsistency in public attitudes was forced to the fore. An example of charismatic personality *par excellence* is Shabetai Zevi or Rabbi Israel of the Good Name; an example of inconsistency is that of the hostility to politics exhibited by traditional Rabbis and the need to rescue the Jews of Europe, which led to the first march on Washington – by 500 orthodox rabbis during the Holocaust. (This march was a precedent, leading to those against race discrimination; the latter vanished when the oppressed minority developed extremist positions borrowed from the oppressors, and were then replaced by different sorts of marches, originated by Bertrand Russell.)

Extremists often gamble on effecting a radical change. A change which is truly radical, however, may transform the social entity, the nation, the church, the workplace, to the point that it becomes different, and the most important characteristic shared by the old and the new, then, is the mere continuity between them. When awareness of the change that has occurred as radical prevails, then the past extremists may have become the main stream of an already seceded body and they may then declare the secession as a *fait accompli*. The paradigm case is, perhaps, the fourth century Council of Nicea which declared the Christian Church not Jewish. A more recent paradigm is the Mei-ji Restoration which was a drastic revolution effected by the leadership designed to open Japan to western influences one century ago. Let us examine the transformation a bit more closely, since it is one which makes some of the old extremists into the new leadership and their extremism into the new centre.

4. THE LOGIC OF CONSENSUS MAINTENANCE AND OF CONSENSUS RESHAPING

The crux of our discussion, then, turns out to be the awareness of certain defining characteristics. The awareness, however, cannot be shared by all members except in the extreme case of an extremely orthodox and cohesive social entity – whether a primitive society or a primitive sect (religious, political, psychoanalytic or intellectual). Such an entity is very stable within narrow confines and very unstable otherwise. This is a great disadvantage. It is replaced in the modern world with a more lax and broad-spectrum kind of awareness which includes the awareness of the very breadth of the spectrum of admitted characteristics. And so the awareness, unless strict or tight, is known as the consensus. And the extremes, then, are those still within the consensus but on the edge of it – or else they do not count at all, though at times they may be tolerated. And so there is no logic of consensus. Unfortunately there is much too little knowledge of the logic of consensus. (This, we remember, is chiefly because of traditional reductionism: the individualist reductionist sees the consensus as the sum-total of the views of members and the collectivist reductionist sees the consensus as determined by the national destiny. Due to Rousseau, the two reductions are often known as the common will and the general will, respectively. The consensus is neither.)

The consensus – national, church, professional or any other – is a distinct characteristic which operates, or rather which the leadership uses, as a means of coordination: the consensus is taken as given, as the mean, as something not lightly to deviate too much from. The consensus is thus a political instrument, not a foolproof mystical divination of a mystical national destiny; and it is thus not an option and need not be an option held by anyone, nor is it the derivative of opinions since opinions about it contribute to the life of the nation (or of any other body). When a political leadership takes a step which may risk the national consensus, whether Israel's invasion into Lebanon or the war on the Falkland Islands, whether the expulsion of Trotsky or the declaration of Luther a heretic, it is done either by irresponsible leaders or by ones who consider carefully the threat to unity. Indeed, the consensus operates very forcefully as a means to determine, one way in stable conditions and differently in times of stress, the point up to which the extreme can distance itself from accepted norms and yet be regarded as fully within the social entity holding the consensus, the point at which the extremists are considered outsiders regardless of whether they leave or stay. (The word 'consensus' thus often means the social body, especially the body politic, holding that consensus.) The question why anybody outside the consensus stays in the body politic, incidentally, may be answered technically – they may have no choice – or by reference to the logic of their situation, especially by reference to their wish or hope to alter the consensus of the body politic or even to capture it – to effect a revolution, or a revolution from within, as the expression goes.

There is thus ample sense in the observation which political extremists repeatedly voice when they say that they see no difference between the diverse parties within the national consensus. The national consensus may include nothing but the adherence to parliamentary democracy. In that case the communists, for example, may be outside the national consensus even while having representation in parliament. Of course, they traditionally oppose parliamentary democracy on the ground that it is a hoax, that it is a mere mask – a mask for the free-enterprise system or for the late-capitalist system, or for any other system. Let us call it X (in allusion to the conspiracy theory of democracy). What this unmasking means is, first of all, that when a majority within the parliament will oppose X, then, though X stands behind parliamentary democracy, X will simply abolish parliamentary democracy rather than give way to the newly elected force, and establish a frank dictatorship of X. (This part of the doctrine was shared by the Chilean Communists and by the Central Intelligence Agency of the United States; it led to a dreadful self-fulfilling prophecy.) The second part of this communist unmasking is that we must rid ourselves of X, which hides behind the mask of parliamentary democracy, and when we successfully get rid of them, then we will have a true democracy. The theory here presented – which contrasts true democracy with the existing parliamentary democracy – may be true. In some instances it is indeed conceivable that some generally received opinion which is a part of the national consensus serves some X which prevents the nation from operating as a parliamentary democracy proper. I shall offer an instance of this later on. Yet the characteristic of the traditional western communist claim against western parliamentary democracy is that parliamentary democracy is not and cannot ever be a proper democracy, that a proper democracy is a populist representation system of an allegedly classless society – akin to the Soviets of Soviet Russia, though perhaps not as powerless. The traditional communist theory leaves no room for politics other than class politics, and so opposes parliamentary democracy on deep and strong philosophic arguments. The view proposed here is its diametrical opposite, not only in its frank advocacy of pluralist participatory democracy but also in its presentation of politics not only in national politics proper but also in every social entity, including churches, work organizations, trade unions, social classes, classless societies, scientific societies, intellectual schools and clubs. So the communist discussion against parliamentarism will here be ignored as one greatly overvalued old utopian dream.

We come back, then, to the party which opposes a significant part of the consensus yet refuses to leave. By definition, then, that party is extremist. Now, under the influence of traditional thinking since the French Revolution, it is customary to view liberalism as the centre and the extremes as illiberal and as anti-democratic. As we saw, this is true for all systems in which liberalism is a significant part-and-parcel of the national consensus. When the national consensus is attacked from a position more liberal than the national consensus, that position is both extreme and liberal, not the illiberal extreme of the traditional kind. For example, the liberals who overthrow the monarchy – for example the *ancien régime* – were once extremists: they were both liberals *par excellence* and extremists *par excellence*. Liberals in countries with weak or fictitious parliamentary systems today are extremists and so their positions are often wildly misunderstood in the truly democratic world, especially by democrats who foolishly also view all their allies as 'the free world'.

The limitation of the discussion concerning consensus to the national consensus of western liberal democracy often leads to the impression that the extremists tend to recommend separatism and/or violence. At times, especially in western liberal democracies, this impression is very sharply illustrated, as, for example, by the Black Panther movement. Yet this impression is easily refuted when taking the consensus within a church, or within a religious order, where violence is simply out of the question, yet where separatism is an option: extremists always have to consider separation or leaving the body in question as a possible political option – even when they staunchly refuse this option. Thus, in particular, the extremists who accuse the national or the church consensus of not being liberal enough are often persons with the option to leave yet with a national or religious identity and attachment strong enough to make them stay and fight – with considerable risk to their personal well-being if their countries or churches are, indeed, far from liberal. Such individuals are, then, the very opposite of the violent anti-liberal extremists, yet extremists they clearly are all the same. The almost forgotten major schism in the Zionist movement at its peak, as well as the recent minute schism in the Church of Rome, with extremist groups seceding at great cost to themselves and to their causes, constitute dramatic illustrations of this general fact.

Now suppose the liberal extreme declares an item in the national consensus of a democracy a factor which renders that democracy fictitious (purely formal, as the Marxist terminology has it). We have then a liberal democratic extremism on hand. The liberals who deem a mixture of theocracy and liberalism otiose, and who demand alteration of the defining characteristic of their nationality so as to omit religious affiliation from the list of defining characteristics, constitute an example of this. Such liberals will not be ejected, as long as their country is even faintly democratic; they will not leave if they are patriots; and they can then only attempt a liberal revolution or retire. The division of politics into liberal centre and extremes, then, is as outdated as the division of the extremes into left-wing and right-wing; it applies only to the French parliament soon after the French Revolution and to similar bodies, but not in general. In German public life today it is commonplace that extreme left and extreme right are not always distinguishable. At least not with ease. In theocracies such as the Islamic Republic of Pakistan they do not even begin to make sense.

5. EXTREMISM AND PLURALISM

Since pluralism is the peaceful coexistence of diverse ways of living, it is by definition liberal. It need not be democratic. Probably some rule of law is inherent in pluralist society, since we can hardly envisage peaceful coexistence utterly without it. Now the extensive rule of law is essential for democracy. Yet non-democratic pluralism exists wherever law-abiding liberal societies live side by side. Since extremism is often viewed as violent, extremism seems obviously incompatible with pluralism. But extremism, we saw, need not be violent. Also, extremism need not be limited to national politics, but may apply to the politics of churches and learned societies and corporations and work organizations and trade unions and clubs, where violence is often not an issue to begin with. Also, extremism may, within a pluralist society, be the mark of any sub-society whose members are all, by definition, extremely orthodox. Such an extreme society, if peaceful, especially if it allows free exit to members, will be tolerated in a pluralist society, but as an outsider it will not really participate in the life of the pluralist society in the midst of which it lives. Examples are those Jewish enclaves in the United Staets which are amazingly cohesive and orthodox and which have the right to be tolerated because they cannot force their members to stay. In Israel these same groups are less liberal not because of any internal differences but because Israel, as an unofficial theocracy, does not jealously guard the right of exit from these groups. What, then, is the general logic of extremes within pluralism? What is the place of democracy within pluralism, what role does it play there and how does it effect the logic of the extremes within it?

Again, the generality of the question forces us to answer it only schematically. And in part the schema is quite clear and may be worded in a testable version. To begin with what was said thus far, clearly, the more democratic a pluralistic society, the more stable it is in that it is able to coordinate responses to outside threats of all sorts. Also, when democratic, it may include extremist subsocieties or parties which threaten its structure more when it is well organized than when it is loosely organized. The logic of this situation is essentially the same as what the celebrated physiologist Claude Bernard characterized as the fixity of the internal environment, noticing that a living body which is better organized withstands external variation better but internal ones less well. (His staple example was the comparison between the cold-blooded and the hot-blooded.) This logic led collectivist reductionist Emile Durkheim (end of his *The Division of Labour*) to an image of an ideal society, a society which is ideally flexible to the external environment yet ideally inflexible to internal pressures. It is a society in which each task is so clearly and sharply specified that a small group of workers not obeying command may threaten the very survival of the nation. This is, fortunately for democracy, sheer fancy – easily refutable by empirical evidence. Yet as a schema it may be thought-provoking: under what conditions is increased flexibility to external disturbances too much of a sensitivity to internal ones? For we want a pluralist society flexible enough to be able to unify under external threat yet allow as much diversity and deviation and division under normal circumstances. In other words, the very factor of awareness may be utilized in times of stress so that society can give up its principles temporarily; thereby exhibiting the kind of flexibility which Claude Bernard ruled out for living organisms. An example may be the strike by the trade union of the nation's sanitary workers. If the nation is inflexible they may cause a total collapse. To that end a part of the nation must strongly side with their cause and a part of the nation must also strongly oppose it. The strike may then be the merely accidental factor which exhibits the nation's readiness to go to civil war. Hence, not the strike but the broken national consensus is the cause of the impending total collapse of the body politic. Here, then the analogy between society and any living organism breaks down. (The breakdown is not only of collectivism, however, its opposite is equally vulnerable to the same criticism. See, e.g., the final chapter of J.M. Keynes's *General Theory*, where the national consensus is endorsed without notice and uncritically.)

Yet this is no solution to any problem; rather, it poses a problem: who is to say what is a time of stress and what is a time in which we may enjoy the luxury of diversity and of democracy? In times of stress the leadership may mobilize the national consensus in favour of suspending democracy. When exactly?

This problem is ancient. It appears already in Pericles' Funeral Oration, in which he says we are not suspicious of our neighbours. It appears in the ancient republic of Rome and in the political philosophy of Hobbes and in that of Spinoza, in the view of a citizen's conduct being considered in accord with the affairs of state unless strong reasons indicate to the contrary. It is, in essence, a difficulty rooted in the ambiguity inherent in the principle of self-defence. Acting in self-defence has a very narrow margin: too late one cannot act and too early one is not allowed to act: self-defence is justified only almost too late. Otherwise all hostilities are justifiable.

Here we see how important awareness is and how important it is that there is a fair degree of consensus of awareness and that awareness be correct: if suspicion is widespread but not in a consensus, then there can be no internal peace; if there is a consensus and if we are suspicious of our neighbours – rightly or not – we cannot live in peace with our neighbours and if our neighbours are aggressive then we ought to be suspicious of them.

This is where the contribution of Watkins looms large. He quotes quite a number of leading democratic thinkers to say that democracy is founded on a broadly-based consensus on a wide variety of issues. He argues that this demand is too strong and leads thus to schism, that parliamentary democracy is rooted in the recognition of significant legitimate dissent. This kind of dissent is within the consensus, as Watkins explains in his essay on revolutionaries and their principles. Extending his idea to pluralism, democratic and other, we consider in each pluralist society the range of significant legitimate variation or difference, the range or kinds of otherness not viewed as intolerable or endangering the body politic as a whole.

Hence, the logic of pluralism is that of a (reasonably but not necessarily fully) justified absence of suspicion given current received opinions. The governments which raise international tension cannot be deemed pluralists in international affairs though they may, perhaps, be pluralists at home. Further considerations will raise doubts about that too, of course: the national consensus is threatened when a government raises international tension unless that government can appeal to all national parties within the national consensus and convince them all of the propriety of its unfriendly acts. The only way to do that is by exhausting all options which are proposed by any party with the national consensus as alternatives to the raising of international tension. The same holds for governments of churches and of other organizations, of course.

This, then, is the logic of extremism within pluralism: to the extent that pluralist society has to act as a unity it must consider the extremist option as the last resort. This, however, does not tell us what option is extreme. The major point of the present essay is just this: there is no general definition of extremes, and this is why the logic of extremes and of the national consensus has not been studied thus far. Yet treating extremes in the abstract it is not difficult to come to certain general observations whose importance, both theoretical and practical, is fairly obvious.

6. CONCLUSION

The opening point of the present essay was an attempt to discuss extremism in general, and so also purity in general. The idea of purity seems to have dropped out of the discussion. Why?

Extremism, we remember, relates to compromise as its opposite pole, or compromise can be the mediation between two or more extreme poles. Compromise can be applied only when we distinguish essentials (on which we may not compromise) from all else. We see here that the primitive political attitude is the purism which declares each characteristic of a given social entity essential. Consider the absence of all deviation; that is to say, consider entities in which any deviation is punishable by death or expulsion or excommunication. This case is considered ideal in the most primitive societies or social organizations and if the ideal includes the concept of a detailed division of labour among its defining characteristics, then it becomes the ideal Durkheimian society. The compromising social entity is much more flexible, both internally and externally, and is thus much more sophisticated. Nevertheless, such a society is still primitive in that it may be swayed by extremists; it thus offers incentives for extremism within the (national) consensus; it thus also offers incentives to sway the (national) consensus as a whole to incorporate increasingly extremist positions. Especially a society which has not many principles and so is more given to compromise, paradoxically encourages extremism, especially in times of stress.

Israel today is a glaring instance of an arena in which there is too much compromise thus becoming the breeding ground for wild extremism. Nor is Israel's excessive readiness to compromise so very surprising: it has no clear awareness of its own national identity and so no sense of purity at all. This becomes intolerable as the national awareness dwindles when the nation loses all sense of priority on the national agenda; even survival, which is always topmost on every national agenda, becomes incomprehensible due to the lack of awareness. Once identity and awareness become clear, the consensus may perhaps be comfortably achieved, perhaps there will be little need to brand any party extremist.

This can be said generally, then: a (national) society or a social entity or an organization which declares a party extremist only as the last resort, and then excludes it from the effective consensus except under extreme duress, such a society is of a special kind – one reasonably open to toleration and pluralism yet reasonably stable. It allows different life-styles, conflicting views and conflicting interests within the clear bounds of a broad consensus; it is a society in which peace and security can be reasonably taken for granted.

Tel Aviv University

and

York University (Canada)

PHILIP PETTIT

DECISION THEORY, POLITICAL THEORY AND THE HATS HYPOTHESIS

In a typically suggestive aside, John Watkins writes:

> Instead of the unified preference-map of normative decision theory, most of us operate with one or other of a number of preference-systems, according to the 'hat' we are currently wearing. When we come home from the office or go out to a party, we may switch easily and unnoticingly from one system to another. (A holiday-maker lazes on the beach in hot sun and hedonic mood. Then he jumps up: a swimmer is in difficulties. His hedonic calculus switches off, moral concern switches on.) But sometimes, of course, the other preference-system stays switched on. If they indicate conflicting decisions the agent will have to make some sort of meta-decision about which system shall now predominate before he can decide what to do.[1]

This paragraph offers a statement of what we may call 'The Hats Hypothesis'. The hypothesis is that human beings are predisposed to act in such a way that we can usually represent their choices as a function of two variables: on the one side, the social context, or at least the perceived social context, of the action; on the other, a utility function, in particular a set of preferences, which that context selects as suitable.[2]

I believe that the hats hypothesis is of great interest and I hope that in this paper I may do something to show why. In the first section I consider what the hypothesis might mean and how it might be defended. In the second I consider the negative significance of the hypothesis for political theory, arguing that it casts doubt on a central assumption of the public choice critique of non-market institutions. And finally in the third section I raise the question of how far it may have positive significance in political theory; I suggest that it gives us a reason for putting faith in an institution that I describe as the intangible hand.

1. THE DEFENCE OF THE HATS HYPOTHESIS

Watkins introduces the hats hypothesis in the context of decision theory. According to that theory, every agent has a more or less complete set of preferences over certain states of affairs, in particular over the outcomes or potential outcomes of choice; this set is expected to satisfy other constraints

Fred D'Agostino & I. C. Jarvie, Freedom and Rationality: Essays in Honor of John Watkins, 23–34.

too but here we may ignore those. Decision theory formulates a requirement of consistency on the states of affairs – in particular, the states of affairs associated with his options – which a rational agent can consistently come to prefer. The different variants of Bayesian theory, for example, require that if an agent's degrees of preferences over relevant outcomes mean, given his probability function, that one option has higher expected utility than alternatives, then in consistency the rational agent must prefer that option.

On one interpretation, what the hats hypothesis says is that every agent has a number of preference-sets over states of affairs, one for each sort of social context, and that as he moves between those contexts he assumes, so to speak, different decision-theoretic identities. That situation would vindicate the claim that we can usually represent an agent's choices in a two-dimensional manner as a function of social context and the contextually relevant preference-set; the agent's probability function will also make a contribution but, assuming it is part of the unchanging background, I do not mention it explicitly.

This interpretation however does not make the hypothesis very interesting. The reason is that the two-dimensional sort of representation envisaged here would seem to ensure the availability, at least in principle, of a one-dimensional counterpart. If I can represent your choices over options A and B as a function of context C^1 and utility function U^1, and your choices over options X and Y as a function of C^2 and U^2, then I can surely construct a single utility function out of the other two, defined over options A-in-C^1, B-in-C^1, X-in-C^2, Y-in-C^2, and I can represent your choices as a function of this alone. Thus what the hats hypothesis would be is simply the claim that for purposes of decision theory options should be more finely individuated than is common; they should be characterised in a way that takes account of the context in which they present themselves.

But this observation need not depress us, for there is a different and more appealing construal of the hypothesis available. In order to motivate this I need to introduce a distinction between preferences over states of affairs – the preferences relevant in decision theory – and preferences over properties that those states of affairs may have.

The best introduction to the distinction may be to consider first that there are two readings possible of many sentences of the form 'I desire that *p*'. The ascription can mean that for the state of affairs expressed by '*p*', I desire it; I prefer it to the relevant alternatives. Alternatively it can mean that while I may not actually desire that state of affairs – my preferences may not yet be formed – I have a tendency to desire it so far as it promises to make it true that *p*; the ascription implies that as between any two possible states of affairs between which I am otherwise indifferent, the fact that only one makes it true that *p* will lead me to prefer it. If I say that I desire to see the sun, that may mean that as between two given alternative states of affairs, say a skiing holiday and one in the tropics, I desire the second. Alternatively however it may just mean that the property of enjoying sunshine is one for which I am liable to prefer one state of affairs to another. Again if I say that I desire to have a job teaching, that may signal a preference for one state of affairs among a fixed set of alternatives or it may simply register that the property of being a teaching job is one that adds in my calculations to the value of any potential career.[3]

This ambiguity may seem to force a primitive distinction on us, as many philosophers have thought, between the desire *simpliciter* for a state of affairs and the *prima facie* desire for that state of affairs. But it is more naturally taken as pointing us to a distinction between the different objects which desire may take, in the one and only sense of the term 'desire'.[4]

I have desires for particular states of affairs. And I have desires for the properties which particular states of affairs may have. I have at the moment the desire to go on writing and the desire to have lunch with a certain colleague; these are desires for particular states of affairs.[5] And I have at the moment, as I have reliably, a variety of other desires that are desires for the realisation of different properties in whatever states of affairs eventuate. I have a desire for the property which a state of affairs might have of bringing me honour or giving me pleasure, of making the world a more just place or of reducing the amount of suffering there. With any such property the fact that it is displayed by a state of affairs disposes me to desire that state of affairs; it will mean that I prefer the state of affairs to a possible object – it will hardly be a real alternative – whose only relevant difference is that it does not have the property.

It is the merest common sense that side by side with our desires for particular states of affairs, we also have such property-desires. The common sense picture is that when I desire a particular state of affairs I always do so on account of some property or properties that it displays. In scholastic terms, the state of affairs is the material object of the desire, the property its formal object. *Quidquid appetitur sub specie boni appetitur*: whatever is desired is desired under the aspect of good.[6]

If we read the ambiguity as pointing us towards this distinction among the objects of desire, we can still draw a divide between desire *simpliciter* and *prima facie* desire. The divide has a derived status, not a primitive one. Under the picture adopted here I come to form a desire for every alternative only so far as I identify it as the bearer of certain attractive properties. I come to desire that *p*, period, only so far as I desire that *p*, *qua F*. And this is just to say, we may take it, that I come to desire that *p simpliciter* only so far as I first have a *prima facie* desire that *p*.

Decision theory inclines us to neglect these observations, since its concerns have to do wholly with how rationality constrains the material objects which someone can consistently desire. Bayesian decision theory, as we have mentioned, lays it down that if a rational agent has preferences among the possible outcomes of two options – and options and outcomes are states of affairs – such that the first option has higher expected utility than the first, then he will choose the first. But it would be a mistake to let decision theory suggest that material objects are the only objects of desire and that the only requirement of consistency that is imposed by rationality is consistency among such objects. Every material desire has a formal as well as a material object and it is just as much a requirement of rationality that an agent be consistent in the sorts of formal objects that move him.

Furnished with the distinction between preferences over states of affairs and preferences over properties, we may return to the hats hypothesis. The reading of the hypothesis which I propose is that it bears on preferences over properties and that what it says is that as we move between different social contexts, different properties become appropriate for determining how we form preferences over states of affairs, in particular over the options that confront us.[7] In a family context the fact that an option promises to help my children's career may be appropriate, giving me a reason for example to spend our spare cash on a home computer. In a political context in which I have to decide on where to award a government contract, it would not be suitable; even if my children work for one of the competing companies, the fact that awarding the contract there would further their careers is not something that it is appropriate for me to consider.

Under the interpretation suggested, the hats hypothesis is a hypothesis about the norms of social life, norms that are relevant to the preferences we come to form. It is a proposition that belongs to the sociology of morals. Whether the hypothesis is borne out or not is an empirical matter but it is surely an intuitive and compelling proposition. Norms are regularities to which most of us conform, and to which most of us approve of anyone else conforming, so that we are naturally authorities on the question of what our norms are.[8] The hats hypothesis, under our interpretation, is intuitively plausible and this in itself suggests that it is probably sound.

Does the interpretation make the hats hypothesis interesting? I think so and the remaining sections of the paper should bear out the claim. But it may be useful here to say why the construal does not fall prey to the objection raised against the earlier interpretation. The objection to that interpretation was that it reduces the hats hypothesis to the observation that in a decision-theoretic representation of an agent, options ought generally to be characterised in a context-bound way. A similar objection may seem to carry against our construal. For can't the claim that the properties appropriate for determining preferences vary from context to context be recast in context-free form? Can't the claim that F is an appropriate property in context C^1 be put as the claim that the property of being F-in-C^1 is appropriate regardless of context, and so on for other cases?

It can; but here this is no objection. The problem in the case of the earlier interpretation was that the context-free reconstrual showed that the hats hypothesis added little to decision-theoretic lore. That problem does not come up with the present interpretation. Let the context-free reconstrual go through, if you wish. It doesn't matter, because the hypothesis still has a substantial point to make over and beyond the points acknowledged in decision theory. It means that our preferences over states of affairs, which decision theory discusses, are formed on the basis of preferences ignored in decision theory: preferences over properties whose appropriateness is contextually related.

2. THE NEGATIVE SIGNIFICANCE OF THE HATS HYPOTHESIS

If the hats hypothesis holds, under our interpretation of it, then important consequences follow for political theory. In this section I will look at a negative consequence: viz., that the critique of non-market institutions associated with the public choice school is put in question.[9]

The public choice critique of non-market institutions, coarsely formulated, comes in a sequence of claims.

1. Whether for empirical or methodological reasons, we should agree with economists that individuals are mostly rank egoists in their behaviour within the market.
2. If we are prepared to make this assumption of individuals in the market, then we should be consistent and apply it also to agents in the polity, the bureaucracy, and every sphere of social life.[10]
3. But if we do apply the assumption of rank egoism, then we ought to be

complacent about non-market institutions only so far as they, like the market, involve an invisible hand whereby private interests are exploited to produce public benefit.

The punch comes in the last proposition. A feature of the competitive market, at least as idealised in neo-classical economics, is that it is meant to deliver the relevant public good – competitive prices – despite the fact, as it is assumed, that individual agents pursue their personal advantage at every point. Private 'vice' is public 'virtue'. This feature does not look likely to hold with most political and bureaucratic structures. Such institutions are not designed to snatch a relevant public benefit from the jaws of private greed. On the contrary, they look like institutions where individuals will put that public benefit at risk if they always pursue their own personal advantage. The politician will pander excessively to the electorally most threatening groups, the bureaucrat will try to expand the size of his personal empire, and so on.

The upshot of the critique is a deep scepticism about political and bureaucratic structures. The following quotation is a good example of the line.

> The market is usually characterised as the private sector, with government agencies and officials occupying the public sector. But what can this possibly mean? It surely doesn't mean that consumers and the managers of business firms pursue private interests, whereas everyone who works for government pursues the public interest. The senator who claims that 'the public interest' guides all his decisions is in fact guided by a personal interpretation of the public interest, filtered through all sorts of private interests: re-election, influence with colleagues, relations with the press, popular image, and a place in the history books. Senators may be less interested than business executives in maximising their private monetary income, but they're probably more interested on average in acquiring prestige and power. The same kind of analysis applies to any employee of a government agency, whether it is a high appointed official on a regulatory commission or someone just starting a job at the lowest civil-service rank. However lofty, noble, or impartial the stated objectives of a government agency its day-to-day activities will be the consequence of decisions made by ordinary mortals, subject to the pull and push of incentives remarkably similar to those that operate in the private sector.[11]

We are now in a position to say what the negative significance of the hats hypothesis is. The hypothesis raises a serious question about the public choice critique of non-market institutions. It means that the fact that we assume rank egoism of agents in the market does not entail that we ought equally to assume such rank egoism of agents in, for example, political and bureaucratic contexts.[12]

To assume rank egoism in an agent, plausibly, is to assume that the properties by reference to which he forms his preferences over states of affairs, in particular over the options he faces, are all self-regarding features. He forms his material preferences by reference only to formal objects such as his own pleasure or success or honour; he is insensitive to any effects on others or on the world more generally. The hats hypothesis means that we may assume rank egoism among agents in economic contexts without thinking that consistency requires us to impose the assumption elsewhere. Economic, political and bureaucratic contexts may make different properties appropriate for determining preferences over states of affairs; they may encourage the wearing of different hats.

The hats hypothesis shows that the consistency claim put forward in Proposition 2 above is overstated. That we assume that market agents are rank egoists does not necessarily entail that we must regard agents in other institutional contexts as rank egoists. It may be that we assume that market agents are rank egoists, for example, so far as we assume that all agents behave according to local institutional norms and that the norms of the market are egoistic. Thus the assumption that market agents are rank egoists may be entirely consistent with different assumptions about how people behave in other contexts. The transition from Proposition 1 to Proposition 2, as the latter Proposition stands, is a *non sequitur*.

3. THE POSITIVE SIGNIFICANCE OF THE HATS HYPOTHESIS

The hats hypothesis reveals a lacuna then in the usual case for the public choice critique. That is some consolation for those who put their faith in non-market institutions. But it is not much since, for all that has been said so far, non-market institutions may still be subject to the sort of malaise alleged by public choice theorists.

The hats hypothesis holds that different properties are appropriate in different contexts for determining agents' preferences over states of affairs. Let it be granted, in particular, that the properties appropriate in political and bureaucratic contexts are indeed different from the self-interested considerations alleged to be suitable in economic settings. Even then the hypothesis offers little consolation for those who look with optimism to non-market institutions. For so far we have been given no reason to think that people will form their preferences on the basis only of contextually appropriate properties. The hypothesis itself does not rule out the possibility, for example, that in awarding a contract in a competition between rival firms, a minister in government may think like a father and award the contract to a firm which employs his son or may think like an entrepreneur and award the contract to the firm that offers the largest bribe.

But after the bad news comes the good. It turns out that if we add a further plausible assumption to the hats hypothesis, all of this changes and the hypothesis assumes positive as well as negative significance in political theory; it becomes significant other than for the negative feature of raising a doubt about the public choice critique of non-market institutions. The extra assumption required is that people generally desire honour. They want others not to think badly of them and, if possible, they want others to think well. In particular they want others to see them as forming their preferences over states of affairs on the basis of contextually suitable properties and only contextually suitable properties.

This assumption is intuitively acceptable, fitting for example into that long tradition of European thought in which the love of esteem is hailed as one of the great human passions.[13] Adam Smith gives forceful expression to it as follows.

> Nature, when she formed man for society, endowed him with an original desire to please, and an original aversion to offend his brethren. She taught him to feel pleasure in their favourable, and pain in their unfavourable regard. She rendered their approbation most flattering and most agreeable to him for its own sake; and their disapprobation most mortifying and most offensive.[14]

The assumption should also recommend itself nowadays, for it is built into at least two major schools of contemporary social theory. Clearly it should be acceptable to those in the sociological tradition of theory within which the desire for status ranks with the desire for wealth and for power as one of the basic human desires. And it should be congenial even to those who work in the more economic tradition of rational choice. The motivational postulate of that theory is almost canonically formulated by John Harsanyi. "People's behaviour can be largely explained in terms of two dominant interests: economic gain and social acceptance".[15] The desire for social acceptance is precisely the sort of thing postulated in our assumption.

If we add this assumption to the hats hypothesis, then we begin to find reason for thinking that people in political and bureaucratic contexts will form their preferences on the basis of contextually appropriate properties – say, properties relevant to the public interest – rather than in the fashion of rank egoism. So far as such agents run a risk of exposure if they form their preferences on any other basis, their desire for esteem will encourage them to form their preferences on this. Any significant chance of exposure, even a low one, may be all that is needed, since even a single disclosure – a single scandal – may destroy the agent's reputation with the relevant audience.

But here's an objection. What the desire for esteem will encourage, so it may be said, is not sensitivity to the contextually appropriate properties, only sensitivity to the property of seeming to be sensitive to those properties. The observation however is not damaging. First, even if true this wouldn't matter from the point of view of how people behave – and that is the final evaluative viewpoint – so long as there are few opportunities where it is rational to reject the dictate of the contextually appropriate properties in the hope of not being detected. And secondly, since it will be costly and difficult actually to form one's preferences with a view to honour, while seeming to do so with a view only to contextually appropriate properties, the rational strategy may well be to look normally only to those properties, safe in the knowledge that this is probably the best way to promote one's honour and other relevant goods. The rational strategy may be to go restrictive, making one's decisions by criteria other than reference to the goods – in this case honour – with which one is ultimately most concerned.[16]

If we follow the line of thought suggested, ignoring the objection, then we begin to connect with a tradition of thinking which has been neglected by recent liberals, particularly by those of the public choice bent. That is the tradition of republicanism which, derived from ancient Roman sources, came to dominate progressive political thinking from the Renaissance, in particular from Machiavelli, down to the end of the 18th century.[17] One of the main assumptions in that tradition is that if we cannot rely on nature to produce contextually appropriate deliberation and action, we can at least rely on pride.[18] One of the distinctive contributions of the tradition was to emphasise that the way to exploit pride for the collective good of society is to subject all holders of public office to the glare or threat of publicity.[19]

The institutions which exploit pride and publicity in this way, many of them the product of the republican tradition, are various. Consider for example the institution of the jury. Jurors are encouraged to meet suitable standards of deliberation and action – ultimately to form their preferences on the basis of appropriate properties – by a number of stratagems. First, those with such a special interest in the outcome that the threat of publicity might not inhibit them are filtered out in jury selection. Secondly, jurors are protected by confidentiality from any threats to their welfare that might be made from outside. And thirdly, jurors are exposed to internal pressure to behave virtuously by being put in a situation where they have to answer to other members of the jury for how they vote. The institution of the jury exemplifies a general pattern. It is replicated for example in the appointment committee, as that is usually established in the bureaucracy and related areas, including universities. Members of such committees are vetted for a special interest, are protected from external pressure, and are subjected to the internal pressure of having to answer for how they vote.

Moving to higher levels of public office, judges, parliamentarians and members of government are also subjected to publicity constraints which, when they work well, ought to produce behaviour that is consonant with the context. Such officers are forced to answer for how they deliberate and act, in particular to show that they form their preferences over options in a contextually suitable manner: by reference to the appropriate properties. Judges are at least informally required to make statements in defense of their rulings and politicians are expected to justify themselves to parliament and to the public. And apart from such direct pressures, judges and politicians are also faced with the prospect of appeals and inquiries which are designed to inhibit any tendency to deviate from the standards we would wish to obtain.

These are brisk observations but I hope they may be sufficient to show that the hats hypothesis with which we began leads in directions that are of great positive significance for political theory. In conclusion I would like to introduce a nice metaphor which serves to catch the lesson of our observations.

Public choice enthusiasts, and liberals more generally, rely on one sort of mechanism in particular – the invisible hand – to achieve the social alchemy we all desire: that is, to ensure that public benefits will be produced by agents who are not mainly concerned with those benefits and who may even be rank egoists. The invisible hand does its work by suitably aggregating whatever actions are forthcoming from the individuals involved. Thus, to take the case of the idealised competitive market, it puts the producers of different goods and services in predicaments such that by each rationally pursuing their personal advantage – undercutting one another and undermining potential cartels – they ensure that prices fall to the competitive level.[20]

What we have identified here is a different sort of mechanism for achieving a similar magic: a mechanism that I describe, in parallel metaphor, as the intangible hand. The intangible hand produces the public benefit sought, not by suitably aggregating whatever actions are forthcoming from individuals, but by influencing individuals in such a way that they act as if they were predominantly concerned with the benefit in question. It is a formative rather than an aggregative mechanism.

There is much more to be said about the nature of these two institutional mechanisms and about their distinctive advantages and disadvantages. But it must wait for another place. I have been concerned in this paper only to make a connection between acceptance of the hats hypothesis and recognition of the intangible hand. I hope enough has been said to forge that link.[21]

Australian National University

NOTES

[1] John Watkins, 'Imperfect Rationality', in Robert Borger and Frank Cioffi, Eds., *Explanation in the Behavioural Sciences,* Cambridge University Press, 1970, p. 207.
[2] Watkins would be impatient of the idealisation involved in ascribing a utility function, even a contextually specific utility function, to an agent; but the idealisation is convenient for my purposes and nothing of substance is affected by it.
[3] On this ambiguity see Frank Jackson, 'Internal Conflicts in Desires and Morals', *American Philosophical Quarterly*, Vol. 22 (1985) and 'Davidson on Moral Conflicts' in E. LePore and B. McLaughlin, (Eds.), *Actions and Events*, Blackwells, Oxford, 1985.
[4] On this point see Jackson, 'Internal Conflicts in Desires and Morals'.
[5] That such a state of affairs is conceived by me as a particular does not mean of course that it may not turn out in different ways, as different possible worlds are realised; it is a more abstract particular, an expectation of such particulars.
[6] The picture is well entrenched in the Aristotelian tradition. See Joseph Raz (Ed.), *Practical Reasoning*, Oxford University Press, 1978. See too David Milligan, *Reasoning and the Explanation of Actions*, Harvester Press, Brighton, 1980, Chapter 3. It is invoked in Jackson's 'Internal Conflicts in Desires and Morals'. The Gorman Lancaster translation of commodities into characteristics represents a parallel in economics. See Amartya Sen, *Choice, Welfare, and Measurement*, Blackwells, Oxford, 1982, p. 30.
[7] See Philip Pettit, 'The Life-World and Role-Theory', in Edo Picevic (Ed.), *Phenomenology and Philosophical Understanding*, Cambridge University Press, 1975.
[8] See my paper 'The Inevitability of Norms', *Ethics*, forthcoming.
[9] For an overview of the tradition see Dennis Mueller, *Public Choice*, Cambridge University Press, 1979 and Iain McLean, *Public Choice: An Introduction*, Blackwells, Oxford, 1987.
[10] On this consistency requirement see Geoffrey Brennan and James Buchanan, 'The Normative Purpose of Economic Science: Rediscovery of an Eighteenth Century Method', *International Review of Law & Economics*, Vol. 1, (1981), and 'Predictive Power and the Choice Among Regimes', *The Economic Journal*, Vol. 93, (1983). Brennan and Buchanan are themselves sophisticated enough in their account of

consistency not to let it force them to endorse Proposition 2. But the same does not hold of others.

[11] Paul Heyne, *The Economic Way of Thinking*, 4th Edition, Science Research Associates, Chicago, 1983, pp. 271–72.

[12] See in this connection my paper 'Towards a Social Democratic Theory of the State', *Political Studies*, Vol. 35 (1987), 42–55.

[13] See for example Arthur O. Lovejoy, *Reflections on Human Nature*, Johns Hopkins Press, Baltimore, 1961, Lecture V.

[14] *The Theory of Moral Sentiments*, D.D. Raphael and A.L. Macfie (Eds.), Liberty Classics, Indianapolis, 1982, p. 116.

[15] 'Rational Choice Models of Behavior versus Functionalist and Conformist Theories', *World Politics*, Vol. 22, (1969), 513–38. The postulate is quoted with approval in Michael Taylor, 'Rationality and Revolutionary Collective Action', in Taylor (Ed.), *Rationality and Revolution*, Cambridge University Press, 1967, p. 66.

[16] See Philip Pettit and Geoffrey Brennan, 'Restrictive Consequentialism', *Australasian Journal of Philosophy*, Vol. 64, (1986), 438–55.

[17] See J.G.A. Pocock, *The Machiavellian Moment: Florentine Political Thought and the Atlantic Republic Tradition*, Princeton University Press, 1975.

[18] See Arthur O. Lovejoy, *Reflections on Human Nature*, Johns Hopkins Press, Baltimore, 1961, Lecture V.

[19] For a fuller development of this theme see Philip Pettit, 'The Freedom of the City: A Republican Ideal', in Alan Hamlin and Philip Pettit (Eds.), *The Good Polity: Essays in Normative Analysis of the State*, Blackwells, Oxford, 1988. That paper draws in particular on two articles by Quentin Skinner: 'Machiavelli on the Maintenance of Liberty', *Politics*, Vol. 18, (1983); and 'The Idea of Negative Liberty: Philosophical and Historical Perspectives', in Richard Rorty, J.B. Schneewind and Quentin Skinner (Eds.), *Philosophy in History*, Cambridge University Press, 1984. See also his more recent paper 'The Paradoxes of Political Liberty,' in S.M. McMurrin (Ed.), *The Tanner Lectures on Human Values*, Vol. 7, Cambridge University Press, 1986.

[20] The predicaments are all prisoner's dilemmas. See Philip Pettit 'Free Riding and Foul Dealing' *Journal of Philosophy*, Vol. 83 (1986); reprinted in *The Philosopher's Annual*, Vol. 9, (1987).

[21] I am much in the debt of John Braithwaite and Geoffrey Brennan, both close colleagues for discussions over a long period on matters related to the distinction between the invisible hand and the intangible hand. The distinction is central to a book which I hope to publish with Braithwaite entitled *Not Just Deserts: A Republican Theory of Criminal Justice*. It is the topic of a paper that I hope to co-author with Brennan. I am also indebted to Fred D'Agostino and Barry Hindess for comments on an earlier draft.

J.J.C. SMART

NEGATIVE UTILITARIANISM

I

One of John Watkins's many notable contributions to philosophy is his paper 'Negative Utilitarianism', which is the second part of a symposium of that title, the other symposiast being H.B. Action.[1] Both symposiasts consider and reject a form of utilitarianism that had been extracted by my brother Ninian Smart from some remarks in Popper's *Open Society and Its Enemies*.[2] However the interpretative principle of charity should prevent us from interpreting Popper as a negative utilitarian in my brother's sense: as my brother indeed remarks, if Popper did have the negative utilitarian principle as a fundamental axiom of his ethics, he also had at least two other principles as well. Certainly if Popper had been a negative utilitarian at all he would have been a singularly muddle headed one.

Indeed in an addendum to the 5th edition, p. 296, Popper makes it clear that he did not regard himself as a utilitarian of any sort. He says that he did not put up the negative utilitarian principle as a *criterion* of right action, and that he would object to the idea of such a criterion just as much as he would object to the idea of a criterion of truth. I am not sure that this by itself makes him a non-utilitarian, since the utilitarian should not put up his or her principle as a *criterion* of right action, if this is meant as definition of right action. Certainly as a utilitarian myself I put up the principle as the expression of an attitude, and not as definitional. (If it were definitional it would be ethically trivial.) As a matter of fact I regard the word 'criterion' as it has been used in philosophy as a bit of a weasel word: I believe that in the hands of Wittgenstein at least it has led to obscurity and fence sitting. Be that as it may, Popper makes it clear that minimization of suffering is only one of several moral principles that he accepts. Moreover in the addendum Popper asserts that he put up the principle of minimization as a matter for public policy rather than for private morality.

The negative utilitarian principle is that we should minimize the amount of suffering and unhappiness in the world. Classical utilitarianism is concerned with the maximization of happiness. In classical utilitarianism unhappiness is conceived as negative happiness, and if we considered happiness as negative

Fred D'Agostino & I. C. Jarvie, Freedom and Rationality: Essays in Honor of John Watkins, 35–46.

unhappiness, then the positive utilitarian principle could be stated as that of the minimization of unhappiness. But this clearly would not be negative utilitarianism, in my brother's sense, that is in the sense he extracted from the passage in Popper. In negative utilitarianism positive goods such as enjoyment do not come into the calculation at all. Negative utilitarianism enjoins us to build hospitals for the relief of pain and other miseries, but it gives us no warrant for supporting the study of metaphysics (for example), however much are the pleasures of metaphysical discussion, unless it can be shown that metaphysical discussions indirectly bring about better hospitals, more food for the starving or other techniques for relieving miseries. If so it is an odd doctrine. We often choose pain and hardship for ourselves because we think that these lead to enjoyments that outweigh them. Why should we not make a similar judgement in considering the effects of our actions on others?

Part of the value of Watkins' abovementioned article is his brief critique of negative utilitarianism. The major part of his paper can, however, be seen as a propaedeutic to this critique, since he gives an interesting and at least initially attractive proposal for the methodology of ethics. Thus, besides discussing negative utilitarianism, I wish in what follows to sketch a different account of the methodology of ethics. Indeed I believe that it is necessary to do so if I am to give any sort of defence of classical utilitarianism against some of the objections that Watkins and others have brought up against it.

In brief, my brother Ninian's main objection to negative utilitarianism is that it contains the unpalatable consequence that it implies that a negative utilitarian who was able to destroy all life on earth ought to exercise this ability, i.e. should act as a benevolent world exploder. Whether or not misery continued to exist on planets of distant stars, all future misery on this earth would be prevented. And yet, says my brother, this is an unpalatable conclusion. Of course a pessimist who thought that in fact the condition of life on earth was such that misery would always outweigh happiness might think that the conclusion that we should explode our planet was quite a palatable one, but in so thinking he could be doing so on the basis of classical, no less than negative, utilitarianism.

If we take the passages in Popper to which my brother referred as giving a critique of classical utilitarianism, they seem to be easily answered. Perhaps they should not be taken in this way though despite the clarifications in the aforementioned *addendum* I do not find Popper's intentions here quite clear. So without wanting to pin any accusations of error on to Popper, let me give a utilitarian reaction to Popper's remarks, even if I have to take them, possibly wrongly, *au pied de la lettre*. It will at least help to clarify the relation of the classical utilitarian's position to that of Popper. Possibly, if Popper's remarks are taken in a different way, it will turn out that there is not quite so much disagreement between him and the classical utilitarian.

The utilitarian may be moved by Popper's remark that "human suffering makes a direct moral appeal, while there is no similar call to increase the happiness of a man who is doing well anyway." This is surely because the relief of suffering is an urgent matter so that the increase of happiness can wait. Thus if I can increase your happiness by urging you to read *Middlemarch* I can do this as well next week as I can this week. If you have a toothache your dentist needs to do something about it *now* or you will have a greater total amount of the unhappiness of toothache. Suppose that the alternatives would be (1) to continue having the toothache for three days while you read *Middlemarch* or (2) to have the toothache relieved at once. You will prefer (2), since you can read *Middlemarch* at another time. It is not clear what the verdict would be if this present time were the only time in your life that (for some strange reason) you would be able to read *Middlemarch*. The verdict might well depend on the intensity of the toothache and the enjoyment you would get from *Middlemarch*. (I neglect here considerations arising from the *fecundity*, to use Bentham's expression, of the pleasure of reading *Middlemarch*.) Of course if the toothache were so great that you were unable to read *Middlemarch* anyway, alternative (1) would not arise.

It will be noted that Popper talks of *human* suffering. I take this to be a slip of the pen, and that he would regard suffering of a sheep or a pig to be morally more urgent than some positive happiness on the part of sheep, pig, human or other sentient being.

In Chapter 11, note 22, Popper advocates the principle 'minimize pain' because if we try to apply the principle 'maximize pleasure', this will lead to totalitarianism. This does not look like negative utilitarianism, nor is it a criticism of classical utilitarianism. Utilitarians presumably object to totalitarianism precisely because it does not make for happiness. If totalitarianism *did* maximize happiness presumably they would be all for it. There are good empirical reasons why totalitarianism does not maximize happiness. Many of these are brought out by Popper himself in his book. The classical utilitarian may surely agree with Popper that on the whole it is a good thing to look for ways of minimizing pain and unpleasure. But not all the way. Otherwise we would have more hospitals and no universities, or at least no faculties of arts. (Or almost so. Perhaps arts faculties help to humanize the medical profession, philosophy departments may beneficially affect scientists and so on. Without a cultural background of the humanities sciences might not flourish so well. But this sort of consideration should not be overemphasized in the present connection.)

One reason why a utilitarian will regard the moral urgency of relieving misery as greater than that of increasing joy is suggested by a remark made by H.B. Acton, in the symposium with Watkins. Following Popper, he remarks that "when we are moved by someone's suffering we feel called upon to help, whereas when we are moved by someone's joy there is no such practical urgency".[3] Besides the consideration in my *Middlemarch* example, there is also the fact that there are practical ways of relieving at least some sorts of misery, while if a person is very happy it is probably impossible to make him or her *more* happy, at least to the extent that would outweigh the pain of someone else's toothache, for example. We must not forget, of course, that the ideas of relieving misery and of increasing positive happiness are by no means incompatible: perhaps one can do both. Indeed the connection between mental well being and bodily health suggests that one can often promote the one by promoting the other.

In Chapter 5, n. 6, of the *Open Society*, to which my brother refers, Popper seems to be concerned to put forth certain rules of thumb for public policy. These are in brief: (1) tolerance to all who are not intolerant, (2) minimization of suffering, (3) fight against tyranny. They do not seem to be put forward as ultimate very high-level moral principles. If they were they would strike me as a rather arbitrary and *ad hoc* set. Yet a utilitarian could well advocate them as useful rules of public policy. Surely tolerance makes for happiness, either fairly directly or for its instrumental effect in improving science and human institutions. Tyranny makes for unhappiness and so should be fought. And on the whole, or to a great extent, governments will do best for the happiness of their subjects by concentrating on the removal of obvious evils. Popper worries that the principle 'maximize happiness' is apt to produce a benevolent dictatorship. It might be thought that the utilitarian answer to this is 'What would be wrong with a *benevolent* dictatorship?' but of course this would not do. Dictatorship is not an efficient form of government, and so would not be beneficient even if the motives of the dictator were benevolent. Moreover power corrupts and a benevolent dictatorship would be unlikely to remain benevolent for very long. Thus a utilitarian would oppose even a benevolent dictatorship, and if it were really true that propagating the utilitarian principle among the public would lead to harm, he or she would keep his or her utilitarianism private, or within a trustworthy elite.[4] Mind you, I think it far fetched that the injunction to maximize happiness *would* lead to dictatorship: the empirical arguments for the beneficence of democracy are too well known. What we perhaps may be opposed to are certain pieces of social engineering: but why should a utilitarian not also oppose certain sorts of 'do gooding' social engineering if these do not in fact on the whole lead to good results?

II

So far I have been saying that Popper was not giving ultimate moral principles. As he confirmed in an edition of his book published later than when my brother wrote, he was no sort of utilitarian, whether positive or negative. I have tried to suggest that a utilitarian could accept all or many of Popper's prescriptions if these are taken as useful rules of thumb for political policy. John Watkins, however, at the beginning of his paper on Negative Utilitarianism, does describe himself as a 'sort of negative utilitarian'. He says that he regards all suffering as bad, though he also regards some things as bad that do not involve suffering. The most interesting parts of his paper are concerned with a discussion of how argument over conflicting moral principles is possible and with his embedding of negative utilitarianism (in a modified version) into a wider moral negativism. The wider moral negativism is methodological in nature and is compared with Popper's negativism in scientific methodology, the idea of falsificationism as opposed to justificationism.

In the present section of the paper I shall argue that Watkins's methodology in ethics must lead to the rejection of utilitarianism whether positive or negative. No real person who accepted it could remain a utilitarian. I shall sketch an alternative way of looking at meta-ethics which might explain how a real person *could* remain a utilitarian, and shall suggest that this meta-ethical approach is perhaps more appropriate than Watkins's ethical methodology, which comes close to John Rawls's method of reflective equilibrium, in Rawls's *A Theory of Justice*[5] which was published subsequently to Watkins's paper. Perhaps the main difference comes from Watkins's stress on criticism and refutation as opposed to justification.

Watkins compares the way in which normative ethical principles can be criticized with the way in which scientific theories can be tested. In science we assume certain basic statements, which are statements of observation and also of initial conditions. Observations statements are not themselves bedrock. For Popper there is no bedrock, and an observation statement taken as basic in order to test a certain hypothesis could itself be criticized by reference to yet other such statements (together with theories) on some other occasion. Thus as Lakatos was quick to point out, theories can no more be conclusively falsified than they can be verified or justified. Nevertheless we can agree with Popper that scientific practice does to a considerable extent bear out Popper's stress on the importance of falsification, and also that it ought to do so.

We criticize a scientific theory by testing its observational consequences. (Actually these are the consequences of the theory together with a statement of boundary conditions. Or we can take the consequences of the theory that are observation conditionals, of the form $O_i \supset O_k$ and test these.)[6] We can say that a theory has unsatisfactory implications if it implies false basic statements (observation conditionals).

Watkins allows a modicum of scientific conservatism in his remark (p. 104) that "after falsification by basic statements a hypothesis is not automatically condemned without right of appeal". The basic statements are accepted only provisionally, and we may test these. If one or more of them is falsified the original hypothesis may survive.[7]

Similarly, Watkins holds, a moral theory or policy can be criticized by showing that it has unsatisfactory implications. My concern here is to consider the application of this idea to utilitarianism (whether positive or negative). There is no doubt that the utilitarian principle has unsatisfactory implications. Thus, for example, the utilitarian will almost certainly feel distress when he or she draws the well-known conclusion from utilitarianism that there are conceivable circumstances in which it would be right to punish (or 'telish'[8]) an innocent person. The word 'telish' is used to pacify those quibblers who (rightly or wrongly) complain that it is part of the meaning of the word 'punish' that the punished person be guilty or thought to be guilty. The utilitarian will doubtless comfort himself or herself by reflecting that he or she has a vanishingly small probability of being in the position where it would be his or her utilitarian duty to do this sort of thing. Nevertheless the implication will still be felt as worrying. (It is true that on many other moral systems the same implication may well follow, if for example 'telishing' is the only way of preventing a world nuclear war, but the stories to make such a conclusion plausible will doubtless be more strange and fantastic than in the utilitarian case.)

This consideration should worry the negative utilitarian too. Will he or she be happy about telishing an innocent person even though this will reduce the *total* amount of misery in the world? This will be felt to be unfair, and so unsatisfactory, but it would be a consequence of negative utilitarianism. So I think that Watkins's methodology (or Rawls's method of reflective equilibrium) would imply that no real person could be even a negative utilitarian. It is true that Watkins allows a difference between ethics and science in that while there is a true theory of the world, even though we do not know what it is, *all* possible ethical theories may have unsatisfactory implications.[9] So it is presumably a matter of choosing the least unsatisfactory. If we are to go by Watkins's methodology it would appear that a modification of negative utilitarianism to take account of certain common sense deontological principles would have fewer unsatisfactory implications.

A scientific hypothesis or a set of moral principles of course implies itself. (In the case of utilitarianism the set will be a one-membered set.) Rejection of a set of moral principles may be felt as unsatisfactory. In the scientific case we can try testing the observation statements originally taken as basic. In the ethical case we can do no such thing. We can of course ask ourselves whether we dont't feel *worse* about a negation of an ethical precept or conjunction of precepts. But generally feelings are not altered by further thought. No amount of further thought would make a normal person who claimed to be a utilitarian feel happy about telishing an innocent person, or about shooting one prisoner to save the other nineteen. (Cf. the by now well-known case of Jim and Pedro put forward by Bernard Williams.[10])

I shall concede that if we accept Watkins's ethical methodology we must say that no normal person who claims to be utilitarian can really be one. A utilitarian does have feelings about fairness of distribution of happiness and unhappiness. These may in part have a sociobiological basis: notice how small children come to say quite naturally 'It's not *fair*'. Retributive feelings are very natural to us and are reinforced by moral training. Moral training also ensures that a utilitarian feels guilty if he or she has, on good utilitarian grounds, to break a promise or tell a lie. I am myself a rather compulsive rule obeyer and this sits oddly with my utilitarianism, so that, to take a trivial example, I feel uneasy when I am in the car with my wife and she parks in the 'residents only' car park at University House, even when there is plenty of room to spare. My wife, who I don't suppose is a utilitarian, is more rational than me and feels no guilt or embarrassment. As a utilitarian I backslide and she acts as a good utilitarian would do.

I propose a different method of ethical thinking and feeling. I conceive of ethical principles as expressions of *ultimate* attitudes. They are not open to testing, though they may be advocated rhetorically. Thus I am even more of a non-cognitivist than is Watkins. I wish to advocate the view, for which I am indebted to the fine book by D.H. Monro, *Empiricism and Ethics*,[11] that ethical principles are expressions of *over-riding* attitudes. The fact that an ethical principle has unsatisfactory implications need not vitiate the principle if the unsatisfactoriness is felt only as a result of a lower order attitude. A higher order attitude is in part a tendency to act in such ways as to strengthen or to weaken certain lower order attitudes.

One needs a hierarchical theory of moral psychology. Without such a theory it would seem that no moralist, whether utilitarian or deontological, could be sincere, since everyone backslides from his or her moral principles on occasion. Thus most of us have selfish attitudes, nationalistic attitudes and all sorts of moral principles inculcated in us in our childhood. Of course, over-riding attitudes will very likely have arisen as a result of early moral and intellectual influences too. But the lower order attitudes do not just disappear. If Jim can bring himself to appease Pedro by shooting one prisoner in order to save the other nineteen, he will surely feel guilt and remorse. On the other hand if he refuses and consequently Pedro executes all twenty, Jim will feel guilt and remorse too.

Why does a person become a utilitarian? In my case a youthful reading of G.E. Moore's *Principia Ethica* was influential: its consequentialism appealed. Other reflections, such as 'What could be important, in the end, other than maximization of happiness?' helped. A generalizing tendency, on the analogy of science, helps too. We all probably have an innate tendency, explicable sociobiologically, to a *restricted* benevolence. We may be driven intellectually to extend this in space and in time, and thus form a higher order attitude of caring for all sentient creatures, everywhere and everywhen, even far into the future.

I have gone into these matters at slightly greater length elsewhere,[12] and so I shall not go further into them here. I hope that I have done enough to hint at a different sort of moral psychology. Watkins's model puts our attitudes on a level, while I have a hierarchical picture. Whether he or I is right is a nice question in moral psychology. According to me, the analogy between science and ethics is less close than it is on Watkins's view. We agree, I think, in both of us being some sort of non-cognitivist. The difference between us is over whether we need give weight to moral feelings that we cannot shed. On my hierarchical view we need not. Folk morality persists in, for example, the most utilitarian psyche. On the other hand, folk science is easily shed. We can easily give up the view that the Earth is flat or that space is exactly Euclidean.

Watkins's method or Rawls's method of reflective equilibrium seems to me to weight the scales too much in favour of 'folk morality'. I do not think that Watkins' negative utilitarianism could be supported by means of his own methodology. There is certainly a negativism at a meta-level, which parallels Popper's negativist scientific methodology of falsifiability. I think that neither is quite as negative as it purports to be, in the moral case because a moral principle implies itself and so rejection of the principle will be one of the relevant things that are felt to be unsatisfactory, and in the scientific case because the basic statements are themselves open to test on other occasions, and so hypotheses are never conclusively refuted. This does not mean that we should lose sight of the great interest and importance of Watkins's methodology of morals and Popper's methodology of science.

Ethics seems to me to be fundamentally different from science. Science is a matter of what the world is like. Ethics is a matter of what we want. Ultimate ethical principles cannot be intellectually criticized, unless they lack clarity or are inconsistent with one another. They can be modified by rhetoric, which for example can help to extend people's sympathies, from the family, to the nation, to the human species, to all sentient creatures on earth, to all sentient creatures anywhere, and from here and now to the far future. I used to think that a person could fail to be a utilitarian only if he or she was either muddle headed or non-benevolent. I no longer think this. A person who supplements the utilitarian principle by a principle of fairness or equal distribution of happiness need not lack benevolence or clarity of thought. A utilitarian thinks of people as buckets for containing happiness, and does not mind whether or not all the buckets are the same size. How we feel about this issue may depend to some extent on our views about the unity of personality.[13] I am here talking about equal distribution of *happiness*. A utilitarian may well approve of equal distribution of *money*, for example, since this may lead not only to more equal distribution of happiness but to a greater amount of happiness.[14]

III

I shall conclude with some scattered observations on negative utilitarianism itself, and the question of a benevolent world exploder.

A classical utilitarian could be a benevolent world exploder only if he or she were a pessimist who, like Schopenhauer, believed that sentient beings inevitably, or even perhaps for the most part, are more miserable than happy. A negative utilitarian, on the other hand, could be a benevolent world exploder even if he or she believed that on the whole sentient creatures are more happy than unhappy. (Evolutionary considerations might suggest this optimistic belief, since happiness makes for health and efficiency.) *Prima facie* at least, the decision to have a child could be supported neither by a negative utilitarian nor by a Schopenhauer type pessimist. (Unless perhaps the negative utilitarian thought that the child would grow up to be a reducer of other people's miseries, perhaps to be a compassionate doctor or even a benevolent world exploder.)

Acton, in his contribution to the symposium mentioned earlier, argues that the negative principle (of minimization of suffering) 'is not merely a matter of irrational choice, but that there are reasons for it'. I have argued that a positive utilitarian could agree with this if the negative principle were taken as a useful but not inviolable subordinate rule of thumb. I think that Acton would agree with this, because at the end of his paper he contrasts urgency with importance. Sanitation, he suggests, might be said to be more urgent than poetry, though poetry is more important than sanitation. Being a lover of prose rather than of poetry, I am not sure that I can entirely agree with this, but possibly as a positive utilitarian I ought to do so, at least if 'importance' tends to maximization of happiness.

Acton has a curious objection to my brother's example of the benevolent (negative utilitarian) world exploder. Acton wonders whether the world exploder might be making a terrible mistake by ignoring the distinction between eliminating suffering and reducing it.[15] But surely eliminating is a case of reducing – the best case of all, the negative utilitarian would say. In suggesting that eliminating is not reducing, Acton seems to me to be like a person who says that zero is not a number. I am also a little puzzled by Watkins's remark that the pain minimizer would have to destroy the *very possibility* of life.[16] If the sentient forms of life were totally destroyed, it might be that the sentient forms would be most unlikely to evolve. This is on the supposition, held by some experts, that the evolution of higher forms of life on earth depended on a lot of lucky accidents. If this is not the case, then the benevolent world destroyer should ensure that *all* forms of life are destroyed, even bacteria and plants and insects, but should this be impossible the world destroyer might have at least ensured a pain free globe for hundreds of millions of years to come. In any case my brother's example was of a world *exploder*, and I think this would ensure the destruction of *all* life on earth. Of course there might be sentient life on planets of distant stars. No doubt the world exploder can do nothing about this, even with the resources of a future physics, but his or her negative utilitarian duty would not be to do the impossible, but would be to minimize suffering as much as lies within his or her power.

In conclusion, let me reiterate that I am much impressed by Watkins's sketch of a methodology of ethics. If his views on this are correct I do not think that any form of utilitarianism can be sustained. In suggesting a different methodology, I tentatively indicate how a utilitarian might sincerely put forward his or her ethical principle, despite his or her own tendency to backslide into deontological behaviour. This possibility depends on a hierarchical moral psychology, which I have not explored here.[17] Which is the correct moral psychology is still an open question.

Australian National University

NOTES

[1] H.B. Acton, 'Negative Utilitarianism, I', and J.W.N. Watkins, 'Negative Utilitarianism, II', *Aristotelian Society Supplementary Volume* **37** (1963), 83–94 and 95–114.
[2] R.N. Smart, 'Negative Utilitarianism', *Mind*, **67** (1958), 542–3. He refers to K.R. Popper, *The Open Society and its Enemies*, 2nd Ed. revised, London: Routledge and Kegan Paul, 1952, vol. 1, chap. 5, n. 6(2) and chap. 9, n.2.
[3] *Op. cit.*, p. 86.
[4] Sidgwick remarks that "...any particular existing moral rule, though not the ideally best for existing men under existing circumstances, may yet be the best that they can be got to obey", *Methods of Ethics*, 7th Edition with a Foreword by John Rawls, Indianapolis: Hackett Publishing Company, 1981, p. 469. Sidgwick also points out that "on Utilitarian principles, it might be right to do and privately recommend, under certain circumstances, what it would not be right to advocate openly", *ibid.* p. 489.
[5] Oxford University Press, 1973. See pp. 20–1.
[6] On observation conditionals see W.V. Quine, *Theories and Things*, Cambridge, Mass.: Harvard University Press, 1981, p. 26 and pp. 70–1.
[7] I am not sure of the exact relations between Watkins and Popper here. Because of Popper's liking for revised and yet more editions, postscripts and addenda, interpretation of Popper seems to be a full time job, perhaps only to be done by those brought up in the LSE.
[8] The word 'telish' was introduced by John Rawls. See his 'Two Concepts of Rules', *Philosophical Review*, **64** (1955), 3–32.
[9] On this point see also J.J.C. Smart, 'The Methods of Ethics and the Methods of Science', *Journal of Philosophy*, **62** (1965), 334–5.
[10] See J.J.C. Smart and Bernard Williams, *Utilitarianism: For and Against*, Cambridge University Press, 1973, pp. 98–9.
[11] Cambridge University Press, 1967, See especially Chapter 17.
[12] In my paper 'Utilitarianism and Punishment', given at the International Conference on Justice in Punishment, Jerusalem, March 27–31, 1988. See also J.J.C. Smart,

'Benevolence as an Over-Riding Attitude', *Australasian Journal of Philosophy*, **55** (1977), 127–35, reprinted in J.J.C. Smart, *Essays Metaphysical and Moral*, Oxford: Basil Blackwell, 1987.

[13] On this question see Sidgwick, *op. cit.* pp. 418–9, and Derek Parfit, *Reasons and Persons*, Oxford: Clarendon Press, 1984, especially Part 3.

[14] See J.J.C. Smart, 'Distributive Justice and Utilitarianism', in John Arthur and William H. Shaw (Eds.), *Justice and Economic Distribution*, Englewood Cliffs, N.J.: Prentice-Hall, 1978, reprinted in J.J.C. Smart, *Essays Metaphysical and Moral, op. cit.*

[15] Acton, *op. cit.* p. 84.

[16] Watkins, *op. cit.* p. 96.

[17] But see Monro, *op. cit.*

GREGORY CURRIE

METAPHYSICAL INDIVIDUALISM

An essay on the metaphysics of the social sciences is appropriate for a volume that honours John Watkins. For John has contributed significantly both to the programme of methodological individualism and to the conception of metaphysics as continuous with and productive of scientific theorizing.[1]

Like good science, good methodology needs firm metaphysical guidance if it is not to lose its way. If we want to be clear about what counts as a genuinely individualistic mode of explanation we ought to take a closer look at the metaphysics of individualism. We need to ask: What propositions about the relation of mind, action and behaviour to society are fundamental to the individualistic outlook?

One such fundamental proposition takes the form of a supervenience thesis, about which I have had something to say on other occasions.[2] If the thesis is false individualism can have little chance of success, for it would follow that society was in the strongest sense *independent* of the individual.

'Supervenience', as I use it here, is a term with little intuitive content. A rough formulation of the supervenience thesis is this: The facts about society could not be different unless the facts about individuals were different. Equivalently: Where the facts about individuals are the same, the facts about society are the same also. Call this thesis *S*. If *S* is false we should have to allow that two situations (conceived of broadly, so as to take into account all relevant surrounding context) might be the same so far as the facts about individuals go, yet different in some social respect. And the social respect in which they differed could not be thought of as having any explanation in individualistic terms. Facts about individuals could be pressed into service to explain only those social circumstances that are *common* to the two situations.

I shall try to improve on the formulation of *S* just given. But my aim is not merely increased precision. For the rough formulation may seem to beg the question at issue, by presupposing a distinction between facts about individuals and facts about society; a distinction that occupies centre stage in the debate between holists and individualists. In Section 2 I shall lay out the problem in greater detail, and motivate a solution to it. To solve the problem I

Fred D'Agostino & I. C. Jarvie, Freedom and Rationality: Essays in Honor of John Watkins, 47–65.

shall need recourse to an idea of Carnap's about the interpretation of theories (Section 3). The solution will be seen to depend for its acceptability on a thesis – another supervenience thesis in fact – in the philosophy of mind (Section 4). Finally I look at some apparent counterexamples to *S* that require a further reformulation; this time one that introduces counterfactual states into the calculation (Section 5). I begin, however, with a brief introduction to the idea of supervenience.

1. SUPERVENIENCE

The positivists hoped for a science unified by intertheoretic reductions down to the level of fundamental physics. But everything above the chemical level has proved resistant both to reduction and to the positivist's alternative: the revoking of cognitive status. Recently it has been suggested that we can think of science as unified by a weaker, non-reductive relation: supervenience. As the grip of positivism weakened so did the imperative to speak in the formal mode, and supervenience has come to be thought of as a relation capable of holding not merely between theories but also between kinds of events, properties and propositions.[3] In what follows I shall move easily between talk of theories, predicates, and properties. And I shall make use of the possible worlds of modal semantics. This is meant to render perspicuous appeals to the notions of *what might have been, what must be*, etc. Those who see nothing but metaphysical confusion in the idea of possible worlds will, if they are charitable, grant my formulation a heuristic value.

Supervenience theses come in various shapes and sizes, though by taking a sufficiently generous approach to the closure conditions on property kinds we get some surprising equivalences between them, as well as some unlooked for reductions.[4] I shan't try to assess the significance of these results here. Instead I shall focus on one formulation that seems appropriate for the purposes I shall presently introduce and in terms of which I can pose the problem to be solved.[5]

If **A** and **B** are two property-kinds then the claim that **B** supervenes on **A** will be formulated in this way: if two possible worlds are alike in the distribution of **A**-properties they will be alike in the distribution of **B**-properties. (Let us then call the **A**-properties *base* properties and the **B**-properties *supervening* properties.) More formally:

$$(1) \qquad \forall\omega\forall\upsilon\,[\forall x\forall A\varepsilon\mathbf{A}\,(A_{\omega}x \equiv A_{\upsilon}x) \supset \forall y\forall B\varepsilon\mathbf{B}\,(B_{\omega}y \equiv B_{\upsilon}y)].$$

where '$A_w a$' is to be read: 'object a has property A in world w'.[6] If we think of facts and events as property instantiations (1) could equally well be read as asserting a relation between kinds of facts or events. (Throughout I suppress the reference to time, for the sake of simplicity.)

I shall now make some comments on the chosen formulation. (1) is rather restrictive in the sense it accords to the phrase 'alike in the distribution of **A**-properties' (or whatever properties we have in mind). That is, it allows worlds to be alike in this sense only when the same objects have the same **A**-properties in both worlds. And we might wish to allow that likeness is preserved by the permutation of objects; to move, that is, from the *identity structuring* across worlds of (1) to the more permissive *isomorphic structuring* across worlds of

$$\text{(2)} \qquad \forall\omega\forall[\forall x\forall A\varepsilon\mathbf{A}\,(A_\omega x \equiv A_\upsilon f(x)) \supset \forall y\forall B\varepsilon\mathbf{B}\,(B_\omega y \supset B_\upsilon f(y))],$$

which differs from (1) in that objects are not compared with themselves across worlds but rather with their images under some mapping f, where f is 1–1 from the underlying domain onto itself. While (2) is perhaps preferable to (1), it will be altogether simpler to develop the argument in terms of (1), and that is what I shall do.

We could ask, further, how supervenience might be formulated in terms of a relation between worlds that have domains of different cardinalities; a question that naturally arises if we allow that there might have been more, or fewer things than there actually are. But this deep and perplexing issue I shall have to put aside for now.

Notice that our formulation (1) allows for both **A**- and **B**-properties to be restricted in their application. So, to consider a relevant example, it would be natural to say that banks have social but not individual properties, that asocial persons (if there are any) have individual but not social properties, that molecules have neither. Any two things lacking **A**-properties will have the same **A**-properties, and hence will contribute positively to the antecedent of (1). We might on the other hand choose to regard property kinds as closed under negation and to say, consequently, that everything has, of necessity, some **A**-properties, for any property kind **A**.[7] Instead of saying that a molecule has no individual properties, we shall then say that it has no *positive* individual properties but only negative ones. Thus a molecule has the individual property of not believing that today is Thursday, and so on; the distribution of individual properties it gets is what I shall call the *null distribution*. If two objects get the null distribution of **A**-properties they have *the same* **A**-properties and again will contribute positively to the antecedent of (1). So the issue need not concern us here.

Let us now consider a slightly different way of formulating (1). Given a possible world w we can speak of the **A**-history of w; the set $\mathbf{A}(w)$ of temporally determinate propositions made true in that world by (dated) **A**-property instantiations. And we can speak similarly of the **B**-history, $\mathbf{B}(w)$ of w. Now two distinct worlds may have the same **A**-histories (as long as they differ in other ways). (1) says that in that case they have the same **B**-histories. In other words there is a function Φ such that

$$\Phi(A(w_i)) = B(w_i, \text{ for all } i\varepsilon I, \text{ where } I \text{ indexes the class of worlds.}$$

We can now see a sense in which the thesis S is very weak. It asserts the existence of the function Φ; is says nothing about the properties of that function. For example, it might be desirable to impose a condition on Φ corresponding to something like the mathematician's notion of uniform continuity. Suppose that w_1 and w_2 are very similar in their distribution of **A**-properties (but not quite the same). For all the supervenience thesis says, w_1 and w_2 might be very different in their distribution of **B**-properties (in which case Φ would not be uniformly continuous.) Perhaps we need a thesis of uniform continuity to add to our thesis of supervenience: worlds individually close are socially close. Perhaps there are arguments for saying that the relevant function is not uniformly continuous for at least some choices of **A** and **B**; perhaps there are arguments for saying that the question cannot be given a precise formulation, because the relevant notions of similarity are not well defined. I would like to know which of these is true; but here I just want to emphasize that S leaves the question quite open.

However, (1) does require supplementiation of another kind if it is properly to express a supervenience thesis. Suppose now that we are considering the thesis that **B**-properties supervene on **A**-properties, and that w is a world which gets a null distribution of **A**-properties *everywhere* (that is, every object in w gets the null distribution of **A**-properties at all times). If the distribution of **B**-properties depends upon the distribution of **A**-properties we should say that in w the distribution of **B**-properties is also everywhere null. (To consider the example that concerns us here: a world in which there were no acting, thinking persons would be a world without (positive) social characteristics.) But our formulation of supervenience in (1) does not tell us this. It allows w to have any distribution of **B**-properties, as long as that distribution is shared by any other world which also has an everywhere null distribution of **A**-properties. So we must add the stipulation: any world with an everywhere null **A**-distribution is a world with an everywhere null **B**-distribution.

By adding this stipulation we avoid the charge that we have too easily shown that we might call *invariant* properties – properties distributed in the same way in every world – to be dependent on every other property kind. Consider mathematical properties; being two, being prime, being a continuous real function. If mathematical objects are 'necessary beings' these properties are distributed in the same way in every world. So any two worlds automatically satisfy thesis (1) if we take the **B**-properties to be mathematical and the **A**-properties to be any properties whatsoever. The sense in which mathematical properties are *not* dependent is exactly the sense in which they violate the additional stipulation just given; worlds with null distributions of at least some other properties are non-null in their distribution of mathematical properties.

Sometimes supervenience theses are accompanied by identity theses of various kinds. Thus Davidson claims that mental events are not merely supervenient upon physical events but that every mental event (token) is identical to some physical event (token). Should our thesis *S* be accompanied by a thesis of event identity between the individual and the social? Is each event (token) of a social kind identical to some complex of events of an individual kind? I think not. It is a metaphysical fantasy to suppose that it is determinate of every individual event whether or not it is constitutive of, say, the event; the Allies defeating Germany. Exactly which individual act of turpitude initiated the decline of the Roman Empire? The relation of particular individual events to particular social events is going to be indeterminate in some places. But this should give little comfort to social holists. For, as John Haugeland has argued, there is similar indeterminacy within purely physical systems between the micro and the macro levels.[8]

According to *S*, social property distributions supervene on individual property distributions. Probably there are other kinds of properties upon which social properties supervene. Perhaps everything, society included, supervenes on the physical (the post-positivist's thesis of unified science). There may be a hierarchy of supervenience relations; chemistry supervening on physics, biology on chemistry, the mental on biology. The individual level would then be an amalgam of the mental and the physical (behavioural) facts about persons, with society, semantics and value-facts (if there are any) supervening thereon.

We can think of this hierarchy as a structure partially ordered by the relation of *immediately supervening* upon (the ancestral of which is the relation of supervening upon). Intuitively **B** supervenes immediately upon **A** if it is not possible to interpolate another property kind **C** between **A** and **B** such that **C** supervenes on **A** and **B** on **C**. **A** then constitutes what we might call a *minimal supervenience base* for **B**. So an interesting question we can ask concerning any given property kind **B** is this: what is the (or perhaps a) minimal base for **B**? We are concerned with this question as applied to social properties. So I begin with the not implausible hypothesis that the minimal base for this class of properties is the class of what I have called individual properties – properties that enjoy a non-null distribution over rational agents and which are descriptive of their mental and behavioural states. Our problems begin at this point.

2. INDIVIDUAL PROPERTIES

If we substitute 'individual properties' for **A** and 'social properties' for **B** in our formulation (1) we have a precise version of the thesis *S*. (1) now says that a kind of property – social properties – depends for its pattern of instantiation on another kind – individual properties. So we don't want our *supervenience base* (the properties quantified over in the antecedent of (1) to contain any properties that are themselves social; only by excluding them from the base can we establish their dependence on something else. So we need an assurance that the properties that belong to the base – individual properties – are not also social properties. At this point we might ask what being a social property amounts to. But this is a question to which there doesn't seem to be any uncontroversial answer.[9] Perhaps we can agree on something less controversial: that a property is social if (but of course not only if) its instantiation presupposes the instantiation of properties and relations of an intuitively social kind. That which presupposes social arrangements is surely social itself. Granting this, there seems to be trouble for *S*, because many of the properties that are natural candidates for being individual properties turn out to be social by this criterion. Thus methodological individualists sometimes speak of the social as explainable in terms of the *actions* of individuals. But many actions are such that the property of *performing that action* will count as social. To vote, attend a meeting, or to purchase shares will require various social arrangements to be in place. Stalnaker voices this objection to a supervenience thesis of the kind I'm presenting when he says that 'an adequate description of the actions of individuals requires reference to institutions and institutional action'.[10] We might hope to get out of this difficulty by retreating to a 'thin' notion of action describable in terms of mental states and movements of the body, together with causal connections between them. But the appeal to mental states is itself problematic if we have reason to think that any mental properties presuppose the instantiation of social properties. That the mental is, on the contrary, autonomous has been a dogma of philosophy from Descartes to Frege; according to classical epistemology the identity of a subject's thoughts depend entirely on how things are with the subject; it is only when we assess the truth value of his thoughts that we need to look at his environment. Recent challenges to Cartesianism have emphasized the ways in which subjects intrinsically alike (alike, say, in being indistinguishable at the molecular level) might have thoughts with quite different contents.

Thus we are invited to consider cases in which a subject's physical and social environment is varied while his own internal constitution is held constant, and to conclude that in these varied environmental conditions the contents of the subject's thought would vary also.[11] *Singular* thoughts about social entities provide ready examples of this kind. My thought that The Bank of New Zealand is charging too much interest is not a thought shared by my twin on distant Twin Earth; for he has never heard of The Bank of New Zealand. Yet he and I are intrinsically indistinguishable. It is, moreover, a thought I can have only if The Bank of New Zealand exists, and that requires, of course, the instantiation of complex social relations.

One might argue that singular thoughts are special in some way; eliminable, perhaps, in favour of constructions involving only conceptual contents. Or one might argue that, really, there are no social entities that can be the subjects of singular thoughts. I don't see anything promising in either of these lines of defence, but it is worth asking whether social dependency holds also for fully conceptualized content. There are interesting arguments that we might appeal to here; we might begin by asking where a being would *get* a system of complex and interlocking social concepts if not through transaction with a social environment. We might appeal to the thought experiments of Tyler Burge, which are designed to show that community norms concerning things like deference to linguistic experts are among the determinants of a subject's conceptual repertoire.[12] We might even be tempted by a radical anti-realism about content of the kind advocated by Kripke in his 'sceptical solution' to the problem he finds in Wittgenstein.[13] For if we follow Kripke we shall have to say that there are no facts about what a person means by saying something, and hence, presumably, no facts about what he believes or desires; otherwise these facts would surely generate facts about meaning. And if there are no facts about content there can be no facts about content that form part of our supervenience base. We could, in sympathetic response to this argument, say that we need simply to adjust the supervenience thesis to fit an account of content-attributions in terms of assertibility conditions instead of an account in terms of truth conditions; Kripke's 'sceptical solution' involving, as it does, the idea that statements of content are assertible without being true. We should then say that worlds are content-indistinguishable if and only if what is assertible about content in the one is assertible in the other (retaining the truth conditional story for other individual property attributions). But the assertibility conditions for attributions of meaning essentially involve, according to Kripke, reference to the social community; what is assertible about content in *w* will be so only in virtue of facts about the community in *w*. This will make content attributions social by our criterion.

One might not admire Kripke's argument very much; nor any of the arguments that I've referred to so briefly here. But I think its worth bearing two points in mind when considering this issue. First, since I want to defend the supervenience of the social on the individual, I find it instructive to assume a worst case outcome in which content depends strongly on the social, and then to ask whether there is a viable form of individualism that survives the disaster. Needless to say, I think there is; otherwise I wouldn't be taking this heroic course.

Secondly, and more positively, we might ask: what has content to do with it anyway? Consider our community, both as it is and under certain counterfactual circumstances, but keeping fixed behaviour and the ways in which our internal processes function to generate behaviour. Now there are two possibilities; either the contents of mental states can vary across these two situations or they can't. If they can't, the functional–behavioural story determines content and we don't need to specify content independently as part of our supervenience base. Suppose, on the other hand that content can vary across these situations. How could these variations in content, unaccompanied by functional-behavioural variation, affect the social relations that individuals bear to one another or to institutions? Contentful states that are not manifested can hardly have any social impact. If content makes no independent contribution to the social we can dispense with it; and we ought to dispense with it if we are looking for a *minimal* supervenience base.

These remarks are apt to provoke two worries. First, in view of what I have just said about manifestation, wouldn't behaviour alone be a sufficient supervenience base? In that case my talk about the internal-functional states that generate behaviour would be otiose. Secondly, is there not some significance here for the distinction between actual and potential behaviour? I'll deal with each of these issues in the final two sections.

What we immediately need, and what the next section will provide, is a way of dispensing with talk about the content of mental states while leaving us with enough information about the status of rational agents to meet the constraints on a supervenience base for the social.

3. GETTING RID OF CONTENT

I am going to suggest a method by which we can arrive at the kind of properties that will form our supervenience base in the required way. Starting with a detailed psychological account of an arbitrarily chosen person, including a full account of the contents of his mental states, I shall show how we can derive a weaker account that is still psychological but now neutral as to content.

Suppose we focus on a subject Jones, and give a comprehensive account of his mental life. (Since we are doing metaphysics we can adopt an omniscient perspective and ignore awkward questions about how we come to know so much about Jones.) Our account will involve specifying Jones's (dated) propositional attitudes together with a full account of the internal *dynamics* of his mental life. That is, we specify the relations between Jones's beliefs and desires and his perceptual states – how his states change under pressure from other states as they are acquired – and how his mental life feeds into behaviour, which we describe non-intentionally in terms of 'colourless' bodily movements (recall that we may not invoke properties like voting, cashing cheques, etc.). Thus our account is a partly propositional and partly functional account of Jones's mentality; it specifies Jones's propositional attitudes, together with the functional roles that these attitudes play in Jones's life – their relations to each other and to his behaviour. What we end up with is a theory – a true one – about Jones. The objects referred to in the theory are Jones, his mental and behavioural states, and the theory says that these states stand in certain relations to each other and to behavioural states.

What we need to do now is to extract from this theory a purely functional component which, together with comparable theories for all of Jones's peers, will form (a provisional version of) our supervenience base. How are we to perform this extraction?

There is a method, due to Carnap, for extracting a purely empirical component from a scientific theory.[14] The method depends upon a device of Ramsey's,[15] who suggested that we replace all the theoretical terms $t_1 \ldots t_n$ in a scientific theory

$$(T)\ T[t_1 \ldots t_n],$$

by existentially quantified variables, to obtain the corresponding Ramsey Sentence:

$$(R(T))\ \exists x_1 \ldots \exists x_n T[x_1 \ldots x_n],$$

where our original theory T said that the entities $t_1 \ldots t_n$ stand in certain relations to one another and to the observable phenomena, *R(T)* says merely that there are n entities that stand in these relations. The theoretical terms $t_1 \ldots t_n$ we may call the *T*-terms of our language, and the rest of the vocabulary we may call *O*-terms ('*O*'- for 'observational', since Carnap's idea was to formulate, with *R(T)*, a 'purely observational' version of the theory).

T and *R(T)* are related in the following way: *R(T)* is strictly weaker than *T*, being entailed by it but not entailing it. And both entail exactly the same *O*-sentences (sentences containing only *O*-terms). *R(T)* has, therefore, fair claim to embody the observational content of *T*.

R(T) expresses a part of what is expressed by *T*. What expresses the other part? What can we conjoin to *R(T)* to get *T* and no more? Carnap pointed out that if we conjoin *R(T)* with

$$(C(T))\ \exists x_1 \ldots \exists x_n T[x_1 \ldots x_n] \supset T[t_1 \ldots t_n],$$

(the conditional that has as its antecedent the Ramsey sentence of the theory and as consequent the theory itself) called the 'Carnap sentence' of the theory *T*, we have something logically equivalent to *T*. Carnap proposed that we take *R(T)* as the synthetic part of our theory, and *C(T)* as the analytic part; analytic in the sense that it gives a partial specification of the meanings of the *T*-terms. *C(T)* tells us that, in any world where the theory is observationally correct, the *T*-terms refer to those entities which satisfy the open sentence $T[x_1 \ldots x_n]$. For any world where the theory is not observationally correct, *C(T) tells* us nothing about the referents of the *T*-terms.

I wish to take over the Ramsey–Carnap method, with some changes of application and interpretation, for my own purposes. My strategy will be to extract from the theory of Jones I described a few paragraphs back a Ramsified sub-theory which, I claim, is neutral as to content and which, I claim, can serve, along with Ramsified versions of comparable theories of Jones's peers, as our desired supervenience base. Recall that our theory of Jones attributes to him a number of mental states, some of them propositionally individuated, and goes on to describe in full detail their interconnections and their connections with his behaviour. Let our theory be written as

$$(J)\ J[p_1 \ldots p_n],$$

where the p-terms denote propositionally individuated mental states. The rest of the vocabulary that occurs in the theory is either logical, or signifies non-propositional mental states (pains, tickles, etc), or components of Jones's internal functional organization, or behavioural states of Jones. The corresponding Ramsey sentence

$$R(J))\ \exists x_1 \ldots \exists x_n J[x_1 \ldots x_n]$$

gets rid of problematic references to content. Where $J[p_1 \ldots p_n]$ might tell us among other things that Jones believes *that is the cat that bit him* and desires *to kick the cat that bit him*, and that this belief and desire of Jones (in connection with his other beliefs and desires) cause him to move his body in a cat kicking way, *R(J)* tells us that there are two states that Jones is in, connected with each other and with other states and that the result of these connections is that Jones moves his body in a certain way. While *J* identifies these states as particular propositional attitudes, *R(J)* is silent about their content.

(*R(J)* is a theory about Jones. The name 'Jones' appears, we may assume, in its formulation. Let that name be replaced by a free variable. Now we have a predicate that names a property possessed by Jones, but possibly by others as well. When I say that *R(J)* is a theory true of Jones and possibly of others, I mean strictly speaking that it names a property possessed by Jones and possibly by others.)

Because *J* is a theory that specifies the contents of Jones's states it is a theory that is false of Twin-Jones, the well known inhabitant of Twin Earth, who is intrinsically indistinguishable from Jones but who has states with different content, the differences in content being the result of environmental differences. But *R(J)* is true not only of Jones but of his twin also. For Twin-Jones has states (distinct in content from Jones's) that are causally related to each other and to his behaviour just as Jones's are. To put it rather crudely, *R(J)* is the psychological history of Jones, minus the facts about the contents of his mental states.

4. FUNCTIONALISM AND INTELLIGENCE

The methodology of the previous section has obvious affinities with functionalism about the mind; the theory that mental states can be identified in terms of causal role. Thus believing a certain proposition will be, for the functionalist, a matter of being in a state with certain kinds of inputs and outputs. A standard technique of functionalists is the Ramsification of 'folk-psychology', a supposedly uncontroversial and quotidian theory about mental states and their roles in our lives. Each mental state is then defined via the Ramsification.[16]

But the similarities between definitional functionalism and my own proposal do not go very deep. The theory that I Ramsify is not a *general* psychological theory that might be constitutive of psychological concepts; it is a highly contingent theory (really just a description) of the adventitious psychological history of a particular individual; a history that *could* be true of some other individual (Twin-Jones, for example) but which would be so only by some amazing coincidence. Our theory *J* tells us what Jones's total mental state is. It tells us what roles his propositional attitudes occupy; not what roles they occupy in others, or might occupy in Jones under other circumstances. It tells us nothing definitive or essential about those attitudes. So it is quite consistent with the method proposed here to think that mental states do not have any *essential* connections with particular roles. All one has to suppose is that each mental state occupies some role or other (different roles in different worlds perhaps).

Functionalism is usually understood as an attempt to define mental states in terms of their typical causal roles; that is how I have explained it in contrasting it just now with my own programme. But a weaker functionalism would eschew definition in favour of *supervenience*; holding to the principle that creatures indistinguishable in functional organization and behaviour would be mentally indistinguishable as well. Now recall what was said above about the status of the Carnap sentence of a theory. As we saw, Carnap proposed to treat the Ramsey sentence as embodying the synthetic claim of a theory, while the Carnap sentence is a partial specification of the meaning of the theory's 'theoretical-terms', and is thus analytic. What we might call 'narrow supervenient functionalism' is just the view that the Carnap sentence of *J*

$$C(J))\ \exists x_1 \ldots \exists x_n J[x_1 \ldots x_n] \supset J[p_1 \ldots p_n]$$

(and of all comparable theories) is, if not analytic, at least a necessary truth. For narrow functionalists believe that anyone functionally-internally identical to Jones would be identical in all mental respects to Jones.[17] So they hold that anyone, in whatever circumstances, of whom *R(J)* was true, would be someone of whom *J* itself was true, and hence that it is not possible for the Carnap sentence to be false.

Now I make no such claim about the way our theory of Jones breaks down. While our theory *J* decomposes into a Ramsey sentence and a Carnap sentence, I take the Carnap sentence to be as synthetic and as informative a part of the theory as is the Ramsey sentence. It says that if Jones's states are related as *J* says they are then those states are $p_1 \ldots p_n$. This claim could, so far as I'm concerned, be false of Jones; that is, *R(J)* could be true of Jones but made true by his possession of propositionally individuated states other than $p_1 \ldots p_n$. (Recall that I'm assuming, for present purposes, that what propositionally individuated states Jones is in is not determined by his internal states alone.) Thus our decomposition of *J* is a decomposition into propositional and functional parts, not into analytic and synthetic (or necessary and contingent) parts.

The functionalism that says that content supervenes on the sort of functional story summarized in our theory *R(J)* I have called *narrow* functionalism; the states described in *R(J)* are intrinsic to the subject and say nothing about his relations to the environment. *Broad* functionalists will argue that to get mental sameness we must pack a bit more functional story into *R(J)*, including functional connections that extend beyond the subject's sensory surface.[18] They will hold, therefore, that the Carnap sentence of *J* is synthetic; it's possible for beings with distinct mental contents both to satisfy it. But they look forward to finding some more embracing functional account, that involves inputs and outputs outside Jones's body, where this functional account will fix the contents of Jones's states. I am neutral for the purposes of this paper about the truth or falsity of broad functionalism.

(We could recapture something of Kripke's position here by saying that, according to Kripke, the only part of *J* that has fact-stating potential is *R(J)*, and that the Carnap sentence is to be understood as telling us that if *R(J)* is true then *J* is assertible. *C(J)* is, on this view, a partial specification of the assertibility conditions of *J*. Clearly there are parallels here between Kripke's position and scientific instrumentalism.[19])

While, as I hope I have just made clear, I reject the idea that the content of Jones's states is determined by his internal-functional organization (while, in other words, I reject the necessity of *C(J)*), I want to claim that there are important facts of a psychological kind about Jones that are so determined. In fact this claim is an important motivation for the account I have been developing here. Let us see why. The hypothesis currently being considered is that the properties I've called 'individual' properties and which purportedly suffice for our supervenience base are those properties of individuals that we specify when we specify, for each individual, a theory comparable to our theory *R(J)*; 'functional theories' we shall call them. Worlds the same in respect of the functional theories that are true in them will be the same in respect of the distribution of social properties. But isn't there overkill here? Wouldn't a complete specification of *behaviour* for all individuals (something included in the functional theories) be sufficient? I've already said that differences in content unaccompanied by differences in behaviour are socially insignificant. Why, then, isn't behavioural sameness enough to guarantee that these worlds are socially the same? Because behaviour alone does not determine whether the behaving creature is intelligent. Worlds can be behaviourally the same while in one of them the behaving creatures are intelligent and in the other not. And in a world where the creatures are not intelligent – where they are just 'automata' or whatever – there aren't any social objects or relations. Society is a set of relations between intelligent creatures, capable of intentional, rational behaviour. (Ant colonies are social only by courtesy. Or, if there is a sense of 'social' according to which ant colonies can be said to be genuinely social, that is not the sense I'm interested in here.) In that case behaviour alone does not give us a supervenience base. But perhaps behaviour, together with a guarantee that the behaving creatures are fully intelligent beings, does give us a supervenience base. Two worlds in which beings with the same kind and degree of intelligence behave in the same way would be worlds socially indistinguishable from one another. If there were differences between the worlds from the point of view of the actual thoughts had by the creatures in them, they would not be differences that showed up at the social level. Society requires intelligence, so it requires something more than is determined by behaviour alone; but it is insensitive to differences in content not reflected in differences in behaviour.

No doubt intelligence is a matter of degree. Perhaps it is a notion subject to indeterminacies of application. But such variability and blurred edges need not worry us if we hold to the following supervenience thesis:

(3) Creatures indistinguishable as to functional organization are indistinguishable as to intelligence.

Consider Jones. Say what you like about his intelligence – say that he is intelligent to degree *r*, that he is more intelligent than Smith, that he exhibits some intelligence, that there is no fact of the matter as to whether he is intelligent or not – as long as you say the same thing of any creature functionally indistinguishable from Jones; that is, any creature of whom *R(J)* is true.

If (3) is correct our functionally specified supervenience base will be sufficient – so it seems – to give us a distribution of social properties. But is claim (3) true?

I don't claim to have a proof that it is. But it seems to me that the burden of proof lies with someone who wants to deny (3). Thus it seems to me at least very plausible that a creature indistinguishable in behaviour and internal functional organization from an intelligent creature is intelligent. Behaviour alone, even dispositions to behave, are insufficient to guarantee characteristically human mentality, as Ned Block's super robot programmed with a huge but finite set of stimulus–response patterns shows.[20] But Block's robot differs in functional-internal ways from an intelligent human even if its behaviour is the same; it apes intelligence by brute force, searching vast areas of memory for an appropriate response rather than by having sophisticated techniques of learning, problem solving etc. (Since our functional account of Jones is not motivated by definitional functionalism we can make it as detailed as we like, descending to the level of algorithms or heuristics for carrying out sub-tasks in a way that a definitional functionalist could not without incurring the charge of rampant chauvinism. At least at this lower level, we will find differences between Jones and the robot.) If, on the other hand, the robot was so constructed as to have an internal functional organization just like that of Jones, I think we would have every reason to regard it as an intelligent, rational being.

If this line of thought is correct, then in specifying functional theories for our individuals, we specify all that we need to determine the social states of the world.

5. GOING COUNTERFACTUAL

Our aim was to find a genuinely individualistic base for our supervenience thesis; a base of properties each of which can be instantiated without the necessity for social arrangements to be in place. Have we found one now? Not, of course, in our theory of Jones alone. Jones is but one individual; nothing very interesting in the social world supervenes on our Ramsified theory of him. But every cognizing individual comes with a similar theory. Put them all together and we have at least a candidate for what we are looking for – a mind-bogglingly vast set of properties possessed by all the individuals there are or were.

But is all this information going to be sufficient to give us the supervenience base we need? Note that our theory *J* of Jones (from which we derived the Ramsified version, *R(J))* is a theory about Jones's occurrent states. It tells us how things are, as a matter of fact, with Jones and his peers; not how they might have been under other circumstances. There is a much richer story to be told about how things would have been different with Jones if any of his beliefs or desires or experiences had been different. Telling that story would give us a vastly more complex theory of Jones, call it 'J_c'. The complexity of it need not worry us of course; we are doing metaphysics, not epistemology and so we may assume an omniscient perspective.

Do we need to tell the more complicated story in order to get an adequate supervenience base? I think we do. If we confine ourselves to occurrent states of individuals we may fix occurrent social states, but we shan't fix counterfactual ones. We may have determined the fact that at a certain time the Bank rate is 3%; we shall not have determined the fact, if it is one, that if the bank rate had been 10% at that time the mortgage rate would have been 5% higher than it actually was. But this last counterfactual is as much a candidate for being a social fact as any occurrent fact is. Presumably what this counterfactual, along with all others of social significance, supervenes on is the class of true counterfactuals concerning individual states. We had better think, therefore, in terms of Ramsifying J_c instead of the weaker *J*.

There is another reason why we need to go counterfactual. If we considered only occurrent functional states as part of our base, our thesis *S* would tell us that worlds indistinguishable in terms of these occurrent states are socially indistinguishable. But I think that there are world-pairs occurrent-indistinguishable at the individual level, but such that one of them has a social environment while the other has none; it has what I have called an everywhere null distribution of social properties. Consider the actual world

@, and a world *w* indistinguishable from @ at the level of occurrent individual states. In @ we notice a *coherence* between the states of individuals. If one individual speaks, individuals nearby will respond as if they heard and understood; if one individual lashes out with his fist individuals in range will duck, or complain about the injuries they receive if they don't. In *w* we notice the same coherence between individuals since *w* is indistinguishable from @ at this level. Both @ and *w* seem to be worlds that sustain richly social interactions between people. But suppose that this coherence in *w* is entirely accidental. *w* is a world in which, by some means or other, the life histories of individuals unfold quite independently of one another. By some incredible chance, they all unfold in such a way that it looks, for all the world, as if individuals are interacting with each other. In that case the coherence between individuals will not carry over to what is counterfactually true. We can say of @ that if Jones had struck out, Smith *would have* done such and such, where such and such is an appropriate thing to do when you are being attacked. We cannot say the same thing of *w* , since *w* is a world of accidental coherence; it is not true in *w* that if I had acted differently you would have responded differently – in *w* your responses don't 'track' my actions. In *w* people appear to interact, but it is appearance only. Since I think that interaction between people is a necessary condition for society, I hold that *w* is a world without a social environment. Since @ and *w* are occurrent-indistinguishable at the individual level, and socially so dissimilar, it is clear that our supervenience base must go counterfactual in order to distinguish them.

Note that a world in which Berkeley's philosophy is true is not a world like our asocial world *w*. In a Berkeleyan world coherence is not accidental; it is divinely maintained. Thus the sorts of counterfactual relations true in @ are true in the world as Berkeley conceived it to be; the difference is only in the nature of the causal channels. In a Berkeleyan world there is a social environment just as there is in @ (so at least we suppose).

Thus by going counterfactual in our consideration of individual states an additional advantage accrues: we enable ourselves to distinguish between @ and *w*, and so we won't fall into the trap of supposing that @ and *w* are socially indistinguishable.

SUMMARY

In this paper I have suggested that a basic insight into the metaphysics of the social world might be the thought that social properties are supervenient. And I tried to determine what a most minimal, most immediate class of properties it is upon which they supervene. My claim has been that such a class is the class of properties the distribution of which we specify when we specify the (occurrent and counterfactual) functional–behavioural states of individuals.

University of Otago

NOTES

[1] For the first see especially 'Ideal Types and Historical Explanation', *British Journal for the Philosophy of Science,* **3** (1952), 22–43, and 'Historical Explanation in the Social Sciences', *British Journal for the Philosophy of Science,* **8** (1957), 104–117. For the second see 'Between Analytic and Empirical', *Philosophy,* **32** (1957), 112–131, 'Confirmable and Influential Metaphysics', *Mind,* **67** (1958), 344–365, and 'Metaphysics and the Advancement of Science', *British Journal for the Philosophy of Science,* **26** (1975), 91–121.

[2] See my 'Individualism and Global Supervenience', *British Journal for the Philosophy of Science,* **35** (1984), 345–358, and 'Realism in the Social Sciences', in R. Nola (Ed.), *Realism and Relativism in Science,* Dordrecht, Kluwer, 1988, 205–228.

[3] Davidson's claim that the mental supervenes on the physical is given in terms of events ('Mental Events', in *Essays on Actions and Events,* Oxford University Press, 1980, p. 88.) Kim formulates supervenience as a relation between property kinds ('Supervenience and Nomological Incommensurables', *American Philosophical Quarterly,* **15** (1978), 149–165, see especially p. 149; 'Causality, Supervenience and the Mind-Body Problem', *Midwest Studies in Philosophy,* **4** (1979), 31–50, see especially p. 41). Hellman and Thompson give a formulation in terms of elementary equivalence between models ('Physicalist Materialism', *Nous,* **11** (1979), 309–345). John Haugeland gives a formulation in terms of worlds as distinguished by the sentences held true in them ('Weak Supervenience', *American Philosophical Quarterly,* **19** (1982), 93–103).

[4] See e.g. John Bacon, 'Supervenience, Necessary Coextension, and Reducibility', *Philosophical Studies,* **49** (1986), 163–176.

[5] The appropriateness of this formulation is argued for in 'Individualism and Global Supervenience'.

[6] Throughout I use the term 'object' to refer to entities in the underlying domain of first order quantification, reserving the term 'individual' for *persons* – a subclass of that domain.

[7] Kim takes property kinds to be closed under infinitary boolean operations (see e.g. 'Supervenience and Nomological Incommensurables').

[8] For Haugeland's argument see his 'Weak Supervenience'. See also Stalnaker,

'Events, Periods and Institutions in Historians' Language', *History and Theory* (1967), 159–179. Philip Pettit, previously an advocate of supervenience plus identity, now expresses doubts about the identity thesis (see his 'In Defence of "A New Methodological Individualism"', *Ratio*, **26** (1984), 81–87).

[9] For an analysis of social properties that owes much to the Griceans see David-Hillel Ruben, *The Metaphysics of the Social World*, London, Routledge & Kegan Paul, 1985, Chapter 3.

[10] *Inquiry* (1984), p. 162.

[11] The *locus classicus* of this position is Putnam's 'The Meaning of Meaning', in *Philosophical Papers*, Volume 2, Cambridge University Press, 1975. See also various works by Tyler Burge: 'Individualism and the Mental', in *Midwest Studies*, **4**, 1979; 'Other Bodies', in A. Woodfield (Ed.), *Thought and Object*, Clarendon Press, 1982; and 'Intellectual Norms and Foundations of Mind', *Journal of Philosophy*, **83** (1987), 697–720.

[12] See the work of Burge cited in note 11.

[13] Kripke, *Wittgenstein on Rules and Private Language*, Oxford, Basil Blackwell, 1982.

[14] See Rudolph Carnap, *Philosophical Foundations of Physics*, New York, Basic Books, 1966. David Lewis gives an interesting exposition and variation in 'How to Define Theoretical Terms', in *Philosophical Papers*, Oxford University Press, 1983, volume 1.

[15] See F.P. Ramsey, 'Theories', in H. Mellor (Ed.), *Foundations*.

[16] See David Lewis, 'Psychophysical and Theoretical Identifications', *Australasian Journal of Philosophy*, **50** (1972), 249–58.

[17] See e.g. Jerry Fodor, *Psychosemantics*, Cambridge, Massachusetts, Bradford, 1987.

[18] See e.g. Frank Jackson and Philip Pettit, 'Functionalism and Broad Psychology', forthcoming.

[19] As McGinn suggests. See his *Wittgenstein on Meaning*, p. 66.

[20] See Block, 'Psychologism and Behaviourism', *Philosophical Review*, **90** (1981), 5–43. Robert Kirk gives another example of a behaviourally sophisticated but non-sentient robot ('Sentience, Causation and Some Robots', *Australasian Journal of Philosophy*, **64** (1986), 308–321). Kirk's robot, like Block's, differs functionally-internally from a sentient human.

Block's argument has other interesting ramifications. For example it undermines the Wittgensteinian claim that use is the authoritative criterion of understanding; that 'if the use criterion is adequately met then nothing could rationally persuade us that this was not a case of understanding' (McGinn, *op. cit.*, p. 111). Block's robot does as well as any human so far as use is concerned, but it does not understand.

DAVID-HILLEL RUBEN

THE ONTOLOGY OF EXPLANATION*

John Watkins and I seem to be equally out of fashion in our shared views on explanation. There is no shame in this, since fashions change but the truth is forever. I, like him, cannot make such sense of the idea of probabilistic or statistical explanation. I do not share his rejection of inductive argument *per se*, but since I agree with him that "the occurrence at a particular time of a chance event simply defies all explanation",[1] I hold that no such non-deductive argument whose conclusion mentions a particular event could count as an explanation of that particular event, however highly probable, given the premises, that conclusion is.

However, what I want to discuss in this paper is an issue about explanation that is rather neglected by Watkins, and by many others who have written about explanation.[2] It seems to me that it is important to get clear on the ontology of explanation, before developing a theory about it, for many issues that arise in the treatment of the subject depend on the ontological decisions one makes. If we think of explanation as a relation, what sorts of entities are its relata? Are they events tout court, as Watkins seems to suggest when he asserts that, if indeterminism is true, we "cannot provide deductive explanations of single events" (p. 228)? Sometimes, for Watkins, the explanandum is an event's having a specific (objective) probability, which seems to make the explanandum not an event, but an event's having a certain feature. What sort of entity is this? A situation, a state of affairs, a fact, or what? Perhaps there is no exclusive choice involved here; sometimes explananda can be particular events, and sometimes they can be an event's having a specific feature or characteristic. (I use 'explanans' and 'explanandum' ambiguously in what follows, to speak both of the things that do the explaining and get explained, as well as the sentences about those things that stand in deductive, or other logical, relations. The idea of a (set of) sentence(s) explaining a sentence is merely derivative; it can only mean that whatever the first (set of) sentence(s) is about explains whatever the second sentence is about.)

I do not suppose that, if this were all that was at issue, Watkins would find the problem very challenging. Whether explananda are events, or events' having features, the important thing about explanation for him is that explanation is 'non-psychologistic'. For Watkins, whether one event, or an

Fred D'Agostino & I. C. Jarvie, Freedom and Rationality: Essays in Honor of John Watkins, 67–85.

event's having a feature, explains another is *objective*, in the sense that it does not depend on how we humans happen to conceptualise or think about the world. Indeed, on the standard Deductive–Nomological account of explanation, one could explain an explanandum unwittingly, without knowing that one had explained it. If I produced premises meeting all of the requisite D–N conditions, from which the conclusion that the explanandum occurs deductively follows, I have explained the explanandum even if I am not aware that I have done so or even if I think that I have not explained it. The standard D–N account of full explanation takes no notice of the epistemic state of the persons involved in giving or getting the explanation. To borrow a distinction from general epistemology, the D–N account of explanation is an 'externalist account', rather than an 'internalist account'. Whether or not an externalist account of knowledge is possible, is an externalist account of explanation on the cards?[3]

Again, like Watkins, I am drawn to an objective account of explanation, as free as possible from any epistemological or other human-dependent intrusion. What I believe, and hope to show in this discussion of the ontology of explanation, is that we cannot pull off an account of explanation that frees it entirely from reference to how we conceptualise or think about the world. Of course, the account proposed here isn't entirely subjective either; it is not just a matter of whatever we happen to think is explanatory being explanatory. But, in the end, considerations of how we think about things play an important role, in addition to the role played by the various 'objective' features in an account of explanation.

What are events, and in particular what is the criterion for individuating and identifying events? On this topic, I should here like to say as little as possible. But I must say something. I think of an event as the exemplification by an object of a certain kind of property, namely, a dynamic property, at a time. I assume, without argument, that event identity does not require the identity of the properties instantiated. That is, event e can be identical with event f, even when the properties instantiated by e and f differ. I accept the view that the event e, say, o's being P at t, can be identical with event f, say, o's being Q at t, even though the property $P \neq$ the property Q. I note that this view is controversial, and that what I have to say in what follows depends on this view of event identity.

In what follows, questions of extensionality will arise. Let me start by introducing some terminology with which to discuss questions of extensionality. Extensionality is a single idea, but the question of extensionality arises for different expressions. As Haack puts it,[4] "A context is extensional if co-referential expressions – singular terms with the same denotation, predicates with the same extension, or sentences with the same truth-value – are substitutable within it without changing the truth-value of the whole, '*salva veritate*', i.e., if Leibniz's law holds for it; otherwise it is intensional". In what follows, I use 'intensional' and 'non-extensional' synonymously. Let's call the first sort of truth-preserving substitutability of singular terms with the same denotation 'transparency'; the second sort, substitutability of predicates with the same extension, 'predicate extensionality'; the third sort, substitutability of sentences with the same truth value, 'sentence extensionality' (this is the same as truth-functionality).

There is a fourth sort of extensionality that we shall need. Consider two predicates, '*P*' and '*Q*' which express or stand for the same property (whatever the criterion of property identity the reader prefers, perhaps weaker than synonymy but surely stronger than co-extensionality). Call any pair of such predicates 'co-intensive predicates'. A context is 'co-intensive predicate extensional' if there is substitutability *salva veritate* of predicates which are co-intensive. Notice a feature of my use of 'co-intensive predicates'. In my sense, two predicates are co-intensive if they stand for or express the same property in this world, whether or not they stand for or express the same property in all possible worlds, and *a fortiori* whether or not they have the same meaning. Red, I take it, is the colour of fire engines. This is a contingent property identity, true in this world but false in lots of possible worlds, namely in those in which 'the colour of fire engines' does not stand for the colour red. Co-intensive predicates need not express the same property in all worlds. At most, we might argue that if two predicates are both names of properties, then, if the predicates express or stand for the same property in this world, then they stand for the same property in all possible worlds. Even this last claim is controversial, and I do not wish to commit myself concerning it. (In '*x* is red', 'red' does not name at all, and *a fortiori* does not name the property, red. I call 'red' a name of red, to distinguish it from descriptions of red, like 'the colour of fire engines'.)

THE RELATA OF THE EXPLANATION RELATION

Explanation, let us assume, is a dyadic relation (perhaps it is triadic: persons explain things by means of other things. If so, let us treat explanation simply as if it were dyadic, since we are here uninterested in the ontology of persons). It hardly needs saying that I do not, in what follows, try to offer some complete account of explanation. I try to limit my discussion to the question of the relata of the explanation relation. However, even within this reduced ambit, a question that I do not wish to discuss is the place of laws in explanation. When I speak of an event explaining an event, let the idea of 'explaining' therein expressed be understood as 'at least in part explaining'. Perhaps laws, or patterns of regular occurrence, need to be added to the ontology of explanation, I omit any discussion of this issue here. I shall speak of two 'candidates' for the relata of the explanation relation: the first is events; the second, facts. I do not deal separately with the mixed possibilities, on which facts might sometimes explain events and events sometimes explain facts.[5] Moreover, I do not assume that the choice between the candidates is exclusive; both candidates might be acceptable.

If we accept events as the relata, then at least sometimes, explanation is by a particular event tout court of a particular event tout court. Or, if one wishes, explanation is at least sometimes of the occurrence of one particular event by the occurrence of another. These come to the same, there being no difference, as far as I can see, between explaining the fire at 10 Downing Street and explaining the occurrence of the fire at 10 Downing Street. This proposal for the relata of the explanation relation can be somewhat broadened to include states, in addition to events, to cover the sort of case in which we might want to say that some unchange, to borrow Ducasse's apt expression, explains or is explained. Further, one might also wish to add actions, or dispositions, or thoughts, to the list of explaining or explained items. Whatever the precise list of items to be admitted, the important point is that the first candidate should be some sort of particular worldly item, some individual chunk of the world, quite apart from how we conceptualise it. For simplicity, I shall continue speaking only of events explaining events, leaving it to the reader to understand that other wordly particulars or individual chunks might do as well.

Two writers who accept that there are explanations of events by events are David Lewis[6] and James Woodward.[7] David Lewis' theory that all explanation is causal explanation is restricted to the explanation of particular events; in reply to an alleged counterexample, he asserts that "I don't agree that any particular event has been non-causally explained" (p. 223). So there must be explanation of particular events, if Lewis is right.

James Woodward argues that there is a sense of explanation, "...causally explains...", which is transparent. In this transparent sense, the explanation relation does hold between events apart from how they are described. "If what is explained is genuinely an individual occurrence, then the singular term occurring in the effect position will function purely referentially..." (p. 279). Woodward's view is supported by some important distinctions, like that between what an explanation explains and what it presupposes but does not explain. "...[T]here is a clear difference between explaining why some particular event, which has a certain feature, occurred, and explaining why that event has that feature." To take Woodward's example, if the short circuit causally explains the fire, and if the fire was purple and odd-shaped, then the short circuit explains the purple, odd-shaped fire. (Notice that Woodward's point here presupposes a rough-grained criterion of event identity: the event of the house being on fire and the event of the house being on purple and odd-shaped fire can be identical, in spite of the fact that two different properties are involved in the two descriptions). The expression, 'the purple, odd-shaped fire' functions purely referentially, to refer to the fire. The explanation does not explain why the fire was purple and odd-shaped; it merely presupposes that the fire has these characteristics, and reference to the fire is being fixed via this description. Nor, argues Woodward, should we try to escape the transparency point by thinking of the explanation of the fire as really an explanation of the fact that there was one and only one event in the vicinity which was a fire. To explain the latter, I would have to explain why no other fires occurred at the same time in the vicinity. In explaining the fire by the short circuit, it is presupposed that there was only one such fire, but this presupposition is not explained by the short circuit. So this transparent sense of 'explains' is a relation between particular events, occurrences, things in the world, apart from how they are described or referred to. There is no need to import facts, or statements, or anything else into the analysis. In explaining the fire by the short circuit, I explain the occurrence of an event in the world, apart from any linguistic or cognitive considerations about how it is talked about or conceptualised.

The difficulty I have with the Woodward–Lewis view of event explanation is that such explanation sentences simply are not fully transparent. I remind the reader that the topic under discussion is that of explanation, which includes causal explanation, but not the topic of causation itself. Causation might be a transparent relation, even if causal explanation is not. If '*e causes f*' is transparent, it does not follow that the corresponding assertion of explanation, '*e* causally explains *f*', must be.

Consider the following argument:

(1) The hurricane explains the loss of life.

(2) The hurricane = the event reported in *The Times* on Tuesday.
∴ (3) The event reported in *The Times* on Tuesday explains the loss of life.

Or, to pick an example adapted but slightly altered from Mackie:[8]

(1) Oedipus's marrying his mother explains the tragedy that ensued.
(2) Oedipus's marrying his mother = Oedipus's marrying the woman he thought least likely to bring tragedy to Thebes.
∴ (3) Oedipus's marrying the woman he thought least likely to bring tragedy to Thebes explains the tragedy that ensued.

What Mackie says here is this: the first description "helps to explain the tragedy in a way that the ...[latter] does not. What is referentially transparent may be, for that very reason, explanatorily opaque".

If the conclusions of these two arguments are false, the best diagnosis of what has gone wrong in the argument is that, despite appearances, the first premiss of each argument is not to be construed as stating that a relation obtains between two events, as the first stab at the ontology of explanation proposed. It is not enough that some substitutions of co-referring terms preserve truth, as they do in Woodward's examples; all must do so, if the explanation in question is to be taken as a singular event explanation. One should take the falsity of the conclusion to throw doubt on the conception that particular events (or states, or whatever) explain events. Explanation never touches things that directly and immediately, but is mediated through the features or characteristics which are appropriate for explanation. Oedipus's marrying his mother is the same event as Oedipus's marrying the woman he thought least likely to bring tragedy to Thebes, but the event is explanatory as conceptualised in the first way, but not as conceptualised in the second way.

Are the conclusions really false, because the one thing doesn't explain the second at all, or are the conclusions literally true but just poor explanations? This raises a very complicated question about the nature of such a distinction, but let me say here just enough to produce what I hope will be an adequate underpining for the thought that the conclusions of the arguments are literally false.

There are pragmatic considerations in giving explanations. First, it is legitimate to give less than the whole explanatory truth to an audience, given one's knowledge of its interests and existing knowledge. The audience may only want to know about some part of the full cause. Or it may not want a very specific description of that cause, but only a more general description – the audience might need to know only that a plague in tenth century China reduced the population level, but not need to know what the plague was. Or, there may be two levels of relevant description of the cause, a macro-description and a micro-description for example, and one rather than the other of the descriptions might be appropriate for the audience one was addressing.[9] Poor explanations are ones that make these choices between relevant descriptions ('relevance', let us suppose, being cashed out for us by the theory of explanation at hand) or between parts of the relevant description, in the wrong way. But none of these pragmatic considerations covers the case of the introduction of an outright irrelevant description, whose connection to the topic at hand can only be seen in a derivative and parasitic sort of way by someone who happens to know the appropriate identity. To select inappropriately from the relevant information is a poor explanation; to offer explanatorily irrelevant information is to fail to explain at all.

The conclusions of these arguments *are*, then, false. I grant that the person who knows the identities might feel reluctant to count the conclusions as wholly literal falsehoods, since there is a sense in which he can see what the conclusions are trying to get at. The conclusions mention events which, *when differently described*, do explain. But the theory of singular event explanation is, as it stands, false, in the sense that explanation, even causal explanation, can never just be of events by events.

Although, as far as I know, Watkins does not use the expression 'the fact that...' when discussing explanation, other writers slip naturally, and sometimes unconsciously, into that usage. Facts are the second candidate for the relata of the explanation relation: 'the fact that *p* explains the fact that *q*', where typical expressions referring to facts contain whole sentences, here represented by '*p*' and '*q*'. Adopting facts as the relata might still yield a theory of singular explanation, if facts, like events, are a sort of individual entity, in the widest sense. But it is worthwhile marking off a fact theory of explanation from an event theory, for reasons that will emerge in the course of the discussion. If facts are, like events, things that are denizens of the world, then a fact theory would be as objective an account of explanation as the one offered by the event theory.

Other accounts of explanation seem to require that the relata of the explanation relation be linguistic or quasi-linguistic items: statements or propositions being the notable additional candidates. Peter Achinstein, for example, holds that an explanans always includes as a constituent a certain kind of proposition.[10] Davidson's account of explanation is the view that explanation relates *statements*: 'Explanations typically relate statements, not events' (p. 93).[11] Let's call Davidson's theory a 'statement theory of explanation'. Is this a third type of theory about the relata of the explanation relation? For Davidson, an example of an explanation in correct logical form seems to be this:

(8) *The fact that* Jack fell down *explains the fact* that Jack broke his crown.

"In...(8), intensionality reigns, in that similar substitution [i.e., of equivalent sentences or coextensive singular terms or predicates] in or for the contained sentences is not guaranteed to save truth" (p. 86). Note that Davidson's remarks about (8) should not be confused with his better known remarks about (2): "*The fact that* there was a short circuit *caused it to be the case that* there was a fire"; for example: "...substitution of singular terms for others with the same extension in sentences like...(2) does not touch their truth value" (p. 85).

Why does Davidson speak of statements explaining statements, when his (8) appears to speak of facts explaining facts? (8) does not speak of facts explaining facts at all. Davidson does not think of the relation in question as the explanation relation, whose relata are two facts, in this case, the fact that Jack fell down and the fact that Jack broke his crown. Rather, it is a different dyadic relation, the relation, the fact that – explains the fact that –, which relation (presumably for Davidson) relates two statements. The view seems cumbersome. If there are to be statements involved in explanation, why can't they be the relata of the explanation relation itself, rather than the relata of the relation, the fact that – explains the fact that–? Or are these really different relations at all? Moreover, I confess to not really being able to understand how a statement could explain a statement, or how the fact that some statement could explain the fact that another statement, except insofar as I understand that thought in terms of the 'fact' idiom which I develop below, and in which reference to statements drops out from mention altogether.

In a well-known paper, in which he discusses the correspondence theory of truth, Davidson inveighs against facts.[12] The gist of his argument is that if "'*s*' corresponds to the fact that *s*" is taken extensionally, as surely it must by someone trying to find worldly items to match up with bits of language, then if '*s*' corresponds to any fact, it corresponds to them all, indeed to The Great Fact which comprises all the facts that there are.

Unlike fact theories of correspondence, a fact theory of explanation need not take fact locutions as (fully) extensional. Fact theories of explanation can discriminate facts more finely than can an extensional theory which is committed only to the existence of the one Great Fact. Or so I will argue in what follows, as I develop the notion of fact. It is not a notion that Davidson would necessarily reject. Indeed, such facts are close in kind to the statements to which Davidson appears to commit himself in the quoted remark. Whether there is anything more than a terminological difference between a fact and a statement (or proposition) theory of explanation will rather depend on what facts turn out, on such a theory, to be like. My strategy is to start with facts, and move towards language or concepts only if necessary, rather than to start with statements or propositions initially. My thought is to push the idea of fact explanation being non-epistemic, not relative to the ways in which we think about or conceptualise things, as far as it will go, and hence to make it as 'objective' as possible. In the end, it fails to go far enough.

Let's return for a moment to Woodward's distinction. Suppose I want to explain the house's burning with a purple, odd-shaped flame, and not just to explain the house's burning, which burning happened to be a burning with a purple, odd-shaped flame. The explanation of the house's burning with a purple, odd-shaped flame demands something that is not equally demanded by the explanation of the house's burning. If we had agreed to the idea of event explanation, and if we were also prepared to adopt a very fine-grained criterion of event (and state) identity, on which two events (or states) are different if the properties exemplified are different, we could see this second explanation of the burning as an explanation too. On this fine-grained criterion, the house's burning and the house's burning with a purple and odd-shaped flame would count as two different token events, and it would be no surprise that their explanations are different. We could then refuse to accept that an explanation of the burning is an explanation of the burning with a purple and odd-shaped flame. We could say that the short circuit is the explanation of the house's burning; the presence of a certain impurity in the combustible material (say, titanium) is the explanation of the house's burning with a purple and odd-shaped flame, and that both explanations are event explanations, being explanations of (two different) particular events.

But I have rejected this fine-grained analysis of event and state identity, and the view that explanation can be of events. We need some other way to get at the difference between the two explananda, the house's burning and the house's burning with a purple and odd-shaped flame. The explanation of the house's burning with a purple, odd-shaped flame is best understood, according to me, as an explanation of the *fact* that an event, the house's burning, has a specific feature. Any concrete particular has an indefinitely large number of characteristics or features; what we want to explain may be not just the particular's occurrence, and certainly not its having all the features that it does have (an impossible task). What is needed is what I would call 'a feature introducing operator', which introduces a (usually short) list of features of the particular for which an explanation may be required. In 'the fact that the fire is purple and odd-shaped', three features seem to be introduced: that the particular is a fire, that it is purple, and that it is odd-shaped. Not every explanation explains all the features introduced by the fact locution, for context might make it appropriate to explain some feature and to ignore others. Selection of introduced features for explanation is a different matter. But features cannot be explained which are not introduced by the fact locution at all. The fire was, no doubt, a fire that occurred at some time, t, but the fact that the fire occurred at t is a different fact from the fact that it was purple and odd-shaped, and its occurring at t is not a feature mentioned in the latter fact. Aristotle's well-known terminology is designed for a similar purpose: it isn't Polyclitus the man who explains the statue, but Polyclitus *qua* sculptor. '*qua*' is Aristotle's feature introducing operator, 'the fact that...' is mine.

THE NON-EXTENSIONALITY OF FACTS

Are the identity conditions for facts predicate extensional? They are not, as the invalidity of the following argument makes clear:

(1) The fact that x is P = the fact that x is P.
(2) $(x)\ (Px \equiv Qx)$.
∴ (3) The fact that x is P = the fact that x is Q.

Since (3) is false (1) does not permit substitution of co-extensive predicates *salva veritate*. A similar argument shows that the identity conditions for facts are not sentence extensional (truth functional) either. Indeed, I take it that one moral of Davidson's Great Fact argument (not that he would put it in this way) is that, if there were facts, they could not be sentence extensional.

That the identity conditions for facts are not predicate extensional suits them for this feature-introducing role in explanation. The fact that x is cordate is a different fact from the fact that x is renate, even though the properties of being cordate and renate are co-extensional. Happily so, because an explanation of the fact that x is renate will not do as an explanation of why x is cordate. A feature introducer introduces a feature, without thereby introducing all other features co-extensive with the first feature. Fact locutions offer a means for introducing the property and making the property matter, in a way that events and states of affairs do not (or, do so only at best controversially). So one might wish to opt for facts as the relata of the explanation relation, to account for the presence of the features which matter in the explanation. Even Woodward's example of explaining the occurrence of the fire sans phrase can be construed as explaining the fact that the fire occurred, where the fact locution introduces the occurrence, but nothing else, as the feature of the fire which matters in the explanation.

FACT: WORDLY OR WORDY?

Some have argued that facts are a special kind of combined linguistic and non-linguistic item. J.L. Austin claimed, in his well known debate with Strawson on 'truth', that: "and so speaking about 'the fact that' is a compendious way of speaking about a situation involving both words and world".[13] My denial of the sentence and predicate extensionality of the identity conditions for facts does not suggest the involvement of words or conceptualisation in the constitution of facts. My remarks on the feature-introducing character of facts bring properties or features or characteristics onto the scene but do not bring words or concepts to the fore. In what way are facts on the side of the words (or concepts) at all, as Austin says?

The fact that the fire is purple introduces both the fire and the purpleness of the fire, whether for explanation or whatever. It isn't the word, 'purple' or 'fire', that is relevant, or even the concepts of fire and purpleness, but the real fire and its colour themselves, however described or conceptualised. In order to see the 'words' character of facts, I think one needs to look at the question of the co-intensive predicate extensionality and transparency of facts. Unlike Austin, I prefer to speak of the conceptual component rather than the words component of facts. It will turn out that what matters in explanation is how we conceptualise or think about something, not what words we use in the thought or conceptualisation.

THE CO-INTENSIVE PREDICATE EXTENSIONALITY OF FACTS

The identity conditions for facts are, then, neither predicate nor sentence extensional. But the questions of their transparency and co-intensive predicate extensionality are trickier. For each of transparency and co-intensive predicate extensionality, there are the following two separate questions: (1) are the identity conditions for facts transparent (co-intensive predicate extensional)? (2) Given the answer to (1), are facts so conceived suitable to be the relata of the explanation relation? Rather than impose a theory of explanation onto facts to begin with, to insure their suitability for explanation, I try to answer (1) on the basis of intuitions. My claim is that facts, if the deliverance of intuition is as I think it is, won't do the whole job required by explanation. I refer to such facts as 'ordinary facts'. I do not think it is of any great philosophical importance whether or not facts as ordinarily conceived are as I say. The main point is that we need entities as the relata of the explanation relation which will do the work that explanation requires. I continue to call such things 'facts', but 'in the special or epistemicised sense'. Such special facts might better be thought of as ordered pairs of ordinary facts, and concepts (or, conceptualisations). If the reader thinks that the deliverances of my intuition are in error, and what I call 'special facts' are what he thinks ordinary facts are like, I have no great objection.

Let me take the co-intensive predicate extentionality question first. Suppose property P = property Q. Are the identity conditions for facts co-intensive predicate extensional? If either 'P' or 'Q' are definite descriptions of properties, facts are not co-intensive predicate extensional. For example, red is the colour of fire engines. But the fact that x is red is surely a different fact from the fact that x is the colour of fire engines.

But I claim that the identity criteria for facts, on any theory of facts which is true to how facts are ordinarily thought about, *are* co-intensive predicate extensional, if both 'P' and 'Q' are names of properties. I take the relevant property identities under discussion here to be *a posteriori*, if 'P' and 'Q' are different names of the properties. As I mentioned before, it is controversial whether these identities are contingent or necessary, but a decision on this is unimportant for what I have to say here. In this case, if the property P = the property Q, then the fact that x is P is just the same fact as the fact that x is Q. How could the facts be different? There is no property or object or any other item which is a 'constituent' in one that is not in the other. The fact that x is P and the fact that x is Q, in such a case, differ neither over the individuals they are about nor over the features of the individuals that are introduced.

(There is also a question about the identity conditions for facts in those cases in which the property identities are *a priori*; for example, there is a question as to whether to count the fact that Harry is an eye doctor as the same fact as the fact that Harry is an oculist. I do not pursue this question, since it has no interesting ramifications for a theory of explanation.)

But now we must turn to the second question about the co-intensive predicate extensionality of facts. If facts as ordinarily understood are co-intensive predicate extensional with regard to names of properties, can such facts do what explanation requires? I assume that the following is an explanation (of a macrostate by a microstate): an explanation for the fact that a sample of ideal gas *b* has temperature *t* is the fact that its constituent molecules have mean kinetic energy μ. Such an explanation relies on the property identity, temperature = mean kinetic energy. There are *other* explanations, at other levels, for this fact. For example, another explanation of the fact that *b* has temperature *t* is that I stuck *b* in the oven at gas mark 4. But *an* explanation for its having temperature *t*, and the one most appropriate in certain scientific situations, is in terms of the mean kinetic energy of its molecules. In any case, if the reader is inclined to dispute that this is an explanation, let him take as an example whatever case he wishes, of an explanation of a macrostate by an underying microstate, such that the macrostate is reductively identified with the microstate. Reduction of one science to another has often been taken as paradigmatic of explanation; whether paradigmatic or not, such reductions must yield some examples of property identities which are explanatory.

Consider the following four claims:

(1) Nothing explains or even partly explains itself (the explanation relation is irreflexive).
(2) having temperature *t* = being composed of molecules with mean kinetic energy μ. (A statement of property identity, whether it be metaphysically necessary or contingent.)
(3) The fact that *b* has temperature *t* is explained by (or partly explained by) the fact that *b*'s molecules have mean kenetic energy μ.
(4) Facts are co-intensive predicate extensional.

But (1)–(4) are inconsistent, because (2)–(4) jointly imply that the fact that *b* has temperature *t* explains or partly explains itself, which (1) says is never so. A similar argument uses (2)–(4), but replaces (1) with (1*):

(1*) If one thing explains a second, the second cannot also explain the first (the asymmetry of explanation).

I take these two arguments as demonstrating that explanation is not just a relation between facts as worldly things, apart from how they are conceptualised. Nor is the point just that explanation is feature introducing, for if $P = Q$, the fact that x is P and the fact that x is Q introduce the same feature. What matters in explanation isn't only property introduction, but the way in which we conceptualise the property, viz., whether the property P is introduced as property P or as property Q.

If facts are thought of in the ordinary way, Austin was wrong. Ordinary facts as co-intensive property extensional entities are *not* combinations of words (or concepts) and the world. Even ordinary facts are not wholly extensional, it is true, since properties matter to them in a way for which predicate and sentence extensionality cannot account. But their co-intensive predicate extensionality makes ordinary facts unsuitable as the relata of the explanation relation. We need relata for that relation, for which not only properties matter, but the way in which we conceptualise or cognise the properties matters too.

It seems to me easiest to deal with this point by continuing to speak of facts as the relata of the explanation relation, but by remembering that we intend 'fact' in a special, 'epistemicised' sense, which may not accord with the way in which 'fact' is normally understood. Whether a fact as normally understood is explanatory or gets explained depends at least in part on the way in which the properties involved are conceptualised; relative to the conceptualisation of a property in one way, the fact may be explanatory, relative to a different conceptualisation of the same property, the fact is not explanatory. To this extent, it must be admitted that explanations are not fully independent of how we think about things. One might think of the special, 'epistemicised' facts needed for the explanation relation as ordered pairs, consisting of facts, understood as co-intensive predicate extensional, and concepts for, or conceptualisations of, the properties involved. In (1)–(4) above, (4) is true of facts in the ordinary, non-epistemicised sense; (3) is true only for the special facts which include a conceptual component. My special facts perhaps begin to look more like Davidson's statements, or Achinstein's propositions (although not just like them), but even so, I still think that the idea of facts explaining facts gets at whatever sense there is in the idea of statements explaining statements, or propositions explaining propositions, and not *vice versa*.

THE TRANSPARENCY OF FACTS

We still have to deal with the question of the transparency of facts (viz., the transparency of their identity conditions). Do locutions of the form, 'the fact that...', introduce any failures of transparency not already present in what fills the blank? Zeno Vendler argues[14] that if $o = i$, the fact that i is P is the same fact as the fact that o is P: "...facts are referentially transparent; propositions, even true ones, are opaque". One supposes that on Vendler's view, the fact that the hurricane struck at midnight = the fact that the event reported in *The Times* on Tuesday struck at midnight. If fact locutions are referentially transparent in this sense, this would seem to raise an additional difficulty for taking facts as the relata of the explanation relation, since surely, even though the fact that the hurricane occurred at midnight explains the fact that the ensuing tragedy occurred in the early hours of the next morning, the fact that the event reported in *The Times* on Tuesday occurred at midnight does not, even though the hurricane is the event reported in *The Times* on Tuesday. Explanatory weight can be carried by the referring terms as well as by the predicates.

We can see a certain ambiguity in Vendler's assertion. Consider the fact that Cicero denounced Cataline. Since Cicero and Tully are the same person, the fact that Cicero denounced Cataline and the fact that Tully denounced Cataline are one and the same fact. To repeat the argument from above: how could the facts be different? They involve the same individuals, and the same properties or features. There is nothing over which they could differ. Nothing is a constituent of one that is not a constituent of the other. Identity criteria for facts *are transparent, if* the facts use only proper names for the same individual in the referential position.

But the situation is otherwise if the individuals are denoted or referred to via definite descriptions (assuming that the definite descriptions do not use co-intensive predicates). Consider the fact that the greatest Roman orator denounced the greatest Roman scoundrel, I assume that Cicero was the greatest Roman orator; Cataline, the greatest Roman scoundrel. Now, the feature introducing power of 'the fact that...' introduces both Cicero but only *qua* greatest Roman orator and Cataline only *qua* greatest Roman scoundrel. Fact contexts are sensitive to the rigidity or otherwise of singular terms. When there is descriptive content to the term referring to the objects a fact is about, the feature introducing use of 'the fact that...' introduces for purposes of explanation the features in that descriptive content, exactly as it would in the case of features expressed by predicates. Thus, the facts differ, when the same particular is referred to by a proper name and when referred to by a definite description. The properties used in picking out the things referred to matter, in a way in which transparency can't get at. The fact that the greatest Roman orator denounced Cataline is not the same fact as the fact that Cicero denounced Cataline. When (non-synonymous) definite descriptions, or a proper name and a definite description, are used to fix reference to the same individual, then, there is no transparency. The fact that o is P and the fact that i is P, where $o = i$, are different facts if either 'o' or 'i', or both are (non-synonymous) definite descriptions.

How, then, shall we answer the second question for the case of transparency? Are facts so conceived suitable to serve as the relata of the explanation relation? The difficulty arises over the name transparency of facts. We have agreed that the fact that the greatest Roman orator denounced Cataline is a different fact than the fact that Cicero denounced Cataline. On the other hand, we have agreed that Tully denounced Cataline and Cicero denounced Cataline are the same fact. So the criteria for ordinary fact identity are not 'epistemicised' with regard to names of individuals, any more than they were for names of properties. But if facts lack this epistemic element, such facts will not be adequate for the requirements of explanation. Someone who does not know that Cicero is Tully can do nothing more with 'because Tully denounced him' as an explanation for the fact that Cataline fell from power, then he can do with 'because the greatest Roman orator denounced him', if he does not know that Cicero is the greatest Roman orator. So if we want entities suitable as relata of the explanation relation, our special facts should be ordered sets of ordinary facts and conceptualisations *both* of the properties and of the individuals who are the constituents of the facts. Cicero conceptualised as Cicero might explain, whereas Cicero conceptualised as Tully may fail to explain. These remarks perhaps constitute a general reason favouring the inclusion into the facts which will serve as the relata of the explanation relation the concepts under which the constituent individuals are cognised.

Apart from this general reason, is there a compelling argument for introducing this additional conceptual or cognitive element in the facts which will serve as the relata of the explanation relation? We were compelled to pronounce facts which were co-intensive predicate extensional as unsuitable for the explanation relation, because of the two arguments above, which relied on the irreflexivity and asymmetry of explanation. The same sort of argument is decidedly less compelling in the case of facts which are name transparent. We could try to repeat the sort of argument from above, but not using the example of Cicero = Tully. In order to produce a credible analogue of that earlier argument here, we need to find a case in which $x = y$, and the fact that x is P can genuinely explain the fact that y is P. We are unlikely to get cases of this unless we use macro-micro identities for individuals, and Cicero = Tully is not a case of such an identity.

(I don't think that the fact that Cicero was bald can explain the fact that Tully was bald. Of course, someone who is puzzled about Tully's baldness may have his puzzlement dissipated by being told that Cicero was bald, and that Cicero = Tully, but surely that is not to explain why Tully was bald. Not all resolutions of puzzlement are explanations; if I am puzzled about the height of Mont Blanc, my puzzlement vanishes when I am told its height, but I have not had its height explained to me.)

Moreover, since the question now before us is whether there is some further conceptual element required in facts over and above that introduced for properties, in order to make them suitable for explanation, I assume that I am dealing with special facts which have the way in which the property is conceptualised already built into them. Therefore, I cannot construct the case I want using two different names for the same property, as I did before; I need to use the same name of the property twice over.

Let 'b' stand for some table at time t, and let 'a' stand for the aggregate of molecules which make it up at time t. I assume, therefore, that $a = b$. Now, an argument analogous to the one above might look like this:

(1) Nothing explains or even partly explains itself (the explanation relation is irreflexive).
(2) $a = b$.
(3) For some property P, the fact a is P explains the fact that b is P.
(4) Facts are name transparent.

My hesitation about any such argument is that I cannot really think of any plausible property with which to replace 'P' in (3). But even if no such argument compels us to build into special facts the concept under which the individual is cognised, in addition to the concept under which the property is cognised, the general remarks above still suggest the plausibility of so doing.

Are the (at least partly) epistemicised facts required for explanation, to borrow an apt term from Stephen Schiffer, 'pleonastic' or 'non-pleonastic'?[15] So far, it might seem that they are to be taken as non-pleonastic. They are non-pleonastic, because I take the expression, 'the fact that p explains the fact that q; even when 'fact' is employed in the epistemicised sense, to have a relational analysis, and I take 'the fact that *p*' and 'the fact that *q*' as singular terms which refer to facts. (If one thinks of the fact that *p* and the fact that *q* as two intensional objects, and calls them '*a*' and '*b*', then the sentence '*a* explains *b*' is transparent, since any singular term that designates the same fact (understood in my special way) is substitutable *salva veritate*.) On the other hand, Schiffer connects up the idea of ontological commitment with that of existence which is "language-independent" (p. 145). That seems to me a conflation. The special facts required as the relata of the explanation relation may not be language-independent (or anyway, not conceptualisation-independent), any more than the existence of words is language-independent or the existence of concepts is concept-independent. So, this commitment to facts does not entail realism about them, in one obvious sense of that word. To this extent, since the facts I require for explanation turn out to be entities dependent on human conceptualisation and thought, any non-pleonastic sense of commitment to them would have to be qualified.

My argument shows that there is an epistemic requirement in explanation; facts explain facts only when the features introduced by the facts, and perhaps also the individuals the facts are about, are appropriately conceptualised. What counts as an appropriate conceptualisation? This introduces a new topic, which I cannot embark on here, except to say that there may be quite objective criteria for appropriateness in making this choice. To this extent, that explanation has an epistemic component is not by itself to allow any unwanted subjective or psychologistic features to creep into a theory of explanation. Perhaps on this basis, if no other, I might hope that Watkins would agree with the general thrust, even if not the details, of my argument.[16]

The London School of Economics

NOTES

* Thanks to Mark Sainsbury, Peter Milne, and Graham Macdonald, this paper is somewhat less confused than it might have been.

1 J.W.N. Watkins, *Science and Scepticism*, Princeton University Press, Princeton, U.S.A., 1984, p. 228.

2 A notable exception to this claim is Peter Achinstein, especially in his 'The Object of Explanation', in *Explanation*, S. Körner (Ed.), Blackwell, Oxford, 1975, pp. 1–45; and in his *The Nature of Explanation*, Oxford University Press, Oxford, 1983. I have learned more from Achinstein than from any other writer on explanation.

3 Jonathan Dancy, *Introduction to Contemporary Epistemology*, Blackwell, Oxford,

1985, pp. 46–49.

4 Susan Haack, *Philosophy of Logic*, Cambridge University Press, Cambridge, 1978, p. 246.

5 Zeno Vendler discusses the mixed case of facts and events as the relata of the causal relation. See Vendler, 'Causal Relations', *The Journal of Philosophy*, **LXIV** (21) (1967), 704–713.

6 David Lewis, 'Causal Explanation', *Philosophical Papers*, Vol. II, Oxford University Press, Oxford, 1986, pp. 214–40.

7 James Woodward, 'A Theory of Singular Causal Explanation', *Erkenntnis*, **21** (1984), 231–262; 'Are Singular Causal Explanations Implicit Covering Law Explanations', *Canadian Journal of Philosophy*, **16** (2) (June 1986), 253–279.

8 John Mackie, *The Cement of the Universe*, Oxford University Press, 1974, p. 260.

9 Hilary Putnam discusses a case in which there is both a geometric 'macroexplanation' and a 'microexplanation' in terms of the laws of particle physics, for the fact that a peg 1 inch square goes through a 1 inch square hole and not through a 1 inch round hole, in his *Meaning and the Moral Sciences*, Routledge & Kegan Paul, London, 1978, pp. 42–43.

10 Peter Achinstein, *The Nature of Explanation*, Chapters 2 and 3, *passim*.

11 Donald Davidson, *The Journal of Philosophy*, **LXVI** (21) (1967), 691–703; reprinted in *Causation and Conditionals*, Ernest Sosa (Ed.), Oxford University Press, Oxford, 1975, pp. 82–94. My page references are to the Sosa collection. Davidson actually holds that "the fact that – explains the fact that – " is a non-truth-functional sentential connective, but there is no need to enter into these refinements here.

12 Donald Davidson, 'True to the Facts', *The Journal of Philosophy*, **LXVI** (1969), 748–764.

13 J.L. Austin, 'Truth', *Proceedings of the Aristotelian Society*, Supp. Vol. **XXIV**, (1950), reprinted in *Truth*, George Pitcher (Ed.), Prentice-Hall, Englewood Cliffs, N.J., 1964, pp. 18–31. Quote from p. 24.

14 Zeno Vendler, *op. cit.*, p. 710.

15 Stephen Schiffer, *Remnants of Meaning*, MIT Press, Cambridge, Mass., 1987. See especially p. 51, Chapter 6 (pp. 139–78), and pp. 234–39.

16 Since writing this paper, I have had Barry Taylor's 'States of Affairs' drawn to my attention (in *Truth and Meaning: Essays in Semantics*, Gareth Evans and John McDowell (Eds.), Oxford University Press, 1976, pp. 263–284). For Taylor, facts are states of affairs which obtain. Taylor uses intensions as the predicative element he requires in constructing facts; whereas I use properties (for ordinary facts) and properties as conceptualised (for special or epistemicised facts). Clearly, there are similarities in our approaches. Taylor's facts are useless for a theory of truth (see p. 280). Taylor mentions other possible uses for states of affairs (and facts), but he does not mention their employment in a theory of explanation.

LEON J. GOLDSTEIN

REFLECTIONS ON CONCEPTUAL OPENNESS AND CONCEPTUAL TENSION

I

Readers of a *Festschrift* honoring J.W.N. Watkins will surely know that once upon a time there raged a hotly-debated argument over whether the concepts used in the social sciences are individualistic or not. Having not long ago re-read that debate as it was preserved by John O'Neill,[1] I find myself thinking that the non-individualist papers are hardly the worse for wear, and that the individualist ones no more persuasive now than then. But I have had my say on all that in the terms in which I understood the debate and have no desire to repeat again what I said all those years ago. I would, however, like to indicate something about the nature of the problem from a perspective that was very far from my mind in those earlier years when the debate was taking place, that is, from a perspective that I have come to think of as conceptual tension.

It is not necessary to insist that much of the philosophical tradition has tended to agree with Frege's saying that "a concept that is not sharply defined is wrongly termed a concept,"[2] but in recent times, apparently in no small measure owing to the influence of Wittgenstein, attempts have been made to say something about concepts which do not seem to be sharply defined, yet, for all that, seem to be concepts nevertheless. In particular, some writers have noticed that it is a feature of moral concepts to be flexible so that new applications in new circumstances bring out aspects of them that were not noticed heretofore.[3] Moral philosophy has never been one of my interests, and I cannot presume to speak with even the least degree of authority about it, but even I have noticed one thing about the way in which its discussions are carried on that may suggest something about why it should seem to be so suggestive a source of concepts which do not satisfy Frege's stricture, that are open rather than closed as we seem increasingly inclined to say.

What I have in mind is that in ethics both sides squabble over the same examples. One writer would cook up a case and present it in such a way as to underscore the soundness of the position to which he was committed: utilitarianism, say. His opponent would then take up the same case and show, at least to his own satisfaction, that it clearly pointed to the soundness of

Fred D'Agostino & I. C. Jarvie, Freedom and Rationality: Essays in Honor of John Watkins, 87–110.

deontology. How can this be done? I would suppose that what happens in the different treatments of what may seem to be the same example is that there are changes in emphasis that result from changes of context. One of the presumed features of open concepts is that they are affected by their contexts, and new contexts make it possible to be explicit about what might be conceived to be still unrealized potentialities of a concept; unrealized, that is, until it has been made to function in the new context. I do not wish to sound realistic about such potentialities, as if to suggest that there is some sense in which the concept actually contains potentialities which may yet – under appropriate circumstances – be realized, even though some of them might well never be. First of all, I don't think that such an idea has the least bit of intelligibility. And, second, if anyone could make sense of it, I suppose he would emerge with a view which would tend to subvert the idea of open concepts altogether. An open concept is said to be one which cannot be characterized definitively – with fixed and definitive criteria of application – but is, rather, one which changes with new applications. To suggest that all of its new faces were already contained in the concept as still unrealized potentialities is precisely to deny its openness to genuinely new situations. In principle an analyst with sufficient time and imagination could provide in advance accounts of all the possible situations, and thus provide a complete account of the concept to which nothing further could be added.

However, all this is really quite unintelligble. There are concepts which clearly reveal hitherto-unexpected aspects when applied to new circumstances, and it is that, I think, which emerges from the case-method of moral-philosophical discourse. The theorist attempts to make his point by considering new situations, and the moral concept applied to changing circumstances does, in fact, reveal its multiple character, a character that cannot be exhausted by some limited list of criteria necessary and sufficient for its application. Although I have no desire to pursue further the nature of ethical discourse, there is one thing that is suggested by the case-method of argument in that field that is worth indicating, particularly since it has some relevance to what I am trying to do here. And that is that there can be no pure explication of a concept in the analytical manner, as if, given the concept, it can be unpacked in order to bring forth the meaning it contains. Rather, the circumstances in which the concept is made to function – circumstances which are contingent with respect to the concept at that point – affect the character that it comes to have. One may suppose that one reason – I don't say that there is nothing else to it – for the disagreement among moral philosophers is that the different circumstances to which the concept in question is applied, affect the character of the concept. Thinking more broadly about concepts and their meanings, I incline to suspect that the appropriate question to ask concerning their analysis is not "How are they to be explicated?" but, rather "What are the elements that enter into their having the shapes they have?" That means that experience and history are constantly relevant to the analyses that interest us.

To say that history and experience enter into the formation of the concepts we use to characterize our world is hardly to say anything revolutionary. If Kant thought that the categories in terms of which we order our world are essentially unhistorical – being grounded in the rationality which is common to us all – we have long been able to counter that with alternatives, the formulation of C.I. Lewis of the way Kant's insight into the apriority of the structure of the world may be combined with sensitivity to its socio-historical character being perhaps unsurpassed.[4] Yet it might still be argued that however the apriori comes into being, once our conceptual tools come into being it must be possible to analyze our concepts in ways which make them clear, that is, in ways which make it possible to specify the conditions necessary and sufficient for their application. It is precisely this belief – deeply rooted in the practice of philosophy and science – that I find increasingly hard to credit. In this, of course, I am not alone. I have already noticed in passing the interest of moral philosophers in what is increasingly being called open concepts. W.B. Gallie has attempted to write about them as "essentially contested,"[5] and his paper on the subject has attracted the serious attention of some writers in the field of political science.[6] The late Morris Weitz has written at length on the matter,[7] and I must admit my awe at his seeming belief that while we are virtually at the beginning[8] of serious interest in such matters, it is already possible to classify them and to conclude that open concepts of specifically determinate sort belong to specifically determinate modes of expression. It is surely too soon to come to conclusions of that sort. I do not have any idea of the varieties of open concept there will prove to be, but some surely have features of the sort I shall be dealing with here, and that will have to be justification enough for what I am trying to do. I should like to suggest that there are experiences which are themselves so inherently tension laden that there is no possibility of doing justice to them unless we can introduce concepts with uneliminable tensions. I want to say two things, one descriptive and one prescriptive. The first is that there are in historical fact tension-laden concepts and that any attempt to relieve the tensions by means of philosophical analysis must result in the loss of the concept. The second is to say – or at least to suggest – that there may be some areas of interest which could not be dealt with except by means of tension-laden concepts, so we ought to be open to introducing such concepts.

In order to exemplify what I am trying to say, I want to offer a brief – no doubt inadequate – sketch of an example: the Old Testament concept of covenant.[9] There are some difficulties in doing this. If I want to say that it is the sort of concept the understanding of which requires that attention be paid to its history, one must immediately face the fact that Biblical scholarship doesn't make this very easy. Prevailing conceptions concerning the chronological order of the Biblical texts seem always to be changing, and so we do not have knowledge of what seems to be an obvious temporal sequence, the way, for example, we can hardly doubt that the Homeric epics preceded the odes of Pindar which, in turn, preceded the tragedies of Aeschylus and the dialogues of Plato. And matters are complicated by the recovery over recent decades of all manner of ancient-near-eastern treaties the terminology of which is quite similar to the terminology of Biblical covenant texts. In fact, more than terminology is involved, for in many cases the treaties and the relevant biblical texts are quite alike so far as structure is concerned: a brief introduction, a statement of the past relations of the parties to the treaty or covenant, the terms of the agreement, and, finally, an awesome list of curses to which the junior member of the agreement is subject were the agreement to be violated. It would certainly complicate matters were it to be the case that substantive elements of the Biblical covenant idea were imported together with these formal or stylistic aspects. I cannot claim to be an expert on the matter, but I shall, nevertheless, risk the wrath of those who know by suggesting that what I have read of the scholarly literature leads me to doubt that this is the case. I have no doubt that there are conceptual similarities between the covenantal relationship of the God of Israel to his people and the treaty relations of the Hittite emperor and his vassals, and one may suspect that it was sensitivity to this that led the Biblical writers to see the point of expressing the covenant relation in language drawn from the treaties. Yet I take it that fundamentally the two ideas are different, as different as one might expect ideas to be that are drawn from such different spheres as the religious experience of God and the political experience of power. In Joshua 24, Joshua seeks to determine whether or not the people of Israel will commit themselves to the covenant with God. As I read the text, the choice seems quite genuine, and Joshua goes as far as to suggest that owing to the difficulty of the terms they might well prefer to forget about the whole thing and to follow other gods. Surely, it is not to be believed that the vassals of the Hittite emperor had such a choice. Either they would sign and give tokens of their loyalty, or they would face consequences too horrible to contemplate.

There are quite a number of covenant texts in the Old Testament, but I shall not deal with them separately as if I were preparing a catalogue of them. They differ among themselves, yet I think there is developed in the course of Old Testament history a covenant idea to which they all contribute. For the present, it will suffice to make some general remarks about the idea; after that I may permit myself some speculation on how such an idea gets to be formed. A covenant is essentially an agreement between two parties. Each party is expected to do something, is expected to uphold its end of the bargain as it were, though there are texts which emphasize only the obligation of one side or the other. But in time, the covenant between Israel and its God comes to have two features which are rather unusual for a covenant: the covenant comes to be seen as a gift from God, an expression of his Grace toward the covenant people; and after a time – I cannot say when – the covenant becomes not subject to being abrogated. A covenant which is a gift would seem to be a conceptual monstrosity, and one may wonder whether or not we might apply suitable techniques of analysis in order to relieve the tension it seems to create. And an agreement that continues in effect regardless of the extent to which either or both of the parties to it seem to violate its terms is likewise conceptually odd, to say only that.

Perhaps there are some who would suggest that these conceptual tensions and oddities might be removed if only appropriate techniques of analysis might be applied. Perhaps some way might be found to separate the element of covenant from that of grace, and so emerge with two distinct ideas rather than retain one which seems inherently unstable. There is, however, one thing wrong with that proposal. Even if it were possible to carry it out, we would in the process lose that historically determinate idea which is the Old Testament concept of covenant. It would seem to me that the idea we are talking about takes the shape it takes in the course of the history and experience of Biblical Israel over a period of centuries. In some way, the concept as it matures comes increasingly to represent or encapsulate relevant aspects of Old Testament religious experience. I cannot characterize the most primitive form that the concept had, but, surely, from the start, that concept was sufficiently open so that the historical course of Israel's religious experience led to the slow shaping over the centuries of the Biblical concept of covenant.

In order to be somewhat clearer about what is happening to the concept in the course of time, let us go back to a brief consideration of the kind of concept that interests moral philosophers, and pay particular attention to what it is that brings out the openness of open concepts. If one thinks that there are no open concepts, one, presumably, approaches the analysis of what conecpts there are in a way which is designed to make explicit what the concept contains. What you say about its meaning is either contained in it, or it is not relevant. One who had doubts about the correctness of the analysis would, I suppose, seek to show that it contained points of irrelevancy, that meanings were being extracted which were not logically part of or contained in the concept at all. Defenders of the claim that there are open concepts defend the view that new circumstances reveal about open concepts facets hitherto undisclosed, which, in fact, could not have been disclosed prior to the use of the concept in the new circumstances. My own suggestion would be that the very shape of the concept is changed by the new circumstances, which circumstances are logically contingent relative to the concept. The history of a concept is not, then, the working out of what, at the start, it contains implicitly, but, rather, the consequence of the effects upon a concept of ever-changing circumstances which are logically contingent to the concept. In dealing with the openness of moral concepts, the moral philosophers seem not to attend to the histories of the ideas with which they deal, and perhaps there is no reason for them to do so. But in the case before us, we have no choice at all. I venture to suggest that the tensions and other oddities that seem to lurk within the Biblical concept of covenant are owing – entirely owing – to the specifically determinate character of the religious experience from which it arose, and, in consequence, there can be no analytical refinement of the concept that does not destroy it entirely.

II

There can be no doubting that the question of individualism has long been on the agenda of modern thought, and it is not really difficult to discern reasons for it. One knows, for example, that early modern science, prior to the development of field concepts, was atomistic, with explanations tending to be in terms of matter in motion. And natural science has tended to provide models for advanced thinking, with those who work in fields other than natural science seeking to make their activities accord with what is done in the physical sciences. Thus, it occasions no surprise to discover in Hume's *Treatise*, a work which purports to apply scientific method to the moral sphere, a thorough-going atomism. And if we turn our attention to the sociopolitical sphere, it is not unexpected to discover that it takes the form of construing the region to which attention is directed as being constituted out of individual human beings, the point being that individual human beings are taken to be the elemental components out of which societies are built up. One need only to think of the methodology or, indeed, the social ontology of Locke's *Second Treatise* or Hobbes's *Leviathan* to recognize that this is the case. The matter is further complicated by the slow growth during the course of modern times of the advocacy of political liberty. This is not the place to review the various forms that such advocacy has taken, but one persistent feature of it is the insistence that individuals have rights and that those rights are to be insisted upon in spite of claims made on behalf of the divine right of kings or the superior claims and interests of states and bodies politic. Again, one may think of Locke in this connection, and it is clear that the individualism of the *Second Treatise* is dual in character containing not only a methodological aspect as to the proper way to study society, but, even more insistently, a prescriptive claim in favor of the natural rights of individuals.

In the course of his brief sketch of individualism in modern thought,[10] Steven Lukes put forward the claim that the idea of individualism is actually an amalgam of ideas, that 'individualism' has different senses and nuances as it is used to discuss and advocate human dignity, autonomy or privacy, and as it functions in discussions of economic individualism, political individualism or religious individualism, among others. I do not wish to discuss these individualisms since they are not relevant to the theme of this study. But I do wish to examine a point of method that Lukes, influenced by A.O. Lovejoy, proposes as appropriate to the study of the idea of individualism. Though there have been exceptions, particularly in the present century, for the most part those who receive an education in philosophy are required to devote no small part of their time to the history of the subject. Thus, most of us who have been so educated come to think of ourselves as having some knowledge of the history of our subject, some sense of how the major thinkers – and the systems they wrought – followed one after another in time, being influenced by predecessors and, in turn, influencing successors. But most of us are not historians of philosophy, do not approach the subject systematically – the way we do the theoretical problems on which we actually work – and so there seems to develop what might be thought of as the professional tradition about the history of philosophy, rather more resembling the common-sense view of a subject in contrast to what disciplined research comes to tell us about it. I cannot speak for others, but I know of myself the sense of surprise and discovery when I learned that Descartes could not have invented the method of doubt inasmuch as he philosophized at a time when Pyrhonnism was a philosophical staple,[11] and that Locke could not have written the *Second Treatise of Government* to justify the Glorious Revolution because it was first drafted – though not published – before that revolution took place.[12]

One of the pioneers of the effort to define systematically the methodology of the history of ideas was Arthur O. Lovejoy, and before proceeding to the substantive task of his own book, Lukes seeks to justify what he does by reference to that scholar. Lovejoy's view is that though a doctrine is easily referred to by a particular term or name, hence creating the impression that it is a unity of a sort, in point of fact it is a compound, and the historian of ideas needs to develop techniques of analysis that will enable him to determine what the units are out of which it is synthesized.[13] Should there be any doubt that this is the case for the idea of 'individualism', Lukes calls upon the testimony of Max Weber: "The term 'individualism' embraces the utmost heterogeneity of meanings...a thorough, historically-oriented conceptual analysis would at the present time be of the highest value to scholarship."[14] It is clear that Lukes takes it that individualism is a pervasive tendency of thought in the modern west, but what he is not inclined to allow is that it is a large amorphous tendency. Rather, following the method of Lovejoy, he wants to claim that individualism as a doctrine or a tendency of thought is made up of elements, units of thought, and that the historian of ideas must separate out the different units from the whole and from one another in order to determine out of what the tendency is constructed.

I have no doubt that this seems to be a reasonable procedure, but we may well wonder why. This is particularly so inasmuch as the specifically determinate idea in its actual historical expression or expressions is the complex unity, not the sum of units. I shall return to this a bit further – at which point I shall have shifted away from the idea of individualism somewhat and toward the tension-laden notion of the unity of the individual and the non-individual – but first I want to make some remarks about the plausibility, or seeming plausibility, of the Lovejoy approach.

It requires no profound insight to recognize that Lovejoy's proposal that we analyze doctrines into the unit ideas out of which they are constructed is simply another instance of the atomism which is so pervasive in the history of thought and to which reference has been made above. The fact is that atomism is a pervasive feature of our intellectual life, part of the ideology of at least the English-speaking part of the world, at any rate among those in that world who think of themselves as tough-minded. I do not, of course, mean by 'ideology' political rationalization in the way that we have been taught to use the term by Marx, but, rather, a basic, ontological, if you like, idea as we find in the usage of such anthropologists as P.H. Gulliver.[15] So far-reaching is the principle of atomism, and so committed are so many of us to the primacy of particulars, that for its sake we come to deny what we know through experience and purport to believe in patent absurdities. Why, to give but one instance, should anyone ever have believed that material objects, such as tables and chairs, are not the substantial *wholes* they seem to be, but actually are little bits of ultimate stuff surrounded by lots of unexperienceable empty space? Furthermore, anyone who purports to think of something in holistic, non-reducible terms, tends to be looked at askance and dismissed as a mystic – to use a term that the tough minded find handy as a term of abuse. Thus, it does not matter how frequently the attempt to establish the reality of the other – the so-called other mind – by means of the argument from analogy fails, those who incline to favor it simply never understand – much less take seriously – those who see the presentedness of the other as not being the givenness of sensible parts, but, rather, something given as a whole. The fact that despite the keenness of his intuition in such matters, the tough-minded literature of the problem of other minds completely ignores Martin Buber is a case in point.[16]

Doctrines are complex and convoluted, and it isn't at all difficult to understand that scholars whose intellects are formed in an environment which virtually takes it for granted that the complexities of the world are to be rendered intelligible by techniques that discover their simple components – be they atoms or genes, electrons or chromosomes – should presume to explore the possibilities that that sort of procedure might offer to the attempt to render intelligible the history of doctrine. But for all the value of procedures of that sort, it is surely not the case that they have enabled us to explain all of the things that interest us. Atomic science does not make sense of the beauty of a landscape, or of a painting in which an artist attempts to capture that beauty. And Bergson is surely correct in observing that there is no possible collection of sketches of Paris that would provide for one who looks at them the experience of being in Paris.[17] Nor is it all that obvious – or at least as being beyond challenge – that atomic intellectual history will get us what we want when we confront the complexity of doctrines. On the face of it, the Old Testament idea of covenant seems rather an easy object upon which to apply the methodology we are considering, but we have seen that to analyze the Biblical idea of covenant in the suggested way is not to emerge with the unit ideas in terms of which the idea may be compounded again, but is, rather, to lose the idea entirely. It is Lukes's apparent agreement with the Lovejoy methodology – and his at least tacit acceptance of the atomism upon which it rests – that leads him to the task of disentangling the elements out of which the doctrine of individualism is said to be compounded, although it is interesting to note, in passing, that in that part of his book which he devotes to an analysis of the controversy over methodological individualism,[18] a doctrine which is motivated in part by the same philosophical commitment as I have been attributing to the Lovejoy methodology, he does not come out on the side of the individualists. In any event, individualism is not the theme of this study. The separability or not of individualism *and* non-individualism – whatever this proves to mean – is.

Thus, I do not wish to examine further the nature of individualism and the adequacy of Lukes's treatment of the various kinds of individualisms that in his view make up the doctrine as a whole. Rather I want to look briefly at an example in which individualism and non-individualism are intertwined in order to wonder or speculate a bit about what happens should we attempt to detach the two from one another on the assumption that clarity of analysis requires that we do so.

The focus of attention at the beginning of Rousseau's *The Social Contract* is beyond any question the individual. We are in the state of nature – where nothing social or societal is presumed to exist – and Rousseau tries to tell us what things were like before the constraints of society, or the sociopolitical order, were imposed upon men, thus placing constraints upon their freedom. In the state of nature, I suppose, there are no social predicates that are applicable to the individuals who are to be found there, and that is not to be wondered at inasmuch as Rousseau's concern seems to be the conditions of freedom and the circumstances that lead to its erosion.[19] And it is, of course, individual men and women who are or are not free. It would appear that as long as we are satisfied to remain in the state of nature we continue to enjoy the freedom that obtains in that state. But, for whatever reason, the details are of no interest here, the decision is made to abandon that state and to enter civil society. The decision, of course, is freely made, and it would make no sense at all for those who freely enter into the new arrangement to accept a circumstance that would leave them worse off than before. In particular, people who are free in the state of nature would hardly of their own volition agree to new circumstances which left them worse off in that regard than they were at the start. I suppose one ought to be wary of exaggerating this point, for it is clearly the case in Hobbes's account of the passing from the state of nature to that of political society that the contracting parties are, indeed, less well off with respect to freedom than they were before. But Hobbes's account is complicated by his conception of human nature which leads to his conclusion that though they are less free than at the start, they are, on balance, better off in a political state, be it as authoritarian or tyrannical as it may be. But that is not Rousseau's view. Whatever the advantages of living in a political order, they do not compensate for the loss of freedom should that occur. In the state of nature, each individual is free – freedom is clearly something one ascribes to individuals and not to collectives of whatever sort – thus any individual who is less than free under the new arrangement is getting less than he bargained for. But how are we to treat policy decisions arrived at by less than unanimous accord, let the majority be as large as you like? Those who do not form the majority are, at least with respect to the point of the policy decision, less than free. Yet to insist that only unanimous decisions may have the force of law is in effect, to dissolve the social compact before it gets started. It is this that leads Rousseau to what may well be the most logically incoherent conceptual monstrosity in the history of political thought, the General Will. It will be recalled that in Rousseau's view while the General Will is known through the vote of the majority it is in fact the will of all and not a mere sum of individual wills.

I think that the General Will has two functions in *The Social Contract*, and I shall state what they are without discussion of the text. Rousseau began his account with some observations about the state of nature, that non-social, pre-social state in which human beings are individuals only. The formation of the sociopolitical realm by way of a contract among individuals makes it clear that this procedure – at least at the start – of Rousseauian 'social science' is individualistic, that in a later age Rousseau might have been called a methodological individualist. Yet for all that, Rousseau has an intuition to the effect that societies cannot be analyzed fully in individualistic terms; there are individuals, to be sure, but their characteristics, aspirations and dispositions do not exhaust the reality of the social realm. I think that the idea of the General Will is intended to point to that non-individual aspect of social reality, an aspect that cannot be articulated by means of the individualist methodology with which Rousseau begins. In addition, the General Will functions to assure at least the possibility that society – and its political order – can exist without compromising the freedom of the individuals who make it up. We have seen that the only alternative seems to be majority rule, which compromises the freedom of the minority, or a demand for unanimity, which is unrealistic and compromises the integrity of the polity. The General Will enables Rousseau to avoid both of these undesirable alternatives.

I have made it clear that I do not think that this is the place to deal with Rousseau's text in detail, and I do not wish to make the effort to search for some more conceptually intelligible resolution of Rousseau's problem. But I do want to point out that within Rousseau's account we do find a pervasive tension between the individual and the non-individual, the sort of tension that for the mainstream of philosophical method – assuming that its practitioners would want to bother at all with *The Social Contract* – simply cries out for distinction and separation. We have societies which are constituted out of individual contractors yet the natures of which are not exhausted by those of the individuals in question. And we guarantee the freedom of individuals in society by means of a will which in one way is supposed to absorb the individualities into itself yet at the same time leave those individualities intact. Surely, one may expect to hear, there is much confusion here and the way to proceed, if we have any hope of finding something of value amidst it all, is to separate the conceptual parts out of which Rousseau's concoction was formed, keeping apart the individual and the non-individual. But even if we could accomplish this, it seems to me that there would be a price to pay: the specifically determinate and historical vision that is Rousseau's view of man, society and freedom in modern times prior to the impact of the industrial revolution would be lost. I want to make clear what I am saying and what I am not when I make this observation. I am saying that there is no way that what Rousseau is trying to say in *The Social Contract* could survive the kind of analysis that would relieve the tensions that we have found in it, particularly those centered on the General Will. I doubt that anyone who thought to revise Rousseau in that way could then claim that he has said what Rousseau wanted all along to say and that he has said it with greater clarity and precision. It is obvious enough what the response would be to what I have just said: so much the worse for Rousseau and *The Social Contract*. Let that be. What I have not said is more to the point, surely more to the point of the present study; I have not said anything about the utility or not of conceptual tension for the systematic study of society and to what extent a philosophy of social science must take it seriously. I dare say that writers whose inclination it would be to reject *The Social Contract*, particularly in view of what I have been saying about it, would be equally inclined to reject conceptual tension, would, that is to say, suspect that if I am right about Rousseau it is precisely because logical analysis cannot relieve the tension of the General Will that renders it so useless as a cognitive instrument for sociopolitical analysis. We shall have to see about that. But I should want to say in advance of that seeing, that if the concepts we need to make sense of our experience as social beings – and of the systematic disciplines which have arisen for the purpose of studying that experience – take the shapes they have owing to the nature of that experience, we cannot expect that the predilections of philosophical analysts, in abstraction from that experience, can be expected to have the last word.

III

What I do *not* want to do in what follows is to review the details of the controversy between individualists and non-individualists, microsociologists and macrosociologists, in order to discover weaknesses in arguments, failures of logical consistency, or instances of question-begging. Those who engage in intellectual activities may not face such occupational hazards as black-lung disease, but they have their own occupational ailments, and the list just made contains several of them. In addition, not infrequently the mistakes we discover in others, and which we cannot persuade them to see as mistakes, are owing to the fact that each side begins with different commitments. If I say that I can apprehend immediately that I am confronted by another person and need no argument-from-analogy or other Humean devices, the Humean thinks that I beg the question on behalf of some mysterious mystic-like way of knowing, but when I say that I apprehend immediately that the color of the wall is green he has no problem. The difference, of course, is that he is committed to the view that all empirical knowledge is rooted in the impressions of sense and I am not. Far from begging the question in the way suggested or requiring an argument on pain of being charged with dogmatism, apprehending others as other is something I do everyday, and so does he, and only a philosophical commitment keeps him from admitting it.[20] I believe that descriptions of social action are not possible if we do not have non-reducible social concepts,[21] but I know that anyone committed strongly to the individualist view will insist that when he looks at what are supposed to be instances of social interaction all he sees are individuals. And he might be tempted to drag in an ontological red herring by adding that he certainly does not see super-human entities hovering over the actors and moving things along. One replies, to be sure, that one doesn't believe in such things either, but that an account or description of the interaction qua social is not possible without the use of non-reducible social predicates. But to what do they refer? comes the retort. And so it goes. Anyone who enjoys anatomizing such discussions would have a field day with the O'Neill collection. And the same sort of thing may be found in a volume edited by Knorr-Cetina and Cicourel.[22] The fundamental difference,[23] between the two is that in the

O'Neill debate one side seeks to defend a genuine atomism of individuals, whereas the corresponding side in the other volume takes it that the minimum unit of social action is not the individual: indeed it is not an action at all. It is, rather, an interaction of small numbers, but not less than two, with the tendency of micro-theorists to think that macro-groups are constituted out of micro-groups. And so the issue is taken to be that of whether or not disparate methodologies introduced for the purpose of dealing with micro- and macro-groups can be integrated so that one may look forward to the emergence of a unified sociology. I do not wish to suggest that every contributor shares this conception of the issue, and my impression is that the micro-theorists are less interested in integration than in the reduction of macrosociology to microsociology.

If I think that to a significant degree the issues underlying the controversy or controversies are conceptual, it must be that the problem concerns the way we talk or ought to talk about the social world or the things known by means of the social sciences. We might find ourselves in the warehouse of a furniture company and ask, What have we got here? – What sort of reality is contained in this place? – and the answer might include tables and chairs, beds, sofas and such like. Of course, tables and chairs don't reveal their presence to us unless we come already equipped with the concepts table and chair. We can analyze each of these concepts in terms of a suitable range of material-structural possibilities, not forgetting to include function or purpose in our account. And we might allow for a bit of openness to the concept by recognizing that some modernistic designs may now be included within the extension, of the concepts, objects having shapes we weren't quite used to. But I don't think that we'd have any difficulty in deciding what the tables and chairs are and in terms of what they are tables and chairs.

We can pick out tables and chairs on sight, as it were, and I suppose it is that which leads so many to think that we can analyze them in perceptual terms or that it is not unreasonable to think of them as being logical constructions out of sense data. We take the function or purpose for granted, and given what that is, and the shapes chairs tend to have, it would not be all that easy to think of chairs as serving some other sort of function, even though we may occasionally use them for purposes other than sitting. There is no reason to go about this. The term "aunt", however, clearly unlike "chair," is not one the referents of which we can pick out by perception alone. The annual convention of a women's organization may seem, in some way, like the warehouse mentioned above in that it contains a variety of entities of a certain sort or class, namely women, but without making inquiries one could not pick out the aunts from the non-aunts in the same way that a visitor to the warehouse may pick out the chairs from the non-chairs. But once a technique is devised for determining which of the women are aunts – some simple questionnaire designed to ascertain whether the woman and/or her husband has siblings and whether any of the siblings have children – one supposes that we have a concept with the same degree of stability as the concept chair. But what is conceptual stability? I really have no formulated answer to that question, but I should suppose that with the increasing seriousness with which philosophy is taking the notion of conceptual openness it may well soon become useful – if not necessary – to come up with an analysis of it. I would like to talk about the two so as to make it seem that I am right about it.

I cannot really imagine that there could be an argument about whether or not the term 'chair' is being properly applied. Even if someone thought that some modernistically-designed chair was actually ugly, it is not hard to believe that such a one would understand why it could be called a chair, could see the chair-like use to which it could be put. A certain widower of my acquaintance married again, and all of the children of his first wife's siblings refer to and address his second wife as 'aunt'. Is she really their aunt? Within the family in question, the issue is never raised. To open the question is to allow for the possibility of a negative answer, and that would hardly be politic, particularly inasmuch as the successor wife had the responsibility of raising her predecessor's children, unquestionable cousins with whom relations were to be maintained. When I first planned to write the preceding sentences, I intended to move on to present a genuine clash of opinions on the question of how to analyze 'aunt', but it now occurs to me that contained in those sentences is already material relevant to such a clash. Far from being as stable as our concept of chair, that of aunt already reveals ways of being treated that would have no real parallel in the chair case. Thus, our decision to use or withhold use of the term may be made in terms of the requirements of family politics – or family peace. Given our intuitions with respect to the relations of the intention and extension of a concept, this may seem odd.[24] Another analysis implicit in the example is 'an aunt is a woman who raises one's first cousins', so that so far as aunthood is concerned no distinction is made between the mother or the step-mother of those first cousins. I admit that no one typically thinks about 'aunt' in either of these ways, but I venture to suggest that no small amount of actual usage accords with these ways, and that the reason we do not notice it is because we tend to think of the term 'aunt' – and of kinship terms in general – in other ways.

The popular discussion of kinship can generate no little heat, rather like the famous case of the squirrel discussed by James at the beginning of the second lecture of his *Pragmatism*.[25] The issue concerned a live squirrel clinging to the side of the tree and which kept moving around the trunk of the tree always keeping the trunk between itself and a man who kept going around it trying to look at the squirrel. The question arose among people present as to whether or not the man actually goes around the squirrel, and if we may believe James, the question, trivial as it may sound, generated a considerable amount of heat. James proposed that it all depended upon what is meant by 'going around'. If it means being first to the east of the squirrel, then, successively, to the north, west, south, and finally again to the east, the answer is 'yes'. If it means being first in front of it, then successively to its right, behind it, to its left, and, finally, in front of it again, the answer is 'no'. James reports that for the most part his good sense was appreciated, but for the diehard minority his solution was just quibbling and evasive. I dare say that there are any number of us for whom there is only one answer possible to the question 'Is she their aunt?' and that is because they have no doubt that kinship terms are grounded in consanguinity and affinity, and nothing else is relevant. Hence, in the case in question, she can't be their aunt since she is neither a sister of one of their parents nor the wife of a parent's brother. Nor would they be moved by the observation that she is indeed an aunt in view of being married to an uncle, for they would have no doubt that with the death of the first wife he ceased to be uncle of her nephews and nieces. I do not want to make too much about what are only my unsystematic observations, but I rather suspect that large numbers of ordinary speakers of American English would incline toward the consanguinity-and-affinity view and would tend to think of kinship terms as narrowly construable.[26] But there are people – people affected in their view to varying degrees by exposure to anthropological writings – who might be open to the construal of kinship terms – or the concept of kinship – in broader social terms. Thus, in the case in question, they might be open to the view that inasmuch as the relations of social intercourse which come to obtain between the unquestioned nieces and nephews of the dead first wife and the latter's successor are, on the whole, rather like those that would have been expected to obtain between them and the deceased had she survived, if we think of those behavioral relations as definitory of the concept, then we treat the concept as having a broader scope and as more flexible and open than on the view of the strict consanguinity-and-affinity analysts. We use the concept over a wider range of instances – hence, construe its extension as larger – and this, presumably, because we construe or understand the concept differently – hence, treat its intention differently.

Let us follow the career of 'aunt' somewhat further. There are instances in which the term is used to designate or address women who are friends of one's family, particularly of one's parents, but who are not conceived to be kin in any ordinary acceptance of the idea of kinship that obtains in our society. How shall we understand such usage? Obviously to the consanguinity-and-affinity point of view, such women are not properly aunts, at all. The use of the term in contexts such as these is at best somewhat metaphorical: the 'aunt' is an aunt *honoris causa*. But if one is able to follow the more social construal of kin terms along the lines of the second position sketched above, it is not at all hard to accommodate examples of this kind. In the family system that we find in urban American communities, the family unit consists of parents and children. Siblings of parents, their spouses and children live in their own residences. Face-to-face encounters are relatively infrequent, though they may be supplemented by telephone communication, and tend to result in children not being all that closely involved with their aunts and uncles. If one attends entirely to the character of the social relations involved, without noticing anything about relationships in the consanguinity-and-affinity sense, it would in many instances be difficult to discern serious ways in which relations with aunts of the usual sort and of the metaphorical sort were different.

There is no need to work up further permutations on this theme in which social factors come to be increasingly significant, and relationships more narrowly construed less so, in the understanding of what kin terms mean.[27] I want, rather, to make some comparisons of the concept aunt and the concept chair, comparisons which will point to the character of the difference between them yet will make clear that the differences are not differences in principle. I think it is clear enough that the range of flexibility which we have discovered – even in the brief treatment of the concept aunt that I have presented – does not seem to have a counterpart with respect to the concept chair. Withal the stylistic variability that clever designers may come up with, what chairs are is rather clear cut, and it is not likely that there could be heated disagreement over the matter. Why should there be? At first, this question itself may occasion surprise. Why should there not be differences in the features that concepts of different sorts display? While I do not wish to challenge the principle which may underlie this second question, I do think that it is possible to deal with the first one. And I would suggest that the difference with respect to degree of flexibility – or openness – that we find in the two concepts has to do with the degree of stability we find in their respective areas of experience.[28] The function of chairs continues to be what it has been for as long as we know about chairs, and that, presumably, is owing to the way in which that function is tied to stable features of the human anatomy. If we could imagine the anatomical structure of the human body undergoing change, not the imperceivably slow change that we learn about through evolutionary research, but rather some rapid change of the sort we could know about over the course of small numbers of generations, then we could imagine the development of new kinds of chairs, to satisfy the chair-function in constantly changing bodies, and we could easily imagine that there could be quarrels as to what really were and were not chairs. This would be particularly the case during periods of transition. It would seem, then, that the relative stability of one concept in contrast to another is owing to the relative stability of the realm of experience from which it emerges and to which it is applied. In addition, one must keep in mind that in so far as our minimal comparison of only two concepts is concerned, the elements which enter into consideration seem fewer for the more stable – less open, yet not entirely closed or sharply defined – than for the other. Function or purpose and structure of the human anatomy seem to exhaust what we need consider in trying to understand how the concept of chair takes the shape it has, with the range of what may serve as a chair being limited by these two conditions. The elements that enter into the formation and change of the concept aunt are, clearly, more varied. (And I have purposely not widened the discussion here so as to take into account what experiences relevant to this matter anthropologists have, and how those more extensive experiences enter into the formation of anthropological thoughts on kinship.)

IV

If so seemingly simple a concept as aunt is seen to involve the comparative complexity of experience as we have seen it to have, what shall we say about the more complex concepts with which the social sciences carry on their activities? I do not wish to select any of them in order to treat them or speculate about them in the way I did with the concepts of chair and aunt in the previous section. I am certain that readers will be willing to agree that the notions of social class, determinate kinds of kinship system or rules of residence, are complex and rest upon complex experiences implicating varieties of elements. Thus, I am sure, that all will agree that the concepts required for even a description of social things – or things known which are social – would be complex. And given the ways in which social experience is given to change, one would expect that the concepts used would express the flexibility of open concepts rather than the stability of closed ones. And this, in turn, would lead to the expectation that there would be considerable contestation over what such concepts mean, how they are to be analyzed and how applied. In that respect, we may think of them as essentially-contested concepts, though their character as essentially contested is not restricted to any role they may have in evaluation, which is what W.B. Gallie, to whom we owe the term, seems to emphasize.[29]

We have clearly seen that to describe situations as social requires that we use concepts that are essentially social and not something other. This is entirely like the claim made by J.M. Brennan with respect to morality. We require to use moral concepts if we are to discover or experience moral issues, and there is no way to find moral problems if we characterize the situation in descriptive or morally-neutral terms.[30] And Mandelbaum, in a well-known example, made the point that to describe something as a banking situation is not possible unless we make use of banking concepts.[31] To be sure, the event described, or present to those who witness it, is one in which individuals are doing the things they are doing. But those individuals have to be presented not in terms of their biopsychological characteristics but as occupying socially-construable roles: depositor, teller, and so forth. Mandelbaum's example was presented in the context of the old debate over methodological individualism, with the opposition to a purely social construal of social-science concepts coming from people who took it that the social world is made up of individuals, and believed therefore, that the concepts or terms used to characterize the world must be analyzable in individualistic terms. But even those who see the building blocks of the social world not in individualist terms but in microsociological ones, say nothing that would offset the point that I am trying to make. Thus, Orvis and June Collins develop the point that the smallest unit of social analysis is the interaction.[32] Much of their book is taken up with the problem of determining what such interactions are in presumed independence of non-interactional sociocultural elements which determine what the interaction is. If the interaction of two or three or more individuals is the smallest unit of social study, then the social realm is not constituted out of individuals, but that is not the point I wish to emphasize. The Collinses seem to be saying that the interaction can be picked out and its character studied before we allow for the introduction of sociocultural elements. In actuality, of course, all are co-present, yet some degree of independence is being claimed for the interaction.

One may wonder just how much of this opinion is consequent upon the fact that the interactions that are reported in their book are artificially put together, as typically they are in small-group research. The interaction is first isolated, and then the attempt is made to discover its features and the features of different kinds of interaction. But in the social world at large where people are coming and going, how would one pick out the interactions from those juxtaposings of people not in interaction? One, surely, must know something other than that people are near or not to one another to know that here is an interaction and here is not. I am reminded of Handelman's and Kapferer's attempts to describe joking relations among workers in factory settings in Israel and Africa.[33] Their method is to focus on the actors in interaction and to claim that the relations emerge out of that. Yet the fact that the actors are fellow workers in the same factory surely provides a delimiting setting which is other than the interaction, and a careful reading of the paper, which I do not wish to discuss in further detail, makes evident how much of the socio-cultural world in which the actors participate provides the background for and enters socially into their interaction.

This all points to the complexity of any small situation we might wish to describe. And we have already had reason to believe that the more complex the experience or classes of experience we are interested in prove to be, the more likely the concepts we need in order to deal with them will be flexible or unstable or open. But the history of philosophy – as of science – makes it clear that the kinds of concept to which attention is paid are those that are supposed to be sharply defined. I suppose that a historical sketch of the matter would be appropriate at this point in order to show that this is the case, but even if I were capable of offering one it would require rather considerable space for what would be essentially a digression. Thus, I can only ask that the reader think himself of what the outcomes are of conceptual analyses with which he is familiar, paying particular attention to the fact that in both science and philosophy, it is deductive arguments that are most highly esteemed, which means, of course, results derived from assumptions precisely formulated in terms of concepts or terms held to be unambiguous: sharply defined. It is, I would suggest, the ideal of the sharply-defined concept that both stands in the way of the development of conceptual tools needed for adequate accounts of social reality and leads to quarrels where none are called for.

What we have to deal with in the social field is highly complex, having both individual – or interactional, likely both – and non-individual aspects. Too, institutions are both stable and changing; both at the same time. There is the comparative stability of social systems or social structures, the sort of thing which has attracted the attention of such social anthropologists as Fortes[34] and Evans-Pritchard.[35] And there is the constant instability and change generated by the interactions of individuals within the framework of stable structures, an aspect that has been the focus of attention of Barth[36] and the ethnomethodologists.[37] From a strictly logical point of view, stability and change are antithetical, and it seems difficult to say of the same thing at the same time that it is both stable and changing. And, so, different social scientists have opted to say one or the other, but not both – even when they know that the object of their study is, indeed, both. A remark by the Collinses points to the issue: "Once a person has learned to think in such static terms as role, status, values, and so on, he seems rather incapable of seeing the social phenomenon as moving, changing, and altering in time."[38] To frame one's analysis, the Collinses tell us, in terms like the ones they cite is to close off the possibility of studying social change. One chooses such terms because one is concerned to characterize and explain conditions of social stability.[39] Why not stability-in-change? Because, I suspect, the conceptual tension that this would entail would be hard to reconcile with the – at least tacit – commitment to sharply defined concepts that most scientists and analysts bring with them to their work. It is not that I have specified two conceptual-tension inducing aspects of social reality that are independent of each other and each of which contributes to the problem. They are inter-related. If one is drawn to the stability side of the stability-change opposition, then one is drawn to the non-individual side of the individual-non-individual opposition. The attempt to avoid conceptual tension would, I propose, be a major motive in the generation of the philosophical-methodological controversy we have been considering, and, at the same time, a pressure toward one-sided abstraction from an area of investigation in such a way as to produce a series of skewed visions.[40,41]

State University of New York, Binghamton

NOTES

[1] John O'Neill (Ed.), *Modes of Individualism and Collectivism*, London, Heinemann Educational Books, 1973.
[2] Quoted in Morris Weitz, *The Opening Mind*, Chicago and London, University of Chicago Press, 1977, p. xi.
[3] See J.M. Brennan, *The Open-Texture of Moral Concepts*, New York, Barnes and

Noble, 1976; and Julius Kovesi, *Moral Notions*, London, Routledge & Kegan Paul, 1967.

[4] Clarence Irving Lewis, *Mind and the World-Order,* New York, Charles Scribner's Sons, 129.

[5] W.B. Gallie, *Philosophy and Historical Understanding*, London, Chatto & Windus, 1964, ch. 8.

[6] See William E. Connolly, *The Terms of Political Discourse*, 2nd ed., Princeton, Princeton University Press, 1983, and John Gray 'Political Power, Social Theory, and Essential Contestability', in D. Miller and L. Siedentop (Eds.), *The Nature of Political Theory*, Oxford, Clarendon Press, 1983, pp. 75–101.

[7] Weitz, *op. cit.*

[8] Perhaps not entirely. I believe that R.G. Collingwood's account of the scale of forms of overlapping classes in an attempt to deal with some form of open concepts *(An Essay on Philosophical Method*, Oxford, Clarendon Press, 1933, chaps. 2 and 3), and that Hegelian dialectic is a technique for dealing with the sorts of open concepts that carry their histories with them. On Hegel, see my 'Dialectic and Necessity in Hegel's *Philosophy of History*', in L. Pompa and W.H. Dray (Eds.), *Substance and Form in History*, Edinburgh, Edinburgh University Press, 1981, pp. 42–57, and 'Force and the Inverted World in Dialectical Reprospection', *International Studies in Philosophy*, **XX** (3) (1988), pp. 13–28.

[9] See George E. Mendenhall, 'Covenant Forms in Israelite Tradition', in E.F. Campbell, Jr., and D.N. Freedman (Eds.), *The Biblical Archaeologist Reader*, Vol. 3, New York, Doubleday & Co., 1970, pp. 25–53; Delbert R. Hillers, *Covenant: The History of a Biblical Idea*, Baltimore and London, The Johns Hopkins Press, 1969, and D.J. McCarthy, *Old Testament Covenant*, Oxford, Basil Blackwell, 1973.

[10] Steven Lukes, *Individualism*, Oxford, Basil Blackwell, 1973.

[11] Richard H. Popkin, *A History of Skepticism from Erasmus to Descartes*, New York, Harper & Row, 1964.

[12] See Peter Laslett's 'Introduction' to his edition of Locke's *Two Treatises of Government*, Cambridge, Cambridge University Press, 1960.

[13] Lukes, *op. cit.*, p. 43; for the passage from Lovejoy, see his *The Great Chain of Being*, Cambridge, Mass., Harvard University Press, 1936, pp. 5–6.

[14] Lukes, *op. cit.*, p. 43; the passage from Weber is from *The Protestant Ethic and the Spirit of Capitalism*, London, George Allen & Unwin, 1930, p. 222, the translation being somewhat amended by Lukes.

[15] P.H. Gulliver, *Neighbours and Networks*, Berkeley, University of California Press, 1971.

[16] See my 'The Problem of the Given in Buber's Conception of the Interpersonal', in Arthur A. Chiel (Ed.), *Perspectives on Jews and Judaism*, New York, Rabbinical Assembly, 1979, pp. 135–143.

[17] Henri Bergson, *Introduction to Metaphysics*, New York, Liberal Arts Press, 1949, p. 32 f.

[18] Lukes, *op. cit.*, Ch. 17.

[19] I am taking Rousseau at his word for this, and cannot, in the present context, take up the question of whether, in the end, *The Social Contract* gives solace to democrats or totalitarians. I think one can defend the view that Rousseau is to be taken at his word, but I remember being present at a meeting of the Israel Society for the Political

Sciences, sometimes during the first half of 1970, at which two scholars – one was the late Jacob Talmon; I no longer remember the name of the other – took opposite views on the issue. Clearly, a technique – hermeneutic – for reading *The Social Contract* which would yield results to which most would assent has yet to be devised.

[20] See my 'Why the Problem of Other Minds?', *Philosophical Forum*, **2** (n.s.) (Winter 1971), 271–277.

[21] See Georg Simmel, *The Sociology of Georg Simmel*, tr., ed., with an Introduction by K.H. Wolff, Glencoe, Ill., The Free Press, 1982, p. 8 f.

[22] K. Knorr-Cetina and A.V. Cicourel, (Eds.), *Advances in Social Theory and Methodology*, Boston, Routledge & Kegan Paul, 1981.

[23] There are many important differences, but for the purpose in hand this is the only one I wish to emphasize here.

[24] Does it in any way resemble: "If this is a chair anything is/can be a chair"? I suspect one would say this only in the context of reviewing or commenting on a new style of furniture, that, here, it would be an esthetic judgment rather than one about the limits of the extension of "chair." In the "aunt" case, once the decision is made – in the case I have in mind, no decision was made, the practice was simply taken for granted – the term functions as usual, and the woman was treated as an aunt in every way.

[25] William James, *Pragmatism*, New York, Longmans, Green, Co., 1916, p. 43 ff.

[26] For an account of the variations of kinship conceptions which are said to be found in the United States, see David M. Schneider, *American Kinship*, 2nd ed., Chicago, University of Chicago Press, 1980.

[27] For an ethnographic study of a society in which social and economic factors play considerable roles in the application of kinship terms, see P.H. Gulliver, *op. cit.*

[28] I am trying to make this point without overly much expanding the range of the present discussion. Clearly, the range of experience relative to the concept aunt is different for ordinary people living in our society and for anthropologists whose experience – through field work and reading – extends to all manner of analogous experiences of ordinary people in other societies.

[29] Gallie, *op. cit.*

[30] Brennan, *op. cit.*

[31] Maurice Mandelbaum, in O'Neill, *op. cit.* p. 224 f.

[32] Orvis Collins and June M. Collins, *Interaction and Social Structure*, The Hague, Mouton, 1973.

[33] Don Handelman and Bruce Kapferer, 'Forms of Joking Activity: A Comparative Approach', *American Anthropologist*, **74** (3) (June 1972), 484–517.

[34] Meyer Fortes, *The Dynamics of Clanship Among the Tallensi*, London, Oxford University Press, 1945, and *The Web of Kinship Among The Tallensi*, London, Oxford University Press, 1949.

[35] E.E. Evans-Pritchard, *The Nuer*, Oxford, Clarendon Press, 1940.

[36] Frederick Barth, *Models of Social Organization*, London, Royal Anthropological Institute, Occasional Paper no. 23, 1966.

[37] Harold Garfinkel, *Studies in Ethnomethodology*, Englewood Cliffs, N.J., Prentice-Hall, 1967.

[38] Collins and Collins, *op. cit.*, p. 81 note.

[39] And at the risk of being charged with political motives.

[40] I am aware of the fact that science always abstracts and that its goal is not that of encapsulating wholes as they may really be, along the line, say, of Bergson's *Introduction to Metaphysics*. I am, however, suggesting that sometimes at least the abstracting we get in social science is motivated in the way I say resulting in a one sidedness that is unnecessary. My impression is that one-sidedness leads to one-sidedness of an opposite sort, that over focus on, say, structural or institutional factors will lead to a subsequent over focus on interactional ones. Some may suppose that this shows that in the long run science proves itself to be self-correcting. But what happens over time may perhaps, be characterized as well in the way I have just done: as a sequence of skewed visions.

[41] In my contribution to the original debate – see O'Neill, *op. cit.*, pp. 264–286 – I spoke about theory, not description, and I exemplified my views by an account of the theory of the development and change of systems of kinship nomenclature in George Peter Murdock's *Social Structure* (New York, Macmillan, 1949). I continue to believe that such theories are not individualistic, though they may be suggested by ethnographic material constructed by field workers whose descriptions would require concepts along the lines suggested in the text and would have to be tested by reference to behavior situations which would likewise have to be described in that way. But I did not think it necessary to bring all this up in an account directed less at the logic of social-science theory than at the problem of conceptual tension.

NORETTA KOERTGE

STEREOTYPES, STATISTICS, AND SCHEMATA[1]

> *To be nobody-but-myself – in a world which is doing its best night and day, to make you everybody else – means to fight the hardest battle which any human being can fight, and never stop fighting.*
>
> e.e. cummings

1. INTRODUCTION

Historians of ideas and students of comparative cultures debate the extent to which all people share e.e. cummings's ideal of being "nobody-but-myself". However, within the modern Western European tradition, most of our political philosophies, theologies, and common sense belief systems emphasize the uniqueness and autonomy of individuals.

In the United States there is a particularly strong aversion to being labelled, stereotyped, categorized, type-cast. Part of our objection undoubtedly stems from the inaccuracies and distortions which inevitably accompany these crude classification systems. But I think the intensity of our resentment indicates that we feel that more is at stake. We don't want to be "put-in-a-box", even if the box is only a conceptual one, for fear that this will somehow limit our freedom. One is reminded of cultures in which knowing a person's name is thought to be a source of power over them and hence it is dangerous for an enemy to learn your proper name.

Accompanying this battle to prevent other people from classifying us is an equally intense struggle to find out who-we-are, where-we-belong. In short we seem to feel a need to denominate ourselves, to say what about ourselves is important and enduring. This process of self-naming is sometimes described as a discovery, as in "I finally realized that I was a poet/transvestite/alcoholic". Other times we report having chosen to take on a certain identity: "I was nineteen when I decided to become a Christian/lesbian/vegetarian". But in both cases the application of a rather crude

Fred D'Agostino & I. C. Jarvie, Freedom and Rationality: Essays in Honor of John Watkins, 111–124.

label to ourselves is seen as not only legitimate, but as liberating (even when the designated category is a problematic one).

My project in this paper is to analyze these processes of applying words to people and our reactions to them. How can we take responsibility for the ways in which we conceptualize others as well as ourselves? I will begin with philosophical criticisms of stereotypes in the ordinary sense of categorizing groups of people in a fixed formulaic fashion which ignores individual differences. I will then move on to sample some of the findings of cognitive science about the nature of mental representations. I conclude with some speculations about the personal and political implications of these analyses.

2. A STATISTICAL AND LEGAL CRITIQUE OF STEREOTYPICAL REASONING

We are often warned about the dangers of stereotypical reasoning in a political context. Stereotypes, we are told, are both false and insulting. Feminists have been particularly assiduous in pointing out the ways in which subtle (and some not so subtle) errors infect our conceptual scheme. A riddle which was popular in the early seventies described an elaborate scenario with family gathered in an emergency room. Finally the doctor pulls back the sheet and says, "I can't operate – he's my son!" The key to figuring out how this was possible was to realize the surgeon was a woman. In the early days of contemporary feminism, many listeners took an embarrassingly long time to consider this possibility, even though the universal generalization, "No surgeons are women", was clearly not true.

But sometimes we object to stereotypical reasoning even when it is not obviously based on false generalizations. Consider the following episode. I call up the local travel agent to confirm my ticket, booked as Professor N. Koertge, and to schedule a return flight. "What time of day would he prefer to fly?" says the agent. With some annoyance, I explain that *I* am the party in question and with some confusion, she explains that she thought I was Professor Koertge's secretary.

From a purely statistical viewpoint, the travel agent's assumption was not a bad one. At my university, less than 20% of the professors are female, but more than 90% of the secretaries are. Furthermore, it is quite common for professors to ask their staff to make their travel arrangements and since I had not used any personal pronouns in my conversation with her (was this partly because of female diffidence at admitting to being a professor?), the probability was in fact quite high that the ticket was for a man. However, the travel agent could be faulted for not also considering the cost of being mistaken before she spoke. Since female customers (especially in a liberal college town) are likely to become annoyed at such errors and since it is very easy to speak in a gender-neutral fashion, the agent is well-advised to avoid drawing any conclusion whatsoever, no matter how high the probabilities may be.

So far our examples have all dealt with inferences about a single case. We have made the obvious point that reasoning with false universal generalizations is problematic and the slightly less trivial point that when deciding whether to act on a conclusion, one should consider not only probabilities, but also utilities. And in cases involving humans, sometimes even verbal descriptions can have consequences and hence count as actions. But just because many of the common generalizations about purported religious, racial, ethnic, or sexual differences are false and because even legitimate statistical inferences based on such classifications are often taken to be insulting, are we then to conclude that one should always eschew reasoning based on politically sensitive categories?

A recent Supreme Court case concerning differential retirement benefits for men and women shows how vexed and confusing this issue can become. A class action suit was brought against the State of Arizona because the pension plans offered state employees gave female retirees lower monthly benefits than males who had made the same contributions, on the grounds that women live longer on average and hence their total compensation would be the same. In a rather confusingly written decision the Supreme Court ruled the Arizona plan impermissible, largely because it violated Title VII of the Civil Rights act which "requires employers to treat their employees as *individuals*, not 'as simply components of a racial, religious, sexual, or national class'." (Arizona v. Norris, p. 3498)

However, the dissenting judges argued that insurance companies ought to be able to use sex-based mortality tables if they are actuarially sound (Arizona v. Norris, p. 3504). In effect, these judges who wrote the minority opinion are saying that insurance rates must be set by locating individuals within some reference class and if "suspect classifications" such as race, sex, or national origin turn out to be good predictors then we must not be too "liberal" to use them. If there is in fact a statistically relevant difference between the two groups, we should act on it.

But the judges who wrote the majority opinion which was critical of Arizona's annuity plan pointed out that the tables in question did not incorporate "other factors correlating with longevity such as smoking habits, alcohol consumptions, weight, medical history, or family history." (Arizona v. Norris, p. 3495) They might have gone on to speculate that these other factors might make a *bigger* difference in expected length of life than did sex and a statistician could add that there was even a possibility that if one controlled for differences in smoking, etc., sex might cease to be relevant at all – to use Salmon's (1970) terminology, smoking might screen off sex as a statistically relevant factor.

Since this notion of *screening off* is crucial to our later discussion, let me illustrate it with a simpler example. Suppose I am shopping for a used car. I might well decide to rule out old cars on the grounds that on average old cars will be more apt to break down. However, even though this is true, it may also be the case that when one compares cars which have the same odometer readings, then their age is no longer statistically relevant to their projected usefulness. So although old cars are more likely to have a high mileage and high mileage cars are more likely to break down, I must be careful not to discriminate against the occasional old car which has a low odometer reading.

To return to the case of human stereotyping. We can now add a further caveat for those who would reason from generalizations about groups to the expected properties of individuals. Even when some racial, ethnic, or sexual characteristic is a statistically relevant indicator of some target property (such as life expectancy or suitability as an employee), we should look for more fine-grained analyses (thus taking into account, for example, health or education) which may in fact then render the classification based on stereotypes completely irrelevant. Correct statistical reasoning requires us to use maximally homogeneous reference classes; that is, ideally our projections would take into account all and only factors, which have an effect on the property we wish to infer. However, the dissenting judges in the Arizona case argued that the administrative cost of identifying all relevant attributes would be prohibitive and so the life insurance companies should be allowed to pick criteria which were "readily identifiable, stable, and easily verifiable." (Arizona v. Norris, p. 3509) Unfortunately, many of the characteristics which are easy to identify happen to be markers such as skin color, sex, age, or accent which have all too frequently been implicated in unfair discriminatory practices.

In other cases the U.S. Supreme Court has laid down some rough guidelines as to how homogeneous the reference class should be via the Fourteenth Amendment, which says that "No State shall...deny to any person within its jurisdiction the equal protection of the laws." This so-called *equal protection clause* has in general been interpreted as ruling out any law which treats different groups differently "when it is without any reasonable basis and therefore is purely arbitrary." (220 U.S. 61, 78) Thus a state university may charge non-residents higher tuition on the reasonable grounds that they and their families are unlikely to have paid state taxes. In general, the law is not obligated to specify the fairest possible differential treatment of the two categories of citizens. Instead the courts will uphold the validity of a statute if there is any rational basis whatsoever for it.

However, certain categories are held to be "suspect classifications" and if a regulation singles out people in these categories for special treatment, then the court will subject it to "strict scrutiny" and rule that it violates the equal protection clause unless the state can prove it has a "compelling interest which can *only* be furthered by enactment of the statute in question". (Gifis, p. 71) Classifications are suspect when they involve classes of people who are "saddled with such disabilities, or subjected to such a history of purposeful unequal treatment, or relegated to such a position of political powerlessness as to command extraordinary protection from the majoritarian process." (411 U.S. 1, 28.)

Laws which make distinctions between people on the basis of race or national origin are always subjected to strict scrutiny, but sex has *not* been considered to be an inherently suspect classification, which is the major reason American feminists have felt that an Equal Rights Amendment was necessary and why Congress passed Title VII.

The discussion thus far suggests the following guidelines: Reasoning with stereotypes is a special case of statistical reasoning, namely, we infer an as yet unobserved property of an individual, the so-called *target property* (T), by locating the individual within a *reference class* (R), and noting the frequency (f) with which the target property occurs within the reference class. We may even be able to assign a numerical value to the frequency; otherwise, we will have to settle for rough estimates like "most" or "over half" or "very rarely".

There are two basic statistical errors which can be made. First, one may be grotesquely wrong in the value assigned to f. (It is just not true that most Chinese-Americans run laundries!) Secondly, even if the value of f is correct for the designated reference class (R), one may not have chosen the most appropriate reference class and so f, though correct for R, is nevertheless not a good estimate of the probability of T. (For example, even though it is true that most Americans don't know how to play cricket, we should not conclude that Yankee Jones is uninformed if Jones was a Rhodes Scholar.)

Avoiding the second error, that of using too non-specific a reference class, often poses practical problems because it might require a lot of effort to determine the appropriate reference class. In fact, it might be easier to go out and detect the target property directly instead of trying to infer it! At this point, one has to add in utility considerations and balance off the cost of error versus the cost of getting better information.

The categories we typically think of as stereotypes constitute reference classes which are very easy to identify but which have a history of being used in situations where one or both of the above statistical errors result. Hence, we are cautioned to be suspicious of them.

3. INSIGHTS ON STEREOTYPES FROM COGNITIVE SCIENCE

We have sketched a normative critique of reasoning based on stereotypes. But if we are to improve our cognitive performance, it would be helpful to understand more about how human cognition works. As the story of the travel agent shows, there is not always time to think about the homogeneity of reference classes or the cost of information retrieval before we speak. What we would like to have are some simple rules of thumb or, better yet, well-ingrained reasoning habits which will keep us out of trouble! Out of the explosion of research in cognitive psychology and artificial intelligence I will extract a few results which cast some light on the human problem of understanding our human propensity to use stereotypes.

In their book on intuitive inference strategies, Nisbett and Ross (1980, Chapter 3) summarize a variety of investigations showing how the *vividness* of a piece of information influences the epistemic weight which it is given. (A common example would be the case of a customer who weighs the experience of a neighbor over the test results compiled in a consumer magazine.) Psychologists find that the concreteness, proximity, and emotional interest of data influence the ease and accuracy with which we can retrieve the material from memory and our tendency to rehearse it. Thus vivid information is more apt to be incorporated into the schematic knowledge structures we rely on for ordinary reasoning.

Nisbett and Ross suggest that historically there may have been evolutionary advantages in favoring vivid, immediate experiences over pallid, third-hand reports. For one thing, it may be in the best interest of strangers to deceive us! And until the development of modern science, abstract generalizations would often have been less reliable than anecdotal evidence. However, with the advent of television, this cognitive tendency makes us very vulnerable to well-designed commercials and other vivid images. And as our society gets more heterogeneous and more complex, there is less and less chance of our having knowledge by acquaintance of all of the factors which need to be taken into consideration. Our inclination to rely on a vividness criterion undoubtedly plays a substantial role in the formation of racial, ethnic and gender stereotypes.

Another important short-coming in human inference strategies which helps maintain stereotypes is what Nisbett and Ross call the *fundamental attribution error*, namely, the "tendency to attribute behavior to the actor's dispositions and to ignore powerful situational determinants of the behavior." (p. 31) Studies show that dominance, talkativeness, or honesty behaviors are better predicted by the situation a person is in than by intrinsic personality or character traits. However, naive observers almost always locate the cause of the behavior within the person rather than in the situation. For example, people tend to view Milgram's classic obedience-to-authority experiments as showing something about the special weakness of individual subjects instead of an indication of the strength of that particular set of situational pressures. (Nisbett, 1980, pp. 121–22)

So if we encounter members of some minority group in a disadvantageous situation we are much more likely to explain any non-optimal performance in terms of group inadequacies instead of noting the stresses or constraints present within their situation. Thus, slaves, colonial subjects, women, children, enlisted soldiers, or members of any other dominated group are often thought to be inherently dumb, lazy, passive, aggressive, sneaky, and manipulative.

In an earlier section we analyzed some of the types of mistakes which can result from stereotypical reasoning and the remarks above on the vividness criterion and the fundamental attribution error show some of the psychological mechanisms which contribute to the formation and maintenance of stereotypes. It turns out that codifying information in terms of prototypes, schemata, scripts, or default rules – the various terms correspond to theoretical constructs differing in ways which need not concern us here – appears to be an important aspect of the way we store and access knowledge in every realm. It is not just members of politically and socially disadvantaged groups that we think about by means of rather crude caricatures.

Although I have called these concepts crude (because they neglect all sorts of crucial factors), they can also be remarkably complex. Recent research has started to tease out some of the more remote strands of what Bem (1976) calls gender schemata. A few examples from research on gender will illustrate how these schemata work. It will be useful to begin with a brief look at the history of so-called M/F scales.

The first major attempt to measure what the authors call "mental masculinity and femininity" was published in 1936 by Terman and Miles. During his famous longitudinal studies of people whose IQs fell in the genius range, Terman noticed that gifted boys and girls differed markedly in their preferences for games and amusements. This led him to a full scale investigation of sex differences in what he called "attitudes and interests". The result was a battery of seven different tests, each item carefully selected because males and females on the average respond differently to it. Since the complete test battery contains over 900 questions, this psychological inventory provides us with detailed composite portraits of what young American women and men knew about and liked in the 1930s.

Here are some sample questions from each section of the test with the scoring key also included. (Terman 1936, 485–506) The answers followed by a '+' count as masculinity points; '–' responses subject from the masculinity score, thereby moving one towards the feminine end of the bipolar scale' '0' responses are neutral:

I. Underline the word which goes most naturally with:

POLE:	Barber 0	cat +	North –	telephone +
PURE:	good –	milk +	water –	white 0

[Some of the responses fit in well with our stereotypes, e.g., that girls would associate 'good' with 'pure' and boys would think of a 'pole cat'. But why should boys think of 'pure milk' while girls choose 'pure water'? Does this reflect a difference in farm chores, the boys doing the milking while the girls fetch water? Terman and Miles claims to have used no theory or considerations of 'face validity' while making up the test items, but simply used any question which generated an empirical difference in response.]

II. Underline the word the drawing makes you think of most:

dish –	ring 0	target +	tire 0
birdhouse 0	flagpole +	sword –	torch +

[Again, why are girls more apt to think of swords?]

III. Underline the word that makes the sentence true:

Beam scales illustrate the principle of

buoyancy 0 elasticity 0 leverage + magnetism –

A character in *David Copperfield* is

Betty + Uriah Heep – Sinbad + Oliver Twist +

[The reason for the girl/boy differences in response to the items in this section are depressingly obvious.]

IV. How much anger does each of the following cause you?

Seeing a person laugh at a cripple

very much – much + little 0 none +

Being called a thief

very much 0 much 0 little + none –

V. Show how well you like each book. Skip those you have not read.

Peter Pan and Wendy

Omit – like + dislike + neutral +

Huckleberry Finn

Omit – like + dislike + neutral –

VI. Show how well you like these characters. Skip those you know nothing about.

Lenin

Omit – like + dislike + neutral +

Florence Nightingale

Omit + like – dislike + neutral +

[Of the 28 famous characters 23 are white men, four are white women, one is a black man.]

VII. Answer as truthfully as you can.

Do you ever have the same dream over and over?

Yes – No +

Have you often been punished unjustly?

Yes + No –

This hodgepodge of questions and the ways in which typical young males and females answer them has a certain curiosity value and might even have been valuable for authors of children's books, but what sort of scientifically interesting results did Terman and his associates expect from them? It is clear from the introduction to the book that these researchers believed that an objective measure of the culturally important characteristics as masculinity and femininity would correlate with a variety of other human characteristics, such as body build, age of puberty, popularity with classmates, marital happiness, leadership abilities, social adjustment, etc. However, most of their empirical studies failed to reveal any significant patterns. Particularly disturbing to them was the fact that the distribution of M/F ratings varied with both age and education. As they grew older, *both* men and women became less masculine (i.e., more feminine). Better educated men were less masculine on average while the better educated women were less feminine.

Terman had expected to find that homosexuals were 'cross gendered' and the one group of 'passive' male homosexuals (in an army stockade) which he studied did have scores significantly lower (i.e., were more feminine) than any other male group tested. (One group of female college athletes scored slightly more masculine than they did.) The researchers had also expected that juvenile delinquent girls would be abnormally masculine, but their studies of teenagers in a reformatory showed an unusually wide range of scores. They concluded that perhaps there were two types of delinquents, rebellious tomboys who went looking for trouble and super passive feminine types who were led astray by bad influences.

The story of the proliferation of M/F scales over the next 35 years and their widespread use in vocational counselling and assessments of psychological adjustment is a fascinating one (see Constantinople (1973) for a review of the major methodological critiques which were eventually brought to bear), but for our purposes here we can use this research program to document the pervasiveness and tenacity of the gender system.

As items on the standardized tests reveal, the concepts of masculinity and femininity were bound up with almost every aspect of one's life – the books one read, the games one played, dream contents, free associations to ink blots, one's view of propriety, morality, and justice, aesthetic tastes, fears and delights, preferences in pets, and one's general knowledge of history, science, and geography. There were two cultures within middle-class, white America – the male and the female.

In the 1960s M/F scales were subjected to a variety of criticisms by psychologists. They had limited cross-cultural validity; they had to be restandardized every few years; unlike IQ, the M/F score of an individual changed with age; and there were also internal technical difficulties which seemed to arise from the fact that the scale was bipolar. Any response which raised one's femininity rating automatically made one less masculine.

To remedy these defects, Bem (1976), Spence and Helmrich (1978), and others introduced orthogonal uni-polar scales. They also introduced items based on personality traits instead of using the more concrete, but also more culture-dependent questions about games and books. So, for example, the BSRI (Bem, 1976) asks people to rate themselves on the extent to which attributes such as *self-reliant*, *cheerful*, and *reliable* apply to themselves. All of the items describe desirable personality traits, but according to Bem's standardization studies, some (such as *self-reliant*) are judged to be significantly more important for males than for females; others, like *cheerful*, were thought to be more desirable for women; and then there were items perceived to be neutral, e.g., *reliable*. These serve as fillers to discourage subjects from guessing how the test works.

Respondents then receive two scores, an M-score which reflects how much they feel the masculine adjectives apply to them, and a separate F-score, which indicates how many of the traits judged especially desirable for females, describe them well. Thus it is possible for a subject to score high on both the M and the F scales – Bem labels these people *androgynous*. Others had scores sharply skewed in either the M or the F direction; and a few people (whom we would expect to have a rather low self-esteem!) rated themselves as lacking the positive attributes with either masculinity and femininity.

Using the BSRI score as a measure of one's perceived gender profile, psychologists have gone on to study its influence on a variety of behaviors. Not surprisingly, in experiments requiring subjects to be photographed performing sex-stereotyped activities, such an oiling squeaky hinges on a box or winding yarn into a ball, the androgynous subjects were equally comfortable in each situation while the strongly sex-typed subjects didn't like posing in a cross-sexed role (Bem, 1977).

Evidence that a cognitive schema is playing a major role is provided by experiments which measure a subject's reaction time in a variety of information-processing tasks. To give a very simple example, when an adjective is flashed on the screen and subjects are asked to signal quickly whether or not it applies to them, sex-typed individuals process sex-stereotyped adjectives much more quickly than neutral ones, while androgynous subjects tend to show the same response times for both neutral and gendered adjectives. (Stillings, 1987, p. 35).

These experiments suggest two general features of schemata – they tend to distort our perceptions and limit our choices in undesirable ways (as in the photography experiment), but they also allow us to respond quickly to situations and to arrive at a decision for action, even though it may not be the optimal one.

The concept of gender is probably atypically complex, both cognitively and emotionally, but even the simplest concepts seem to have a complicated structure. Psychologists interested in prototype theory find that a concept such as *bird* orders features and functions according to degrees of typicality. (Cohen, 1984) Thus a robin is a more typical bird than a penguin because it flies. Robins are also more typical in size than eagles, but less typical with respect to color. It appears that people make use of these intuitive assessments of degree of typicality in order to estimate probabilities. Thus subjects may discount rather direct evidence that Jones is a bank teller if Jones fails to match the prototypical concept of bank tellers.

This might not be so bad if our typicality assessments reflected in at least a rough fashion empirical distributions of the high-lighted features, e.g., if our prototype corresponded to characteristics of the population mean. However, this is usually not the case, partly because we aren't that good at intuitive statistics, but more importantly because our prototypes tend to remain centered around the exemplars which are first used to teach us the concept. (Holland, 1987). For example, picture a balloon to yourself. How big is it? What color is it? Does it have a string attached? For most Americans, the paradigm balloon is a bright, saturated red color (as in story books). However, my informal survey of toy stores revealed very few balloons which are truly red, although pink is fairly common.

Thus, the very structure of our concepts tend to reflect our earliest, most parochial experiences. As we encounter penguins and dirigibles we may later refine them, but unless something drastic happens our conceptual default settings or knee-jerk reactions will remain as they were set in childhood. So for most of the readers of this paper, the prototypical human being will be white, a deeply religious person will be Christian, and a surgeon will be male.

4. CONCLUDING REMARKS

Are we then prisoners of our own concepts? Certainly not when we have time to reflect. Using the standard norms of critical thinking we can correct and improve upon our first-order, oversimplified conclusions and decisions. But what about our spontaneous perceptions, assessments, and utterances? Do we have any control over the structure of our concepts or will those same old stereotypes surface whenever we are tired, distracted, or in a hurry?

I think we can take responsibility for those embarrassing or damaging 'slips' which we all find ourselves making on occasion. But the solution is not just to think more before we speak (although that may be a good idea!). Rather what we need to do from time to time is to engage in some conceptual remodeling, to restructure our default assumptions and to change our typicality orderings. And the very properties of our cognitive apparatus which reinforce damaging stereotypes can help us to change them. We can go out of our way to collect *vivid* experiences which instantiate new prototypes. We can rehearse and reinforce the new conceptual structures by correcting our pronouns or introducing new vocabulary. As science students we sometimes succeed in replacing our naive, common-sense pictures of the world with the more accurate representations of physics, chemistry and biology. Skiers practice on overcoming their instincts to lean into the hill and good drivers learn not to jam on the brakes during a skid. As students of social science, and as citizens, surely we are also obligated to improve upon our cultural schemata.

Indiana University

NOTE

[1] Earlier versions of this material were presented to philosophers at Colby College and to psychologists at the University of Maine. I would like to thank the students and faculty for their feed-back on both of those occasions.

I would also like to acknowledge the influence of John Watkins' philosophical work on the general perspective taken in this paper, especially his excellent article, 'Imperfect Rationality', which combines a normative analysis with the pragmatic aspects of human decision making, as well as his spirited defense of methodological individualism.

REFERENCES

Arizona v. Norris: 1983, *Supreme Court Reporter*, **103**, 3492–3512.

Bem, Sandra L.: 1976, 'Probing the Promise of Androgyny', in Kaplan, Alexandra G. and Bean, Joan P., *Beyond Sex-Role Stereotypes*, Boston: Little, Brown, and Co.

Cohen, Benjamin and Murphy, Gregory I.: 1984, 'Models of Concepts', *Cognitive Science*, **8**, 27–58.

Constantinople, Anne: 1973, 'Masculinity–Femininity: An Exception to a Famous Dictum?' *Psychological Bulletin*, **80**, 389–407.

Gifis, Steven H.: 1975, *Law Dictionary*, Woodbury, N.Y.: Barron's Educational Series.

Holland, John H., Holyoak, Keith J., Nisbett, Richard E., and Thagard, Paul R.: 1987, *Induction: Processes of Inference, Learning and Discovery*, Cambridge, Mass.: MIT Press.

Nisbett, Richard and Ross, Lee: 1980, *Human Inference: Strategies and Shortcomings of Social Judgment*, New Jersey: Prentice-Hall.

Reinish, June M., Rosenblum, Leonard A., and Sanders, Stephanie A. (Eds.): 1987, *Masculinity/Femininity: Basic Perspectives*, Oxford University Press.

Salmon, Wesley C.: 1970, *Statistical Explanation and Statistical Relevance*, University of Pittsburgh Press.

Spence, Janet T. and Helmreich, Robert L.: 1978, *Masculinity and Femininity & Their Psychological Dimensions, Correlates, and Antecedents*, Austin: University of Texas Press.

Stillings, Neil A., Feinstein, Mark H., Gardfield, Jay L., Rissland, Edwina L., Rosenbaum, David A., Weisler, Steven E., and Baker-Ward, Lynne: 1987, *Cognitive Science: An Introduction*, Cambridge, Mass.: MIT Press.

Terman, Lewis M. and Miles, Catharine Cox: 1936, *Sex and Personality: Studies in Masculinity and Femininity*, New York: Russell and Russell.

ERNEST GELLNER

HUMAN RIGHTS AND THE NEW CIRCLE OF EQUITY

The issue of human rights is seen in the West in the context of a very distinctive institutional and ideological background. This includes a state which is both strong enough to ensure public order without tolerating coercive self-help on the part of sub-communities, and yet not strong enough to dominate civil society, and which can be checked by it. Moreover, it is an instrumental state, one charged with the performance of a set of tasks which, however, do *not* include responsibility for ultimate human fulfilment, or for the rule of righteousness. The suppression of all corruption on earth is not necessarily the state's business, at least not unless it disturbs the neighbours.

Whatever the ultimate historic roots of the notion of human rights, their content and manner of application has become profoundly secularised. They apply to believers and unbelievers alike, and their validity is not held to depend on any religious premise. Man is claimed to be endowed with inalienable rights as man, and not as a corollary of any theological doctrine. The rights do not stand or fall with the acceptance or rejection of any faith. Neither the validity nor the range of application of basic rights is tied to religious conviction. This premise-less spontaneous generation of human rights may possibly be a typical Western illusion, a characteristic von Munchausen philosophy of the supposition that we can morally lift ourselves by our own braces: critics may well claim that this position is parasitic on a religious tradition which it disavows, and that it sits on a branch which it claims to have sawn off. Be that as it may: as now presented, the doctrine of human rights has disassociated itself from any religious foundation.

The question which one must ask is – what happens to the notion of human rights when these various assumptions do not hold? There is one way of seeking convergence of world-views, by means of scrutinising the texts of a civilisation for pronouncements which seem to converge with ideas which appeal to us. No doubt, isolated statements can often be found which enable one to construct, not one, but many moral and doctrinal overlaps of this kind. I find such a method inappropriate, arbitrary and futile. Instead, I'll try to concentrate on the central moral and political institutions of the civilisation in question, and on a certain concrete tradition of political activity which it embodies. The central idea is that of an eternal, transcendent, unmanipulable

Fred D'Agostino & I. C. Jarvie, Freedom and Rationality: Essays in Honor of John Watkins, 125–140.

and available *Law*. The central tradition is that of a state obliged to implement such Law.

Consider a social order which has the following features: first of all, it is endowed with a central state which is simultaneously too weak to monopolise legitimate violence, yet is absolutist in its pretensions. It is legitimated in terms of its *moral* role, its obligation to endorse good and suppress evil. Its role on earth is to enforce divine law. It is both willing and able to dominate civil society, without sharing power with it.

This characterisation, combining as it does weakness and strength, is not contradictory: the various items in it refer to different parts of the geographical and social landscape. In much of the countryside, especially the more inaccessible, mountainous or near-desert regions, the state's writ fails to run. Though the sovereignty of the ruler is recognised, in written deeds recording allegiance, or in rituals of submission, nevertheless the day-to-day maintenance of order remains almost entirely in the hands of local self-administering, mutual-aid and mutual-insurance groups, commonly known as tribes.

These nested groups practice the *feud*. The feud makes an entire group collectively, and all its members severally, responsible for acts of aggression committed by any one of their number. Thus it obliges them to police their own members, for fear of attracting reprisals directed at all of them. Not only does the maintenance of peace depend on such self-policing by each group: there is simply no other kind of policing. The central state is not locally represented by any kind of militia, and the tribal groups, which are armed, internally cooperative and ideally cohesive communities, have no means of bringing to court an individual from another group, who has or is held to have committed an offence against one of their own number. They cannot normally demand his extradition (though the terms of a peace-settlement will quite often require his exile from his group of origin). They can only put violent pressure on his group as a whole, treating its membership as an undifferentiated and collectively responsible body. The recognised rules of the feud not merely permit them to do so, but positively encourage and endorse it.

Furthermore, over and above the feud provoked by individual acts of aggression, the tribes indulge in raiding as a kind of recognised institution, partly for gain, partly for sport. It is understood that both raiding, and the defence against it, help maintain group cohesion. Performance in this national sport helps select the successful candidates for group leadership. The central state accepts all this, sometimes with better and sometimes with worse grace. It is content, or has perforce to be content, with some symbolic recognition of its own sovereignty, with the display of some respect for its own lines of communication, and – sometimes – with a modest measure of taxation. It seldom dares impose its own nominees as local administrators: it prefers to act through local power-holders endowed with their own social base, and it will try to strengthen some of these and undermine others, aiming more at a balance of power than at outright domination. It is generally unclear whether the payments which flow downwards from the state to local notables (payments which one may consider to be a kind of blend of salary and bribe) are greater than the symbolic taxes which flow upwards from the tribal areas to the centre.

So much for the role, or non-role, of the state in peripheral but extensive tribal regions. In the city, and its surroundings, its role is altogether different. No corporate groups defy its authority. Hardly any corporate groups exist at all. The population it governs is pulverised, and not allowed to combine in any overtly political associations. If families do possess kin links among each other, the state does not allow them to fuse, like their rustic cousins, into order-maintaining, mutual-insurance associations. As Ibn Khaldun elegantly puts it, it is the state which prevents injustice, other than such as it commits itself. It holds, as you might say, a monopoly of injustice, and does not allow others to encroach on it.

So it does not share with others its obligation to enforce good and proscribe evil, and the chief executive is not required to give an account of his trusteeship, or present himself for re-election. Nonetheless, even in theory, his power is not absolute in all respects. Legislation, conspicuously, is in principle outside his competence. It has been preempted by the deity itself. The state is prevented from modifying, not merely some entrenched and constitutional aspect of law, but equally, and perhaps primarily, the law governing the detail of social life.

This written law is in the keeping of a loosely recruited, and more or less open and informal guild of scribes, theologians/jurists. The Law is God's alone, but access to it, given its availability in a corpus of writing, is reserved for the scholars. These scholars cannot, either individually or collectively, defy the state: they possess no social base for so doing. The most that their urban clientele is capable of is a very occasional *fronde*, a riot of protest. It does not possess the clout to overturn the political order or even to menace it very seriously. It is true that the urban scholars confirm a new ruler: but this is an ex post facto ratification of an established power situation, not a free choice.

Nonetheless, though generally quite powerless to determine the occupancy of key positions in the system, the loose guild of scholars does have a remarkable hold on the overall moral climate of the society. This climate in subtly varying ways pervades both the urban and the rustic and tribal segments of the society. The theology communicated by the scholars, with its emphatic monotheism and divine nomocracy, is deeply revered and internalised by the entire society. It confers the ultimate and crucial identity on all its members.

So there is a political threat which does hang over the rulers: the danger of an alliance between the possessors of legitimacy (the scholars), and the monopolists of cohesion (the tribes). This is not something that happens every day, but from time to time it can and does happen. If a highly respected urban scholar condemns state authority for its moral transgressions, say its illegitimate imposition of taxes or its immoral non-Islamic life style or both at once, and if such protestor goes in the wilderness and provides leadership for militarily experienced and cohesive tribal groups, then, and perhaps only then, central authority is in real peril. His authority confers legitimacy on the non-submission, jealousy and rapacity of the tribesmen, and enables them to combine, in the name of lawful divine wrath, in units larger than they would manage in the name of the local shrine. It is in this way that the wheel of political fortune turns, and a new dynasty appears in the name of purification and revival.

This mechanism also provides the answer to the question – where did the state come from in the first place? If the urban population is atomised and politically supine, how did a state organisation ever emerge in its midst? And indeed, the central authority did not spring up as some spontaneous growth from within the city. A cohesive rural group, its cohesion strengthened by religious leadership and enthusiasm, imposed itself on the centre of society, and the erstwhile tribal chief becomes a monarch. He is a kinsman to his tribal army, and an absolute ruler to his urban subjects.

The model I have sketched out is taken almost entirely from Ibn Khaldun, unquestionably the greatest sociologist of Islam, and perhaps the greatest sociologist ever. But is his model applicable to all Muslim societies, even within the arid zone? This has often been contested, largely on the basis of material about the *mamluk* system of government. Ibn Khaldun was fully aware of this phenomenon, and made numerous observations about it, but did not fully incorporate it into his theory. In any case, he lived before the time when this system found its finest expression in the Ottoman Empire.

In the Ottoman/mamluk variant of the model, political cohesion at the centre is forged, not by the vicissitudes of ungoverned tribal life in the desert and the mountains, but by sustained formal training and education from an early age. 'Slaves' are purchased, from the Caucasus or elsewhere, or recruited by means of the famous baby-tax (*devshirme*) from subject populations, and subjected to sustained and testing training, which turns them into the soldiers, administrators and bureaucrats of the state. The state owns them, and they own the state. Their esprit de corps is not that of a kin group, but of a body of men who have jointly undergone drill and indoctrination. Both their recruitment and their training cut them off from other social links and temptations, and their technical lack of property – theoretically they own neither themselves nor anything else – makes it easy for the state to revoke their privileges, and link their position to loyal and efficient service. It can place them wherever they are needed, and order them to do whatever is required.

This is the Platonic formula for social cohesion and political self-perpetuation: a severely trained and disciplined, communistic elite, undistracted from duty by either wealth or kin, committed to a meticulous performance of duty by a deep internalisation of the dominant values. In the end, the system failed to be perfectly self-perpetuating, for reasons Plato himself clearly foresaw. Surreptitiously or otherwise, the elite acquires both wealth and kin, and transmits its position to heirs, in defiance of the original logic of the system. The temptations of honour, of wealth, and pleasure, seduce them from the stern principles of their training.

However, the Ottoman system differed from the Ibn Khaldun model which it replaced, sufficiently at any rate to confer great political stability and longevity on the Empire. It developed its own political theory and economic philosophy, known as the Circle of Equity – attributed to Aristotle, formulated by a Persian, but adopted by the Ottomans. An elaborate version of it runs as follows: no sovereign authority without an army; no army without wealth; no wealth without subjects; no loyal subjects without justice; no justice without harmony on earth; no harmony without a state; no state without law; no enforcement of law without sovereign authority.[1] Its kernel can also be stated far more briefly: the state needs resources if it is to keep the peace. Subjects need peace if they are to produce wealth. If either interferes in the tasks of the other, no good will come of it. Each to their task, and never the twain shall meet. Producers are not to meddle in politics, and rulers are not to meddle in trade and production. Political participation and economic productivity are not meant to be combined in the same persons. Rulers and producers are not to be parts of one community, even if they are parts of one body politic. So once again, civil society is weak and the state strong.

The Ottoman model differed from the Ibn Khaldunian one, not merely in its greater political stability, but also in its greater tendency to deal with socially autonomous communities, defined by religion. It was also endowed with a larger intermediate zone of dominated and exploited peasants, caught between the state on one side and tribesmen on the other, and harassed by both.

It is debatable whether the existence of the Ottoman Empire really demonstrates the inapplicability of the Ibn Khaldun model to post-mediaeval Muslim history. My own guess is that the Ottoman system worked on its own terms only in some central regions – the Balkans, the Nile Valley, Western Anatolia, and perhaps a few other key strategic points. In other extensive areas, the world of Ibn Khaldun persisted unabated under the surface, as the effective local form of organisation. It re-emerged conspicuously and dramatically with movements such as the Wahabis, the Mahdia in the Sudan, or the Sanussiyya in Cyrenaica, or Emir Abd el Kader in Algeria. In regions such as Eastern Anatolia or the Arabian peninsula, it was always conspicuously present.

In either model, *rights* inhere in a man in virtue of his membership of a kin group, or of a religious community. A tribesman has rights primarily as a member of a team, a kin network. A townsman has them as a member of a religious community, and they are linked to his faith. It is less than clear what other rights a man could claim otherwise. In the traditional world, the question of abstract human rights was not frequently posed. What happens to this social order with the coming of the modern world, in which the issue of abstract human rights *is* posed?

The first and perhaps most important point is that the Muslim world defies the Secularisation Thesis. The idea that the modern world tends to become secularised is one of the best diffused truisms of sociology. The precise extent of secularisation no doubt varies from place to place. So does the specific form it takes. Much depends no doubt on what exactly we choose as our criteria of secularisation. If we select something like 'the independent and dominant authority of religious ideas and institutions in society' as the touchstone of *non*-secularisation, there can hardly be any doubt but that, as a rough overall generalisation, the secularisation thesis is indeed valid. In many places, secularisation in this sense is part and parcel of the local belief system: it is held that the state and society are associations which must perform their functions independently of the truth of specific religious convictions or doctrines. They do not depend on a religious consensus for their viability. To wait for it would be a recipe for disaster. Political cohabitation must be established on merit, and without the benefit of consecration. In some places, of course, a positively anti-religious element is part of the official ideology.

None of this is true of the Muslim world. Allowing for one or two exceptions (notably Turkey), the authority of religion over state and society remains unchallenged, and is perhaps enhanced. Significantly, this is as true of conservative regions dominated by traditional elites (Northern Nigeria, Saudi Arabia), as it is of the radical states ruled by new princes (Libya, South Yemen, Algeria). What is the explanation?

A number of elements can figure in a plausible answer.

The strict scriptural monotheism, the nomocratic puritanism and egalitarianism of *High* Islam, are particularly compatible with the requirements of a modern or modernising age. It is indeed 'modern' in most of its features, other than perhaps its firm faith, its lack of doubt. But Islam is relatively free of those embarrassing 'superstitious', magic-oriented, locally rooted items, which function so well in the context of highly localised social structure, but which wither and become hard to sustain in an atomised, mobile society exposed to the chill winds of scientific or scientistic doctrine. A High-Culture Islam had always been genuinely present, even if not much practiced by wider social strata. This genuine continuity enables it to be invoked in self-identification by the newly urbanised masses, and by populations seeking a national image in the conflict with colonial powers or with their own over-Westernised elites (as in Iran). They acclaim it, and disown the erstwhile low-culture variants, repudiating them as corruptions. It ratifies their ascension to urban status, it defines their wider identity with dignity, it expresses their resentment against over-privileged rulers. They can turn to their own, previously half-ignored High Tradition, fairly unburdened as it is by petty magic, and unsullied by the disgrace of the traditional state.

The presence of a religious ideology which is so modern in many of its features – its elegant monotheism, its rule-ethic rather than loyalty-ethic, its scripturalism – but which at the same time is genuinely local, enables Muslim populations to escape the dilemma which otherwise generally faces Third World cultures: is one to ape alien culture (Westernisation), or idealise the local folk culture (populism)? Muslims can opt for the High Culture variant of their own faith, which is genuinely theirs rather than imported, and – except in Turkey perhaps – is relatively untainted by the disgrace of the *political* ancien regime. They need not idealise either an alien or a folk culture.

The Third World once began on the Rhine, and most countries which passed through this condition did face such a dilemma. Muslim reaction to the Western challenge in the main had the form of a sustained purification of the old religion, of *Reformism*. It was very seldom populist in tone. A form of populism was practised only vicariously by Westerners, in what may be called the T.E. Lawrence style.

The second conspicuous feature of the new situation is that the state remains very strong vis à vis civil society. It is now strong even where it had once been weak: it can impose its will on rural populations, even in out-of-the-way locations. Modern military, communications and administrative techniques make this possible. The state also remains strong where it had always been strong, in the cities and amongst settled and easily accessible populations.

There is perhaps nothing distinctively Islamic about this: all this is very common throughout the Third World. There is a certain kind of backwardness, in which a state, propped up by borrowed technology controls, thwarts and inhibits the society it dominates. It is easier, or more tempting, to borrow the technologies of domination than those of production. The result – a relatively advanced system of coercion superimposed on a backward system of agricultural production – may well be self-perpetuating. It leads to a situation in which it is far more advantageous to be a coercer than a producer, and which leads to a severe struggle for positions in the coercive hierarchy. This in turn further inhibits any development of the productive sector. Oil wealth, where it is available, tends to strengthen rather than counteract such a self-perpetuating situation.

The softening of political conflict, when and in as far as it has occurred in the Western world, was conditional on the appeal and attractiveness of rewards within civil society. The life prospects of producers, or some of them, became more attractive than those of coercers, and producers became increasingly protected from depredations by coercers. 'Democracy' and liberalism can work, because the occupancy of top political posts is no longer the most attractive, or only attractive, position in society. On the contrary, the direct rewards of power are relatively meager, and the posts are sought more for the prestige or glory they bring, rather than from material incentives. So the political struggle is no longer carried out on the winner-take-all principle, and its stakes are relatively low. Larger pickings are available in civil society. Under such conditions, a voluntary rotation of power, with respect for the rules, maintained whether or not one wins, becomes feasible. This situation is seldom ever reached in contemporary Muslim politics.

The struggle for power within Muslim societies is admittedly no longer carried out by the old rules. Tribesmen are no longer militarily relevant. The networks which do arise and compete with each other for control of the state are not made up of shepherd tribes, brought together by some Guardian of a Shrine. They are made up of unsymmetrical dyadic patronage relations, which are territorially discontinuous, and are linked by telephone not pilgrimage. These bonds are generally instituted between unequals. Support is traded for protection.

Patronage politics are of course in no way restricted to Muslim lands: they are very common, for instance, throughout the Mediterranean zone, on its northern as well as its southern and eastern shores. But Muslim lands are very prone to them, and the regionally based networks are a kind of shadow of the old kin groupings, which had always used clientship to complement kinship.

Patronage-network politics have many sources, but one of them is this: if a moralistic political order precludes legitimate opposition, and requires social and ideological homogeneity, the only way of resolving disputes is through informal groupings within the nominally monolithic structure. This is something the quasi-theocratic states share with radical-socialist ones. In Libya, one can see the superimposition of both Muslim and socialist/populist fundamentalism.

These then are the main features of the new scene: a strengthened, fervently internalised and purified religion; a strong state superimposed on a weak civil society, haunted by the conflict of rival patronage networks, generally more opportunist than genuinely ideological – though the ideological seriousness and commitment are much more genuine with the fundamentalists. The fundamentalists proper, i.e. under clerical not military leadership, have dramatically secured power in the course of one particular revolution which really brought home to the West the importance of Muslim fundamentalism – the Iranian.

In other Muslim lands one can observe a kind of spectrum, ranging from military leaders with a strong and deeply felt religious commitment, via others whose devotion may be more tepid or instrumental, and ending with the tacit secularism of Tunisia (not to mention the overt but now muted and ambiguous secularism of Turkey). Or perhaps one should say that there are two diapasons not one: a traditional one where the rulers are the heirs of an Ibn-Khaldunian conquering tribal group, as in Saudi Arabia or Northern Nigeria, and a modern one, where they are the new mamluks, fellow-graduates of military and/or technical schools. This distinction cuts across that of whether or not a country has had a colonial period: Morocco and

Northern Nigeria or Malaysia, all with a colonial past, are endowed with non-military traditional political elites, whilst North Yemen, which has only rather briefly been colonised by the fellow-Muslim Ottomans, has experienced military rule.

A certain irony attaches to the fact that Iran constitutes the extreme and purest example of fundamentalist clerical rule. The Persians are defined by their Shi'ism, rather in the way in which Poles are defined by Catholicism or Serbs by Orthodoxy. In most places in which they are found, Shi'ites are the underdogs, but in Iran the sect defines both the rulers and the national identity. Within the *sectarian* prism of Islam, Shi'ism is the furtest removed from nomocracy, from the worship of law, and the one closest to an overt, undisguised anthropolatry and cult of sacred personality. Cult of the community, cult of the Book and cult of Sacred Leadership are the legitimacy-conferring themes intertwined in Muslim theology. Mainstream Sunni Islam blends reverence for the Book with respect of communal consensus, making only moderate concessions to the House of the Prophet, well this side of a cult of personality. The Kharejites (*those-who-left*, i.e. almost literally the Dissenters) constitute a kind of theological Left, with an egalitarian stress on the community and merit. It was the Shi'ites who solved the old Platonic problem of Who Guards the Guardians – which in its theological version reads: who Checks the Interpretations of the Interpreters? – by postulating divine personages, *Imams* in the specific Shi'a sense. Their incorruptible perfection is held to be such that the veridicity of their pronouncements and verdicts are free of possible doubt. In consequence, of course, only such Imams can claim genuine and ultimate political authority.

In the political practice of most though not all Shi'ite sub-segments, the implications of this position have been modified by the absence, the famous Occultation, of the Hidden Imam(s). If the perfect leader and Interpreter is *in hiding*, someone else must assume these tasks *pro tempore*. One of the segments, the Ismailis, have an Imam who is not in hiding, but concretely, physically present. Hence their particular political theology has been most effectively used in the overriding of Law by divine leadership, in the interest of an amazingly rapid and successful commercial adaptation by the Ismaili community to the circumstances of modern trade.

With hindsight, it is obvious that the amazing Iranian Revolution, carried out, not against a militarily defeated and financially bankrupt regime, but against an exceedingly rich one, endowed with a fine and undefeated army, was made possible by its association with a deeply remembered, ritually re-enacted cult of Martyrdom. Shi'ite scribes are not just lawyers, but also professional perpetuators of the memory and life of a Martyr. It is this element which enables them to communicate rather more effectively with the masses, and to arouse their feelings. The more arid Sunni lawyer/theologians are often obliged to leave this task to questionably orthodox Sufi saints. It was the ecstasy of the martyrdom which enabled Shi'ite rioters to break the will of the Shah's armed forces. Khomeini himself has proudly commented on this distinctive Shi'a potential for revolution:

> The Shi'a school of thought ... has had certain distinguishing characteristics from the very beginning. While other schools have preached submission to rulers, even if they are corrupt and oppressive, Shi'ism has preached resistance against them and denounced them as illegitimate. From the outset, Shi'is have opposed oppressive governments.[2]

No doubt the Shi'ite cult of the martyred personality greatly aided the fundamentalist revolution. But *after* the revolution, and in the course of constructing a new regime and formulating its principles of legitimacy, it receded into the background. The ultimate and exclusive authority and legitimacy of the Hidden Imam is of course not denied. That would be unthinkable, and would constitute an intolerable inversion of Shi'ite faith. But, in Khomeini's political writings, in a subtle way it is made irrelevant.

It is *The Law* which is to be enforced, with the same unrelenting severity and rigour, whether or not the Imam has returned from hiding. That is what government is really about, and it does not in any way change either with an Occultation or Second Coming. When the Imam returns, He will assume the burdens of government and administration, which however are quite distinct from his divinity, and a kind of burden imposed on Him: *divinité oblige*. Khomeini makes it clear that the Imam only accepts political responsibility rather reluctantly, as a deplorable chore – somewhat in the spirit in which I undertake administrative duties in the University.

But as long as the Hidden Imam remains absent, the task must be shouldered by those best equipped to handle it. Who else but the Doctors of the Law? They also only take it on as a *duty*. It is a lawyer-led republic which must assume the burdens of Provisional Government. A new Circle of Equity emerges: society needs justice. But justice can only be had if the state is willing and able to enforce it. Only those who *know* the Law – available because Revealed – are able to do this. Khomeini's message to them is that they *must* assume this task and that, at least until the Hidden Imam returns, no one else is capable of discharging the task. Given oil wealth, it is no longer necessary to insist so much on the surplus which civil society must hand over (though Khomeini does insist on the adequacy of orthodox Muslim principles of taxation); what civil society really owes the scribes-rulers is obedience. What it receives in return is an enforced morality and that moral rootlessness and its associated ills, which is the base of a faith-less modern society.

So, with the later thought of Khomeini, a lawyer-led nomocratic republicanism comes to the fore. *Logically*, this had always been implicit in Shi'ism, as a theoretically available option. The pervasive historical circumstances of an earlier age, however, obliged the clerics to accommodate themselves to tribal-leaders-turned-monarchs. The scholars had to be content with merely *guiding* rulers, rather than ruling themselves. But now it turned out to be Shi'ism, which was best equipped to mobilise masses and effect a revolution. At the same time, it was Sunnism, with its emphatic nomocracy, by implication favouring the Doctors of Law, which was best equipped to provide a new political theory once the revolution had succeeded. In Iran, we see a victorious Shi'ism, in effect unavowedly going Sunni in the course of actually governing. The Imams are tacitly pensioned off: the Return of the Imam from Occultation will make no *political* difference. He rather than the scholars will then shoulder the burden of administration: but the way in which it is done will not change one bit. It is implied that the Imams cannot change the Law and they are subject to it. The Law will rule, whoever presides over the social order. The Law is to be enforced one way or the other, by those qualified, whether through their legal expertise or through their ultimate and immutable perfection.

Khomeini's political thought is clear and vigorous, and available in English in a translation by a man who combines outstanding Orientalist qualifications with that of being a convert and an enthusiastic admirer – Hamid Algar, of the Department of Near Eastern Studies at the University of California at Berkeley.[3] All quotations from Khomeini are taken from this source.

> A body of laws alone is not sufficient for a society to be reformed...For this reason, God Almighty, in addition to revealing a body of law...has laid down a particular form of government together with executive and administrative institutions.

> The Most Noble Messenger...undertook the implementation of law and the establishment of the ordinances of Islam, thereby bringing into being the Islamic state. He did not content himself with the promulgation of law; rather, he implemented it at the same time, cutting off hands and administering lashings and stonings. After the most Noble Messenger, his successor has the same function and duty.

No one can say...that the penal code of Islam, with its provision for the payment of blood money and the exacting of requital, should be suspended.

...all non-Islamic systems of government are the systems of *kufr*...and it is our duty to remove from the life of Muslim society all traces of *kufr* and destroy them.

Islamic government does not correspond to any of the existing forms of government...It is not constitutional in the current sense of the word...It is constitutional in the sense that the rulers are subject to a certain set of conditions...that are set forth in the noble Qur'an and the Sunna of the Most Noble Messenger...Islamic government may therefore be defined as the rule of divine law over men.

...in Islam the legislative power and competence to establish laws belongs exclusively to God Almighty. The Sacred Legislator of Islam is the sole legislative power. No one has the right to legislate and no law may be executed except the law of the Divine Legislator.

Islamic government is a government of law. In this form of government, sovereignty belongs to God alone and law is His decree and command. The law of Islam, divine command, has absolute authority over all individuals and the Islamic government. Everyone, including the Most Noble Messenger (peace be upon him) and his successors, is subject to law and will remain so to all eternity – the law that has been revealed by God...and expounded by the tongue of the Qur'an and the Most Noble Messenger.

It is an established principle that "the *faqih* ['theologian/lawyer'] has authority over the ruler." If the ruler adheres to Islam, he must necessarily submit to the *faqih*...This being the case, the true rulers are the *fuqaha* themselves, and rulership ought officially be theirs.

...government is necesary, and...the function of government...is still enjoined upon us by God...The two qualities of knowledge of the law and justice are present in countless *fuqaha* [plural of *faqih*] of the present age. If they would come together, they would establish a government of universal justice in the world. If a worthy individual possessing these two qualities arises and establishes a government, he will possess the same authority as the Most Noble Messenger.

Can there be any distinction in this respect between the Most Noble Messenger, and the Imam, and the *faqih*? Will the *faqih* inflict fewer lashes because his rank is lower? Now the penalty for the fornicator is one hundred lashes. If the prophet applies the penalty, is he to inflict one hundred and fifty lashes, the Commander of the Faithful one hundred, and the *faqih* fifty?

The judicial and governmental functions assigned by the Imams to the *fuqaha* of Islam are retained permanently.

The general thrust of the argument is clear. The Muslim polity is to be governed by divine law, and by divine law alone. This law must, in the absence of the Imams, be both interpreted and implemented by the scholars. Who else? The Shi'ite element in Islam, the transfer of ultimate authority to super-human Imams, is not exactly denied: but it is rendered politically irrelevant. Law not personality is the ultimate fount of rectitude.

One can speculate about what possible role a non-theological conception of human rights could have in such a polity. But Khomeini himself provides one possible answer. He harangued his audience:

> Xenomaniacs, people infatuated with the West...Come to your senses...Look at the West, and see who the people are that present themselves in the West as the champions of human rights and what their aims are. Is it human rights they really care about, or the rights of the superpowers?...Our jurists should not follow or imitate them.

He then proceeds to pour scorn on the Society for the Defense of Human Rights, described by the translator in a footnote as "a body composed chiefly of jurists who objected to the activity of the Revolutionary Courts."

Khomeini does of course have a point when he insists that his conception of the polity is not a tyrannical one. There is something noble about such unwavering commitment to the notion of a Law well beyond the reach of all human manipulation. This had always been *an* element in Muslim civilisation. On its own, it could hardly restrain the rulers: Khomeini's impeccable logic – all power to God's lawyers – was not then implemented, and past Muslim political theorists had to ignore the logic, and accommodate themselves to the reality of monarchical rule. Nonetheless, the reverence for an un-manipulatable Law in the keeping of Scripture and scripturalist scholars, *in conjunction* with the threat of religiously enlisted anger – and greed – of cohesive, aggressive and independent tribesmen, did constitute a kind of check on government. With such ferocious wolves at the gate, and such an unmanipulatable charter of moral/legal rectitude available in unchangeable writing, rulers had to watch their step, lest the guardians of Holy Script and the formidable armed tribesmen make joint cause. Rulers must seek not to give too much offence to potential and respected legitimators of rebellion. Avoid providing occasion and excuse for the fusion of normally fragmented tribal segments!

Tribesmen are no longer of political or military significance, at least in most places. As late as the first decade of this century, Bakhtiari tribesmen could still take part[4] in the Constitutional Revolt (whose aspirations Khomeini repudiates), and influence oil deals; but they are unlikely to do so today. Are there other social forces available, liable to perform the same role at present?

Khomeini is often said to have been brought to power by the bazaar, by the urban petty bourgeoisie: it *is* perhaps capable of constituting some kind of countervailing power? But most of its members tend to be attracted to puritan Reformism. They like the fundamentalist repudiation of Western wisdom, which is outside their reach and which excites their resentment. They like fundamentalism as a form of ratification of their Muslim urban ways, in contrast with their rustic past *and* the cosmopolitanism of the elite. They also like it as the definition of their new anonymous identity. They are reassured by its traditional assumptions about the economic order. This class is not much attracted by Marxism or socialism. It appears that it can shame the technocratic elite into outward and even some inner conformity. What sanctions it possesses is not yet clear.

A recent writer, Talal Asad[5], repudiates the claim that there is an affinity between Islam and Marxism, a claim invoking the fact that both are committed to the imposition of a total moral order in society. His argument purporting to show that there is no similarity between Muslim and Marxist social orders is interesting:

> ...a moment's reflection will show that it is not the literal scope of the sharia [divine law] as text that matters here but the degree to which it informs and regulates social practices, and it is clear that there has never been any Muslim society in which the religious law of Islam has governed more than a fragment of social life.

By contrast, he insists:

> ...in modern states...[t]he administrative and legal regulations...are far more pervasive and effective in controlling the details of people's lives than anything to be found in Islamic history.

Asad certainly has a point: any modern state is in many ways far more intrusive than any traditional state was able to be, however total its aspirations may have been. Asad may be overstating the exiguity of the zone of social life controlled by Islam even in the traditional world. But the crucial point is: can one usefully compare the *modern* socialist state with the *traditional* Muslim one? One must compare a *modern* socialist state with a *modern* Muslim one. Then the comparison becomes more illuminating. In both cases, there is an atomised civil society, a strong and moralistic state, committed to the implementation of an overall vision, is justified as the agency which implements that vision and is charged with leading mankind to a true fulfilment. It has a bias in favour of substantive, rather than merely procedural justice.

By contrast, the notion of human rights, as it is currently understood in the West, presupposes a kind of theological 'veil of ignorance': it respects a shared condition in which ultimate truth is not guaranteed to anyone. Our social contract is based on the assumption that in the public sphere, we share a scepticism, not a conviction. We come to the assembly, not only without arms, but also without dogma. The baseline assumption is that no one is entitled to be sure of possessing the truth, and so may not impose his version of it on others. So basic rights must be rooted, not in revelation, but in its absence. We are all of us entitled to our beliefs, but also to our doubts; and so our obligations to each other are based on the Cartesian notion of man as a legitimately doubting being. (Khomeini is reported to have said that the West is wrong to heed Descartes rather than Aristotle – a perceptive remark.) Man's claims and rights do not on this view hinge on the truth of any one faith, and they transcend them all.

Secondly, the realistic expectation of the enforcement of human rights hinges on the existence of countervailing forces in civil society, capable of guarding the guardians, of checking the state. It precludes the notion of the state as the *obliged* enforcer of an absolutely revealed and unquestionable Faith and Law, accountable to God for the implementation of righteousness – in which capacity the state is not even entitled to brook any opposition, nor to allow the countervailing forces to organise effectively, for to do so would mean betraying its trust. If an important tradition, or one very powerful political implementation of it, is committed to contrary assumptions, the question – what notions of human rights can it share with the liberal/sceptical one – becomes a very deep one.

Cambridge University

NOTES

[1] Cited in Lucette Valensi, *Venise et la Sublime Porte*, Hatchette 1987, p. 105.
[2] *Islam and Revolution, Writings and Declarations of Imam Khomeini*. Translated and Annotated by Hamid Algar. Mizam Press, Berkeley, 1981.
[3] *Ibid.*
[4] Gene R. Garthwaite: *Khans and Shahs*. Cambridge University Press, 1983.
[5] 'The Idea of an Anthropology of Islam'. Occasional Papers series, Center for Contemporary Arab States, Georgetown University, Washington, D.C., March 1986.

ADOLF GRÜNBAUM

THE DEGENERATION OF POPPER'S THEORY OF DEMARCATION*

I. INTRODUCTION

Karl Popper, on the one hand, the descriptive or hermeneutic phenomenologists, on the other, have offered radically different, influential diagnoses of the failure of psychoanalytic theory to pass scientific muster. As Popper would have it "Freud's theory...simply does not have potential falsifiers" (Popper 1974: 1004) and is therefore non-scientific. But, in his view, the time-honored inductivist conception of scientific rationality was unable to detect this fundamental flaw. And, thereby, it purportedly failed to give a correct diagnosis of the scientific bankruptcy of the psychoanalytic enterprise (Popper 1962: 33–38, 255–258; 1974: 984–985; Grünbaum 1984: 103–107). Therefore, Popper (1962: 256) concluded: "Thus there clearly was a need for a different criterion of demarcation" between science and pseudo-science, other than the inductivist one. In this way, psychoanalysis served as the gravamen and benchmark of his case for the superiority of his own falsifiability criterion of demarcation.

In my book (1984: 280), I argued for the following contrary thesis:

> It is ironic that Popper should have pointed to psychoanalytic theory as a prime illustration of his thesis that inductively countenanced confirmations can easily be found for nearly every theory, if we look for them. ...*it is precisely Freud's theory that furnishes poignant evidence that Popper has caricatured the inductivist tradition by his thesis of easy inductive confirmability of nearly every theory!*

But, as he tells us in his (1983) volume, its Chapter 2 contains his first "published ... *detailed* analysis of Freud's method of dealing with falsifying instances and critical suggestions" (p. 164, n. 1; italics supplied). When I was completing my (1984) book on psychoanalysis, Popper's (1983) volume was not yet available to me. And in my later (1986: 266–269) reply to him, I made only a cursory critical remark about his (1983) account of Freud's scientific miscarriage. Here, I shall therefore deal with his most recent and fullest treatment of this issue in adequate detail. And in Section IV, I shall deal with two other flaws in Popper's theory of demarcation that are unrelated to psychoanalysis.

Fred D'Agostino & I. C. Jarvie, Freedom and Rationality: Essays in Honor of John Watkins, 141–161.
Published by Kluwer Academic Publishers. Printed in the Netherlands.

According to Karl Jaspers, Paul Ricoeur and Jürgen Habermas, Freud misunderstood his own theory scientistically by giving far too little weight to 'meaning connections' between mental states, as distinct from causal connections. In their view, hermeneutic victory can be snatched from the jaws of Freud's scientific defeat, once we appreciate that the hermeneutic discernment of so-called meaning connections is at the heart of the psychoanalytic enterprise (Jaspers 1974; Habermas 1971, chap. 11; Ricoeur 1970, 1981). Elsewhere, I have argued that this criticism basically misidentifies both the source and the import of Freud's scientific shortcomings: far from giving too little explanatory weight to 'meaning connections' (thematic affinities), Freud endowed them with much too much explanatory significance, drawing fallacious causal inferences from them (Grünbaum 1989).

Therefore, I can confine the present paper to Popper's (1983) volume. Frank Cioffi (1985, 1988) has published two ill-tempered replies to my earlier criticisms of Popper's charge of pseudo-science against psychoanalysis. It is a mark of Cioffi's mode of disputation that in his (1988), he makes no mention at all of my (1986), where I painstakingly dealt with his objections to my (1984), as he knows. There (Grünbaum, 1986: 271–273), I also documented his exegetical fabrications and straw men. Erwin (forthcoming) has given a careful defense of my views against Cioffi's (1985). And I can now refer the reader to my (1988) for my most detailed treatment of Freud's *Rat Man* case, which undercuts Cioffi's objections to my handling of that case history.

II. POPPER'S 1983 MISDIAGNOSIS OF FREUD'S SCIENTIFIC FIASCO

Let me begin with a retrospect on Popper's treatment of the topic prior to the appearance of his (1983) *Postscript*.

Throughout his career, Popper has repeatedly made two claims: (1) Logically, psychoanalytic theory is irrefutable by any human behavior, and (2) in the face of seemingly adverse evidence, Freud and his followers always dodged refutation by resorting to immunizing maneuvers. According to (1), *none* of the deductive consequences of Freud's hypotheses are refutable by potentially contrary empirical evidence. But clearly, this charge of unfalsifiability against psychoanalytic theory *itself* does not follow from the *sociological* objection that Freudians are not responsive to criticism of their hypotheses. After all, a theory may well be invalidated by known evidence, even as its true believers refuse to acknowledge this refutation. Besides, if

Popper were right that "Freud's theory...simply does not have potential falsifiers" (1974, Book II: 1004), why would it have been necessary at all for Freudians to *dodge* refutations by means of immunizing gambits? Popper's (1) and (2) seem incoherent.

Ironically, it emerges clearly from some of Popper's other doctrines that the recalcitrance of Freudians in the face of falsifying evidence, however scandalous, is not at all tantamount to the irrefutability of their theory. As he tells us, theories, on the one hand, and the intellectual conduct of their protagonists, on the other "belong to *two entirely different 'worlds'*" (Popper 1974, Book I: 144, italics in original). Yet because Popper sometimes discusses them in the same breath, my response to his views on psychoanalysis takes both into account. My principal objection to him pertains to his logical thesis of empirical irrefutability, although I argued in (1984) that he also used gross oversimplification to make his case for his sociological objection to Freud's evasive methodological behavior.

Indeed, contrary to Popper, Freud unflinchingly issued some important theoretical retractions in response to acknowledging the emergence of refuting evidence. A telling case in point is furnished by his 1926 landmark revision of his prior views on the causal relation between repression and anxiety. There Freud wrote:

> It was anxiety which produced repression and not, as I formerly believed, repression which produced anxiety.
>
> It is no use denying the fact, though it is not pleasant to recall it, that I have on many occasions asserted that in repression the instinctual representative is distorted, displaced, and so on, while the libido belonging to the instinctual impulse is transformed into anxiety [footnote omitted]. But now an examination of phobias, which should be best able to provide confirmatory evidence, fails to bear out my assertion; it seems, rather, to contradict it directly. ...It is always the ego's attitude of anxiety which is the primary thing and which sets repression going. Anxiety never arises from repressed libido. ...I believed I had put my finger on a metapsychological process of direct transformation of libido into anxiety. I can no longer maintain this view. [S.E. 1926, 20: 108–109]

In a volume on Popper's philosophy, Popper (1974, Book II) had claimed – once again – that psychoanalysis is an empirically untestable psychological metaphysics, which does "not exclude any physically possible human behaviour" (p. 985). And from this allegation of empirical unfalsifiability, he immediately drew the fallacious inference (p. 985) that psychoanalysis indeed can, in principle, *explain* any actual behavior. Thus, on the heels of saying that Freud's and Adler's theories do not exclude any possible human behavior, Popper tells us that "whatever anybody may do is, in principle, explicable in Freudian or Adlerian terms."

But if a theory, in conjunction with particular initial conditions, *does not exclude* any behavior at all, how can it deductively *explain* any *particular* behavior? To explain deductively is to exclude: as Spinoza emphasized, to assert (entail) *p* is to deny every proposition incompatible with it. Note that in psychoanalytic theory, just as in Newton's physics, for example, law-like statements cannot explain particular behavior without initial conditions: without *suitable* initial velocity specifications, Newton's laws of motion and gravitation do not yield an elliptical orbit for the earth under the gravitational action of the sun. Hence, if no actual behavior could falsify psychoanalysis under given initial conditions, *I*, then this theory, cum *I*, could not explain *any* such behavoir deductively. *A fortiori*, if the theory *T* were unfalsifiable, it could not explain *all* such behavior, as Popper believes. Furthermore, if the conjunction *T* and *I* fails to explain some particular behavior *b* deductively, then *I* and *b* cannot confirm (support) *T* hypothetico-deductively. Thus, if psychoanalysis were unfalsifiable, how could any actual behavior – let alone *all* physically possible behavior – be explained by it so as to confirm it inductively, as Popper claims (1962: 35)? On the contrary, the alleged unfalsifiability would *preclude* such hypothetico-deductive confirmability.

According to psychoanalytic theory, in both genders, repressed homosexual desires are the causal *sine qua non* of paranoia, whatever the variations in the delusional modes of this psychosis. The modes include delusions of persecution, of jealousy, of grandeur, and of heterosexual erotomania (S.E. 1915, 14: 263–272; S.E. 1922, 18: 225–230; earlier Freud gave a *more tentative* statement in S.E. 1911, 12: 59–65). Thus, every sort of paranoiac is an arena for the psychic conflict between homosexual impulses, and the need to banish them from consciousness as objectionable to avoid emotional distress, such as disgust, shame or guilt (S.E. 1905, 7: 127 and 157–159). In brief, his or her delusions (of persecution, etc.) are the *unconscious defense* against the conscious emergence of the forbidden homosexual cravings. And this defense is mediated by two defense mechanisms: (1) Reaction-formation, which converts the homosexual love into hatred toward the love-object, i.e., once a dangerous impulse has been largely repressed, it surfaces in the guise of a more acceptable contrary feeling; (2) projection of the ensuing hatred from the lover by imputation to the beloved.

In 1915, Freud published a paper entitled 'A Case of Paranoia Running Counter to the Psychoanalytic Theory of the Disease' (S.E. 1915, 14:

263–272). The pertinent case is that of a women suffering from persecutory delusions. And, as suggested by the very title of this paper, Freud saw her case history as at least potentially supplying a *refuting instance* or disconfirmation of the particular sexual etiology he had posited for this patient's type of psychosis. In this vein, he notes (p. 265) that the patient's conduct might well supply evidence that she is *not* beset by a conflictual struggle against unconscious homosexual impulses. To take a contemporary putative example, imagine a self-declared lesbian, who is also paranoid, and living in San Francisco as the avowed lover of another woman. Suppose also that she is publicly active in the gay rights movement there. Freud emphasizes that if there are indeed empirical indications of such freedom from repressed homosexuality on the part of a paranoiac, then the person in question would *count against* the psychoanalytic etiology of paranoia. The posited case would be a contrary instance, because this etiology theory declares intense repressed homosexuality to be a *sine qua non* for paranoia.

As Freud points out furthermore (S.E. 1915, 14: 265), the hypothesized etiologic role of repressed homosexuality leads "to the necessary conclusion that the persecutor must be of the same sex as the person persecuted." But, if so, it becomes important to determine empirically to what extent the supposed persecutors are actually of the *same* gender as the persecutees. Since the delusions of persecution are themselves conscious, this empirical determination of gender does not require the techniques of psychoanalytic investigation as such, and is therefore extraclinical in that sense. It is, of course, logically possible that any and every paranoiac feels persecuted *only* by members of the *opposite* sex. For this reason alone, Freud's etiology is falsifiable by a finite number of such instances in the face of Popper's own denial of such falsifiability!

What then are the known findings among paranoiacs as to identity or difference of gender? And does Freud conclude that his etiology of paranoia would simply be refuted by fancied persecutors of the *opposite* sex?

His answer to the first question is unequivocal: "In psychiatric literature there is certainly no lack of cases in which the patient imagines himself persecuted by a person of the opposite sex" (S.E. 1915, 14: 265). Indeed, the paranoid young woman of his 1915 case history sought out a lawyer for protection against the imagined persecutions by a young man with whom she had been having an affair. As Freud admits: In the light of his homosexual etiology of paranoia, "It...seems strange that a woman should protect herself against having a man by means of a paranoic delusion" (S.E. 1915, 14: 268). How then, if at all, does Freud propose to reconcile such cases with the requirement of gender-identity that he himself had deduced from his homosexual etiology of paranoia? His reply takes the form of two claims that are, in principle, empirically testable. First, he reports "as a rule we find [in paranoia] that the victim of persecution remains fixated...to the same sex" (S.E. 1915, 14: 271). Furthermore, he maintains, whenever a supposed persecutor is of the *opposite* sex, then investigation will reveal *this* antagonist to be only a secondary one, whom the paranoiac imagines to be in collusion with an *original primary* persecutor of the *same* sex.

But are there statistics bearing out the first claim that the great majority of paranoiacs believe themselves to be persecuted by members of the *same* sex? Even if there are such statistics, neither Freud nor his followers have provided evidence to support his crucial further thesis that for any presumed persecutor of the opposite sex, there demonstrably exists a person of the same sex who was the *original* object of the paranoiac's persecutory delusions. Indeed, to the extent that, in the case of oppositely sexed persecutors, psychoanalysts *fail* to find the primary, gender-identical persecutor required by their theory, their etiology of paranoia is *disconfirmed.*

There are further reasons for concluding that the psychoanalytic etiology of paranoia may actually be disconfirmed. As Fisher and Greenberg have pointed out (1977: 259), "the appearance of overt homosexual...acting out in the paranoid would represent a contradiction...of Freud's theory of paranoia." How so? That theory makes the following conditional prediction: If a person does not strongly repress homosexual impulses, then this person will *not* be paranoid. But any individual who is openly and publicly a highly active practicing homosexual surely qualifies as a person who does *not* strongly repress his or her homosexuality. And if that same individual also exhibits strong delusions of persecution toward, say, various members of the opposite sex, then this flamboyant homosexual gives a strong observable indication of being paranoid. Yet, according to the prediction derived from Freud's etiology, just such an openly active homosexual should *not* be paranoid.

It should not be unduly difficult for extraclinical researchers to sample the population of avowedly practicing homosexuals of either sex to determine whether there are any paranoiacs among them. In fact, presumably it is not a very risky bet to suppose that there are. And, if so, their existence casts strong doubt on the psychoanalytic claim that all paranoiacs are *repressed* homosexuals. Yet Zamansky (1958) believes to have produced experimental support for it, which I find dubious (Grünbaum, 1986: 269–270).

In sum, the psychoanalytic etiology of paranoia is empirically falsifiable or disconfirmable, and Freud explicitly said so in his 1915 paper. As he noted (pp. 265–266), empirical indicators can bespeak the absence of strongly repressed homosexuality, as well as the presence of paranoid delusions. Thus, empirical indicators can fallibly discredit the stated etiology.

Hence, this example has an important *general* moral: Whenever empirical indicators can warrant the *absence* of a certain theoretical pathogen *P* as well as a *differential diagnosis* of the *presence* of a certain theoretical disorder *D*, then an etiologic hypothesis of the strong form '*P* is causally necessary for *D*' is *fallibly* refutable. The etiology will be falsified or disconfirmed by any victim of *D* who has *not* been subjected to the avowedly required *P*. For the hypothesis *predicts* that anyone not so subjected will be spared the miseries of *D*, a prediction having significant prophylactic import for child-rearing. Equivalently, the etiologic hypothesis *retrodicts* that any instance of *D* was also a case of *P*. Thus, Glymour's (1974) account of Freud's case history of the Rat Man points out that Freud's specific etiology of the Rat Man's obsession with his father's death was fallibly falsified by means of disconfirming the retrodiction that Freud has based on it. (But see Grünbaum 1988, Section 2, B for the several auxiliary hypotheses on which this falsification was predicated.)

Popper undertook to write on the falsifiability of psychoanalysis. And yet he was apparently unaware of Freud's 1915 paper on paranoia, whose mere title would have alerted him to the testability of the Freudian theory of psychopathology, or would at least have made him aware of the fact that Freud had entertained its falsifiability. Indeed, it is a measure of the inadequacy of Popper's psychoanalytic scholarship that, by his own account, he was first made cognizant of such testability by William Bartley. As we learn from an insert in Popper's (1983: 169), in 1980, Bartley pointed out to him that the existence of a homosexually active paranoiac is *ruled out* by Freudian theory, because it demands that the homosexuality of paranoiacs be strongly repressed, rather than active. But Fisher and Greenberg (1977: 259) had already emphasized this prohibition, as we saw.

It so happens that Bartley gave this example to Popper during the discussion of my own paper at the July 1980 London Symposium on Popper's Philosophy. As I had argued in that paper, Freud's theory implies that the decline in the social taboo on homosexuality in communities like San Francisco should, in due course, issue in a decreased incidence of paranoia in both sexes. This quasi-statistical prediction is quite compatible with Freud's allowance for a "multiplicity of mechanisms of repression proper" (S.E. 1911, 12: 65–68), which include autochthonous developmental factors that are independent of exogenous social taboos. Hence I contended that Freud's etiology of paranoia would be discomfirmed epidemiologically by the failure of the quasi-statistical prediction to materialize.

Let us now see how Popper reacted to Bartley's example. In context, the relevant passage in Popper's *Postscript* (1983: 169) reads as follows:

> I cannot think of any conceivable instance of human behaviour which might not be interpreted in terms of either [Freud's or Adler's] theory, and which might not be claimed, by either theory, as a 'verification'.
>
> * Added 1980 [by Popper]. The last sentence of the preceding paragraph is, I now believe, too strong. As Bartley has pointed out to me, there are certain kinds of possible behaviour which are incompatible with Freudian theory – that is, which are excluded by Freudian theory. Thus Freud's explanation of paranoia in terms of repressed homosexuality would seem to exclude the possibility of active homosexuality in a paranoid individual. But this is not part of the basic theory I was criticizing. Besides, Freud could say of any apparently paranoid active homosexual that he is not *really* paranoid, or not *fully* active.

This retort strikes me as unfortunate for the following reasons:

1. To justify Popper's original thesis of non-falsifiability, it was necessary to *show* – not just to assert peremptorily – that there simply are no potentially falsifying empirical instances in the case of psychoanalytic theory, as he claimed anew in his 1974 reply to Lakatos (Popper 1974, Book II: 1004). In that reply, Popper asserts that Newtonian physics does have potential empirical falsifiers, which could be neutralized only by using immunizing strategies, such as modifying some auxiliary hypotheses of the Newtonian corpus. Thus, he admits that Newtonian physics can be immunized against falsification. By the same token, his assertion of its falsifiability presupposes "disregarding the possibility of immunizing stratagems." But in 1974 he then claimed a contrast between Newtonian theory and psychoanalysis as follows: "And this is the heart of the matter, for my criticism of Freud's theory was that it simply does not have potential falsifiers" (1974, Book II: 1004). Furthermore, "I cannot describe any state of affairs concerning Mr. Smith – say about his behaviour – which would need immunization [i.e. which would require neutralization by immunization of the theory] in order not to clash with Freudian theory" (1974, Book II: 1005). But despite Popper's own admitted inability to describe such a case, the documented ability of others to do so shows how wrong he was to have inflated his mere suspicion of the empirical irrefutability of psychoanalysis into a thesis of unfalsifiability. Plainly, his own reported inability to think of contrary instances was, at best, grounds for a mere suspicion of non-falsifiability. Worse, Popper flatly

contradicts his thesis of non-falsifiability: (i) In the *Postscript* (1983: 173), he tells us in italics that "*anxiety dreams constitute a refutation of the general* [Freudian] *formula of wish-fulfillment*", (ii) Popper assumes falsifiability when he complains that Freudians use immunizing gambits to neutralize contrary evidence.

2. It is important to note how Popper tries to practice damage control by qualifying and hedging his more recent admission that the case of a homosexually active paranoiac does contradict his thesis. Thus, he immediately resorts to *two* immunizing maneuvers, which are disingenuous, I submit, besides being demonstrably unsuccessful, as we shall now see.

He admonishes us that Freud's repression etiology of paranoia "is not part of the basic theory I was criticizing." Yet we are left completely in the dark as to just what does count for him as the 'basic' part of psychoanalytic theory, *and why*. In effect, Popper asks us to overlook that, throughout his career, he had leveled the charge of non-falsifiability *tout court*, rather than only against a so-called 'basic' part of the theory, whose identity he strangely fails to specify. Indeed, as we saw (Popper 1974, Book II: 1004), he wrote: "my criticism of Freud's theory was that it simply does not have potential falsifiers."

Vague as it is, even his claim that Freudian etiologic theory is *not* 'basic' to psychoanalysis is simply untenable. As any student of the subject knows, both historically and logically, precisely the theory of psychopathology, which features the repression-etiology of the neuroses, is the *most foundational* part of Freud's edifice. Thus, as we know, its architect had ample reason to declare in his 1914 *History of the Psychoanalytic Movement*: "The theory of repression is the cornerstone on which the whole structure of psycho-analysis rests. It is the most essential part of it" (S.E. 1914, 14: 16).

Not content with being disingenuous, the kettle calls the teapot black, when Popper (1983: 163) writes against Freud:

> I wish to criticize Freud's way of rejecting criticism. Indeed I am convinced that Freud could have vastly improved his theory, had his attitude towards criticism been different...
>
> This self-defensive attitude is of a piece with the attitude of looking for verifications; of finding them everywhere in abundance; of refusing to admit that certain cases do not fit the theory.

In response, I find it hard not to exclaim: "Physician heal thyself."

Moreover, Popper wants to take out an insurance policy in case his immunizing resort to the alleged non-'basicality' of the theory of paranoia is not seen by others as a vindication. And so he comes up with the following (1983: 169): "Besides, Freud could say of any apparently paranoid active homosexual that he is not *really* paranoid, or not *fully* active." In this way, Popper wants to claim that this kind of case does not discredit his lifelong thesis of irrefutability about psychoanalysis after all.

To this I say two things: (i) In regard to Freud's scientific honesty, the question is not whether he '*could*' save his etiology of paranoia in this way, but only whether he *would* try to do so. And his 1915 paper shows explicitly that, in fact, he did not, although his epistemic reasoning there is problematic in other respects, as I have pointed out; (ii) Popper's depiction of what Freud *could* say to neutralize the counterexample does not help to sustain Popper's thesis of irrefutability, as I shall now show.

Assume we grant that Freud could evade falsification by the case of the paranoid active homosexual in the far-fetched manner proposed by Popper. Then surely a like gambit is available to a physicist, *mutatis mutandis*, to neutralize unpalatable contrary evidence, so as to parry the refutation of his theoretical hypothesis, as I shall soon illustrate. Yet Popper claims falsifiability for physics, even though he explicitly *predicates* it on "disregarding the possibility of immunizing stratagems" (1974, Book II: 1004). And what is sauce for the goose is, I submit also sauce for the gander. Thus if, as Popper claims, the physical scientist does *not* forfeit the falsifiability of his hypotheses in the face of immunizing maneuvers that are surely available to that scientist, why then should Freud have to acknowledge irrefutability merely because he too could adopt such stratagems?

As Popper would have it, Freud 'could' evade refutation by pleading that even the most promiscuous homosexual is not 'fully active' after all. But Freud's etiology of paranoia had explicitly declared that *strongly repressed*, intense homosexuality is the *sine qua non* of paranoia (S.E. 1915, 14: 265; 1922, 18: 228). And even an average homosexually active paranoiac violates this declared necessary condition. *A fortiori*, an erotomanically active paranoid homosexual clearly violates it.

Could Freud still say, as Popper (1983: 169) suggests, that the active homosexual is "not *really* paranoid," no matter how delusional his behavior? To deal with this question, let me first recall a relevant moral from a paper by Carnap (1936, 1937). There he acknowledged over fifty years ago that in physics, no less than in somatic medicine or psychology, empirical data are *not sufficient conditions* for the presence of hypothesized states, which they are taken to manifest. Instead of being sufficient conditions for the theoretical states, empirical indicators *underdetermine* such states, and betoken their presence only more or less probabilistically. Thus it is a commonplace, even among laymen, that an x-ray or even the more refined tomogram or cat-scan does not guarantee any one somatic diagnosis. Hence if the repudiation of a psychiatric diagnosis of paranoia is an immunizing option open to Freud, then a like gambit of rediagnosis is available to a radiologist, when expectations based on the original diagnosis fail to materialize.

By the same token, tracks in Wilson cloud chambers *may* actually bespeak the existence of as yet unknown particles, rather than the passage of those postulated entities that physicist now (fallibly!) take them to attest. Hempel has reminded us that even a circular array of iron filings does not guarantee the presence of a magnet deductively. And, as Wesley Salmon noted (1959), the same holds in psychoanalysis. True, even patent delusions of intense persecution are only probabilistic evidence of paranoia. But Popper's appeal to the fallibility of differential diagnosis, and to the possibility of rediagnosis is misleading as a basis for singling out psychoanalysis as remaining irrefutable, even in the face of the example of a homosexually active paranoiac.

In sum, to the detriment of Popper's vaunted contrast between the testability of physical theory and psychoanalysis, the rescuing option that, according to him, is open to Freud, could likewise be exercised, *mutatis mutandis*, in physics so as to preclude refutation of *its* hypotheses. For example, if the measured charge of an electron turned out *not* to accord with theoretical expectations or requirements, a physicist *could* say – in Popper's sense of 'could' – that the particle in question is "not *really* an electron." Besides, as I have illustrated (1984: 121–123), when Freud did retract a differential diagnosis to preserve a specific causal hypothesis, he was sensitive to the need for independent evidence in favor of the alternative diagnosis.

Far from supporting his non-falsifiability thesis, Popper's (1983) account of the case of the homosexual paranoiac evidently relies on just the kinds of immunizing maneuvers he is quick to reject when others engage in them. In any case, as we saw, his account seems incoherent: only a few pages after discussing the paranoia example, he deals with Freud's dream theory and astonishingly declares it to be a "simple fact...that *anxiety dreams constitute a refutation of the general formula of wish-fulfillment*" (Popper 1983: 173; italics in original). Thus, by Popper's own appraisal, the dream theory must be falsifiable, since it had already *been* falsified by anxiety dreams, when Freud first proposed it! And this clear acknowledgement of falsifiability is hardly tempered in the *logical* sense by Popper's detailed complaint against

Freud's own evidential disposal of *prima face wish-contravening* dreams, such as anxiety dreams and nightmares. After all, on Popper's conception, theories are so-called third-world objects that have a logical life of their own, as it were, apart from the attitudinal scientific integrity of their propounders.

As we know, Freud relied on the distinction between the manifest and latent content of a dream to claim that a latent wish engenders even those dreams whose manifest content is anything but wish-fulfilling. And Popper's (1983) critical comment on this distinction is the following:

> Freud repeatedly re-affirms his programme of revealing the latent content of every anxiety dream as a wish-fulfillment. ...*Yet Freud never carries out his programme; and in the end he gives it up altogether* – without, however, explicitly saying so. [p. 165; italics in original]
>
> The reason why Freud does not carry out his original programme of showing (by way of detailed analyses such as he is wont to give) that all anxiety dreams are wish-fulfillments is, clearly, that in the end he no longer believed in it. ... I should be the last to criticize such a change of mind. But the change is not a conscious correction, or the admission of a mistake. [p. 167]

As we learn from Popper's *Postscript* (1983: 164, 172), the psychoanalytic theory of dreams in particular was historically a major inspiration of his falsifiability criterion of demarcation. Moreover, though Popper hails Freud's interpretation of dreams as "a great achievement," he nonetheless considers it to be his own centerpiece for leveling the following reproach: "*But it is a fundamental mistake to believe that, because it is constantly being 'verified', it must be a science, based on experience*" (1983: 172; italics in original).

Yet, as we know from the old inductivist tradition of Francis Bacon and John Stuart Mill, Popper's admonition here just carries coals to Newcastle by echoing one of the salutary injunctions familiar from precisely that legacy of eliminative inductivism. Freud's wish-fulfillment theory of dream-production is clearly a causal hypothesis. And I have argued on earlier occasions (1979 & 1984: 280) that the demands made by that inductivist patrimony for validating such a strong hypothesis are clearly not met by the so-called 'verifications' adduced by Freudians: as I recalled above from my (1984), *it is precisely Freud's theory that furnishes poignant evidence that Popper has caricatured the inductivist tradition by his thesis of easy inductive confirmability of nearly every theory.* Hence inductivistically untutored claims of confirmations made by psychoanalytic zealots cannot serve Popper's advocacy of his falsifiability criterion of demarcation as a replacement for the received Bacon-Mill inductivist criterion, which was in vogue when he came upon the philosophic scene.

Thus, *malgré lui*, the upshot of Popper's account of the psychoanalytic theory of dreams seems to be as follows: Since this part of Freud's corpus is known to be false, it is falsifiable. Hence, I submit, precisely because – qua 'third-world' object – the dream theory itself is false – it is *scientific* by Popper's criterion of demarcation. This conclusion is, of course compatible, with Popper's distinct claim that Freud was aware of the falsity of the dream theory, but was unwilling to admit it explicitly.

In the face of his proclamation of unfalsifiability, Popper has a highly incongruous attitude toward psychoanalytic explanations. If, as he says, psychoanalysis is indeed unfalsifiable, then it cannot make any risky empirical predictions, because such predictions run the risk of turning out to be false. Hence, if Popper is to be believed, psychoanalysis cannot be corroborated observationally by successful risky predictions, as demanded by his standards of corroboration. Yet, in his view, explanations based on a completely uncorroboratable theory are presumably *pseudo*-explanations. Indeed, if psychoanalysis does yield the surfeit of explanations deplored by him, then its explanations cannot be genuine, precisely because they explain too much, as it were.

Unabashed by such apparent corollaries of his account, Popper not only enlists psychoanalytic theory to provide several explanations, but even explicitly declares his belief in its essential substantive correctness. Let me document these very odd twists.

(1) In his (1950: 16–17), Popper invokes Freud's unconscious defense mechanism of 'reaction-formation' to 'explain', as he claims, how Heraclitus embraced two *prima facie* contradictory ideas. But if such Freudian hypotheses as the existence of conflict, actuated by unconscious fear and resistance, are untestable, as Popper thinks, how can they have enough *empirical import* to explain Heraclitus's utterances by Popper's standards of *bona fide* scientific explanations?

(2) In his (1962: 49), Popper wishes to explain the adoption by some people of a 'dogmatic attitude', as distinct from a 'critical attitude'. In the quest for such an explanation, he registers "a [purported] point of agreement with psychoanalysis" when suggesting the following causal hypothesis: "most neuroses may be due to a partially arrested development of the critical attitude...[due] to resistance to demands for the modification and adjustment of certain...responses. This instance in its turn may perhaps be explained, in some cases, as due to an injury or shock, resulting in fear." Yet, according to psychoanalytic theory, dogmatic rigidity may be seen as a manifestation of

neurotic conflict, but hardly as a contributory *cause* of the neurosis itself. To this extent, Popper's purported "agreement with psychoanalysis" here seems spurious.

(3) In his (1962: 275, n. 52), Popper tries to make diagnostic and explanatory use of Freud's theory of the oedipal conflict, which postulates ambivalent feelings toward the father that include hostile, aggressive and even murderous impulses. Thus Popper declares: "One need not believe in the 'scientific' character of psycho-analysis (which, I think, is in a metaphysical phase) in order to diagnose the anti-metaphysical fervour of positivism as a form of Father-killing." But, on his account of psychoanalysis, how can the putative desire for father-killing yield anything but an ad hoc *pseudo*-explanation of the positivist antipathy to metaphysics? What competing explanations, if any, did Popper canvass? He makes no mention of any.

(4) Let us be mindful of Popper's complaints against the spurious verifications of Freud's dream theory. Then one is simply dumbfounded to learn that he has no hesitation to declare his belief in its essential correctness. As he puts it amazingly (1983: 164):

> in spite of severe shortcomings...it [Freud's dream theory] contains, beyond any reasonable doubt, a great discovery. I at least feel convinced that there is a world of the unconscious, and that Freud's analysis of dreams given in his book are fundamentally correct, though no doubt incomplete.

But, if the dream theory – though purportedly falsified by anxiety dreams – is unfalsifiable, then there can be no corroborating evidence for it by Popper's standards. Therefore, we are left to wonder how we can know "beyond any reasonable doubt" *without corroborating evidence*, that Freud's dream theory "contains...a great discovery." And if the distinction between the manifest and latent content of a dream is discredited by anxiety dreams, and Freud's wish-fulfillment hypothesis is false, just what *is* the purported great discovery? Indeed, since Popper believes that (the bulk) of Freud's dream analyses are essentially correct, if incomplete, how can the pivotal distinction between the manifest and latent dream content fail to be sound?

III. THE DEGENERATION OF POPPER'S CRITERION OF DEMARCATION

In the *Postscript* (1983: 163–164), Popper's avowed purpose of treating the dream theory is to show the following: "that the problem of demarcation is not merely one of classifying theories into scientific and non-scientific ones, but that its solution is urgently needed for a critical appraisal of scientific theories, or allegedly scientific theories." Yet ten pages later – at the end of the section on Freud's theory (p. 174) – Popper tells us amazingly enough that "from the beginning," his 'problem of demarcation' ... "certainly was not a problem of classifying or distinguishing some subject matters called 'science' and 'metaphysics.'" But this concluding disclaimer *flatly contradicts* the first conjunct of his initially stated motive for dealing with the psychoanalytic dream theory. By the same token, this concluding disclaimer here is patently incompatible with two of his earlier landmark reports, as we shall now see.

Thus, in his 1953 Personal Report on the development of his entire philosophy of science, which is reprinted in his (1962), he informs us that, beginning in 1919, when he was seventeen years of age, his problem was "to distinguish between science and pseudo-science...or...'metaphysics'" (1962, 33; italics in original). And, in his Reply to Critics (1974, Book II: 984), he explains that his criterion of demarcation

> is more than sharp enough to make a distinction between many physical theories on the one hand, and metaphysical theories, such as psychoanalysis, or Marxism (in its present form), on the other. This is, of course, one of my main theses; and nobody who has not understood it can be said to have understood my theory.

Evidently, in 1983, he insouciantly repudiates just this major, central tenet of his whole philosophy without ado. By the same token, he now denies his repeated other historical accounts of the aim that inspired his criterion of demarcation, starting in 1919. And, as we shall now see, this 1983 retraction occurs on the heels of a qualified, hedged disavowal that plainly serves as an immunizing stratagem for his prior vindication of his notion of demarcation by reference to psychoanalysis and Marxism. Thus, that stratagem in the *Postscript* (1983: 174) runs as follows:

> it hardly matters whether or not I am right concerning the irrefutability of any of these three theories [i.e., psychoanalysis, Adlerian psychology, and Marxism]: here they serve merely as examples, as illustrations. For my purpose is to show that my 'problem of demarcation' was from the beginning the practical problem of assessing theories, and of judging their claims. It [the problem of demarcation] certainly was not a problem of classifying or distinguishing some subject matters called 'science' and 'metaphysics'. It was, rather, an urgent practical problem: under what conditions is a *critical appeal to experience* possible – one that could bear some fruit?

But, if – as Popper tells us here – the stated three theories are not essential to the vindication of his falsifiability criterion of demarcation, and are mere illustrations, then I must ask: what *other* theories for which scientificity has been wrongly claimed can be adduced to furnish such a vindication *vis-à-vis* the much older criterion of evidential support, which he wants to replace as unduly permissive? And as for stating conditions under which a "critical appeal to experience" is possible, Francis Bacon had demanded – three centuries before Popper – that the validation of a given theory *T* requires data that are *contrary* to *T's rivals* or competitors, while being positive instances of *T*.

In an article on psychoanalysis in a German newspaper, the journalist Dieter Zimmer (1982: 18–19) quotes Popper on the status of psychoanalysis. According to that article, Popper characterized it in 1982 as a subject for "weak heads" (Schwachköpfe). If this report is authentic, how does it square with Popper's stated tribute to Freud's dream theory and with his own explanatory invocation of psychoanalysis in the cases I cited?

Several conclusions seem to me to emerge from my scrutiny of Popper's portrayal of psychoanalytic theory *vis-à-vis* his falsifiability criterion of demarcation:

(1) Popper's indictment of the Freudian corpus – or of its unspecified 'basic' portion – as inherently unfalsifiable has fundamentally misdiagnosed its failure as a scientific theory. More often than not, the intellectual defects of psychoanalysis are too subtle to be detected by his criterion of demarcation. For example, there is no systematic published critique by him of Freud's method of free association, *qua* purported method of causal validation. And yet just that method is *the* method of clinical investigation in psychoanalysis. Alas, Popper's myth of non-falsifiability has entrenched itself in current philosophic folklore. Examples of its inveterate enunciation abound. Thus, in a very recent issue of *The New Republic*, we are told that "Freudian explanations" are "beyond the reach of empirical falsificaton" (Wright 1988: 30).

(2) As a perspective on the nature of Freud's theory, Popper's portrayal is not viable. Yet perhaps it has had some value sociologically by putting psychoanalysts on notice to become more accountable scientifically. But it is unclear whether it actually had that effect on their attitudes.

(3) Popper's (1983) notion of demarcation has become a kind of degenerative philosophical research program in Imre Lakatos's sense of that term.

This said, let me gladly make an acknowledgement: My own scrutiny of psychoanalysis was prompted by my initial doubt as to the soundness of Popper's portrayal of it.

IV. OBITER DICTA ON POPPER'S CRITERION OF DEMARCATION

A. As John Watkins (1978: 351) has correctly noted, the actual falsifiability of psychoanalysis does not entail that Popper's "demarcation criterion was in trouble [merely] because it actually included something [psychoanalytic theory] which Popper himself had mistakenly excluded." Nor have I ever suggested such a moral (1984: 105). Instead, I have now furnished further evidence for the falsifiability of psychoanalysis in order to undermine the following pillar of Popper's theory of science: the falsifiability criterion of demarcation is superior to the received inductivist standard, because the former yields a negative verdict on the scientific status of Freud's theory, whereas the latter permissively accords such status to it. As I have argued, it is just the other way around!

But, as for the merits of Popper's criterion itself, it is instructive to appraise his attempt to defend it (1974: Book II: 1038) against Grover Maxwell's objection that it is *too restrictive* (1974: 294). Maxwell had adduced mixed, quantified statements, e.g. 'All men are mortal', as hypotheses that belong to empirical science, but cannot be deductively falsified by basic statements. And, in my contribution to the Lakatos *Festschrift* (Grünbaum 1976: 227), I had countenanced Maxwell's objection. Yet Bartley (1982: 221, n. 125) complained: "Adolf Grünbaum ignores Popper's reply [to Maxwell] and repeats this worn old criticism."

Hence, let me quote Popper's reply and explain why I consider it unavailing. He writes (1974, Book II: 1038):

> Whenever a pure existential statement [or mixedly quantified statement] by being empirically 'confirmed' appears to belong to empirical science, it will in fact do so *not on its own account*, but *by virtue of being a consequence of a corroborated falsifiable theory*.

As in Maxwell's example:

> when existential statements are verified, this is done by means of *stronger falsifiable* (though perhaps still existential) statements. 'Jones dies', for example, might be verified by 'Jones dies before his 150th year'.

Thus,

> the totality of evidence 'confirming' 'All men are mortal' in fact confirms [in the first instance] a much stronger, and *falsifiable*, statement; for example, 'All men die before their 150th year'.

In short, 'All men are mortal' – though not deductively falsifiable – is authenticated as scientific by following deductively from the corroborated falsifiable statement 'All men die before their 150th year', which is logically stronger. Popper is satisfied that this amplified criterion "completely clears up" the scientific status of Maxwell's counterexamples. I claim, however, that it hardly does so, and that contrary to Bartley, 'All men are mortal' is anything but a hoary canard.

Popper (1973: 19–20) had argued that (relative degree of) corroboration is indeed preserved under logical entailment. And, as he would have it, the unfalsifiable statement 'All men are mortal' derives its scientific status from the (degree of) *corroboration* that it inherits deductively from a stronger hypthesis that is *both* falsifiable *and* corroborated. By contrast, there is obviously no deductive inheritance of falsifiability from the stronger hypothesis.

Two serious difficulties, however, vitiate Popper's reply to Maxwell:

(1) The invocation of corroboration to confer scientific status derivatively on an unfalsifiable statement constitutes a *major switch*, rather than an extension, of Popper's demarcation enterprise: He is abandoning his exclusive reliance on potential contrary evidence in favor of the inductivist's dependence on actual supporting evidence. And this abandonment is not lessened by the fact that the corroboration to which Popper appeals pertains, in the first instance, to a *falsifiable*, logically stronger hypothesis.

(2) Popper (1973: 18) tells us that the relative degree of corroboration '*says nothing whatever about future performance...of a theory*.' But, as I have shown elsewhere (1976: 244–247), this anti-inductivist denial is logically incompatible with two other features that he attributes to corroboration (1973: 19–20): (a) degree of corroboration is preserved under logical entailment, (b) "the degree of corroboration is a means of stating preference with respect to truth." Indeed, given (b), (a) contravenes Popper's denial of the corroborability of future-tensed statements, because at least some corroborated theories have *future-tensed entailments* (e.g. theories whose laws are universally quantified with respect to time).

Evidently, by invoking the deductive inheritance of corroboration along with its other avowed properties, Popper's attempt to accommodate the scientific status of 'All men are mortal' runs afoul of his central anti-inductivist plank: degree of corroboration "*says nothing whatever about future performance...of a theory*." Therefore, Popper can neutralize Grover Maxwell's counterexamples to the falsifiability criterion only on pain of fatal damage to his contra-inductivist construal of corroboration or to its vital feature (b). And I conclude, *pace* Popper and Bartley, that Grover Maxwell's counterexamples stand.

B. Popper (1962: 38) offers two illustrations of his contention that "scientific theories originate from myths, and that a myth may contain important anticipations of scientific theories." The myths, though originally unfalsifiable, can become testable. One of his purported examples is

> Parmenides' myth [i.e., *proto*-scientific world picture] of the unchanging block universe in which nothing ever happens and which, if we add another dimension, becomes Einstein's block universe (in which, too, nothing ever happens, since everything is, four-dimensionally speaking, determined and laid down from the beginning).

But here, Popper commits an error to which I called attention twenty years ago (Grünbaum 1968: 22–23): The conflation of an absence of objective coming-*into*-being (becoming), on the one hand, with timelessness or lack of change, on the other. Though Einstein's space-time is devoid of becoming, it hardly features a timeless or unchanging universe. And it is therefore misleading to characterize it as a 'block universe'. By contrast, Parmenides' universe *is* timeless, and – contrary to Popper – does not admit a consistent addition of a time-dimension. To say, for example, that the career of a radioactive atom is not characterizable objectively by the attribution of the coming *into* being of events on its world line, is *not* all to say that this atom does not *change its state* when it emits an α-particle, and thereby loses mass. By the same token, to deny that the events of an Einsteinian space-time themselves change ontologically by becoming present or past is hardly tantamount to declaring that change in such a space-time is illusory in Parmenides' sense. Popper seems to overlook that whereas Parmenides did deny the objective existence of change and hence of becoming, Einstein denied that there is becoming, while asserting that there is indeed change. Precisely by being the *time*-dimension in more than name, Einstein's fourth dimension cannot be 'added' to Parmenides' unchanging world, as Popper would have it, without generating a contradiction.

I conclude that the exclusion of objective becoming from Einstein's space-time does not warrant Popper's belief that this universe originated conceptually in an interesting sense from Parmenides'. Thus, the theories of Parmenides and Einstein apparently do not illustrate the transformation of untestable myths into testable scientific theories, as envisioned by Popper.

University of Pittsburgh

NOTE

* This paper draws on a section of the sixth of my eleven Gifford Lectures, delivered at the University of Saint Andrews, Scotland in early 1985 on the topic 'Psychoanalysis and Science'.

REFERENCES

Bartley, W.W., III: 1982, 'The Philosophy of Karl Popper'. *Philosophia* (Israel) **11**, 121–221.

Carnap, R.: 1936, 'Testability and Meaning', *Philosophy of Science* **3**, 419–471.

Carnap, R.: 1937, 'Testability and Meaning', *Philosophy of Science* **4**, 2–40.

Cioffi, F.: 1985, 'Psychoanalysis, Pseudo-Science and Testability', In: G. Currie and A. Musgrave (Eds.), *Popper and the Human Sciences*, pp. 13–44. Dordrecht, Martinus Nijhoff.

Cioffi, F.: 1988, 'Exegetical Myth-Making in Grünbaum's Indictment of Popper and Exoneration of Freud'. In: P. Clark & C. Wright (Eds.), *Mind, Psychoanalysis and Science*, pp. 61–87. Oxford, Blackwell.

Erwin, E.: forthcoming, *Was Freud Right?: Scientific Realism and Feudian Psychology*.

Fisher, S. and Greenberg, R.P.: 1977, *The Scientific Credibility of Freud's Theory and Therapy*. New York: Basic Books.

Freud, S.: 1953–1974, *Standard Edition of the Complete Psychological Works of Sigmund Freud*, translated by J. Strachey *et al.* London: Hogarth Press.

Glymour, C.: 1974, 'Freud, Kepler and the Clinical Evidence'. In: R. Wollheim (Ed.), *Freud*, pp. 285–304. New York: Anchor Books. An *Afterword* appears in R. Wollheim and J. Hopkins (Eds.), 2nd edition, *Philosophical Essays on Freud*. New York: Cambridge University Press, 1982.

Grünbaum, A.: 1968. *Modern Science and Zeno's Paradoxes*. London: George Allen and Unwin.

Grünbaum, A.: 1976, 'Is Falsifiability the Touchstone of Scientific Rationality? Karl Popper Versus Inductivism'. In: R.S. Cohen, P.K. Feyerabend and M.W. Wartofsky (Eds.), *Essays in Memory of Imre Lakatos*. Boston Studies in the Philosophy of Science, vol. 38, pp. 213–252. Dordrecht and Boston: D. Reidel.

Grünbaum, A.: 1979, 'Is Freudian Psychoanalytic Theory Pseudo-Scientific by Karl Popper's Criterion of Demarcation?' *American Philosophical Quarterly* **16**, 131–141.

Grünbaum, A.: 1984, *The Foundations of Psychoanalysis: A Philosophical Critique*. Berkeley, Los Angeles: University of California Press.

Grünbaum, A.: 1986, 'Author's Response', *Behavioral and Brain Sciences* **9**, 226–281.

Grünbaum, A.: 1988, 'The Role of the Case Study Method in the Foundations of Psychoanalysis'. In: L. Nagl and H. Vetter (Eds.), *Die Philosophen und Freud*. Vienna: R. Oldenbourg Verlag. Reprinted in *Canadian Journal of Philosophy*, **18**, (4), 1988.

Grünbaum, A.: 1989, 'Why Thematic Kinships Between Events Do *Not* Attest Their Causal Linkage'. In: J.R. Brown and J. Mittelstrass (Eds.), *An Intimate Relation: Studies in the History and Philosophy of Science*, a *Festschrift* for Robert E. Butts. Boston Studies in the Philosophy of Science. Dordrecht and Boston: D. Reidel.

Habermas, J.: 1971, *Knowledge and Human Interests*, translated by J.J. Shapiro. Boston: Beacon Press.

Jaspers, K.: 1973, *Allgemeine Psychopathologie*. 9th ed. New York: Springer Verlag.

Maxwell, G.: 1974, 'Corroboration Without Demarcation'. In: P.A. Schilpp (Ed.), *The Philosophy of Karl Popper*, Book II: pp. 292–321. LaSalle, IL: Open Court.

Popper, K.R.: 1950, *The Open Society and Its Enemies*. Princeton: Princeton Unversity Press.

Popper, K.R.: 1962, *Conjectures and Refutations*. New York: Basic Books.

Popper, K.R.: 1973, *Objective Knowledge*. Oxford: Oxford University Press.

Popper, K.R.: 1974, 'Autobiography of Karl Popper', Book I, pp. 2–184 and 'Replies to My Critics', Book II, pp. 961–1197. In: P.A. Schilpp (Ed.), *The Philosophy of Karl Popper*. LaSalle, IL: Open Court.

Popper, K.R.: 1983, *Realism and the Aim of Science*, W.W. Bartley, III (Ed.) Totowa, NJ: Rowman and Littlefield.

Ricoeur, P.: 1970, *Freud and Philosophy*. New Haven: Yale University Press.

Ricoeur, P.: 1981, *Hermeneutics and the Human Sciences*, translated by J.B. Thompson. New York: Cambridge University Press.

Salmon, W.C.: 1959, 'Psychoanalytic Theory and Evidence'. In: S. Hook (Ed.), *Psychoanalysis, Scientific Method and Philosophy*, pp. 252–267. New York: New York University Press.

Watkins, J.W.N.: 1978, 'Corroboration and the Problem of Content-Comparison'. In: G. Radnitzky and G. Andersson (Eds.), *Progress and Rationality in Science*. Boston Studies in the Philosophy of Science, vol. 58, pp. 339–378. Dordrecht and Boston: D. Reidel.

Wright, R.: 1988, 'Why Men Are Still Beasts'. *The New Republic*, Issue No. 3834 (July 11), 27–32.

Zamansky, H.S.: 1958, 'An Investigation of the Psychoanalytic Theory of Paranoid Delusions', *Journal of Personality*, **26**, 410–425.

Zimmer, D.: 1982, *Die Zeit*, No. 45 (November 5), 18–19.

GERARD RADNITZKY

SCIENCE AS A PARTICULAR MODE OF THINKING AND THE 'TAMING OF THE STATE'

John W.N. Watkins in an early seminal paper 'Epistemology and Politics' (1987, originally published in 1957) explored the relationship between epistemological doctrines and political ideas; he defended the thesis that philosophical ideas about the genesis of factual knowledge do have political implications. In various writings he has investigated the problem of the rational preference between competing scientific theories. In a recent paper he contrasts the optimism of many philosophers "in the 'glad confident morning' of modern science, in the early seventeenth century, when it had made only rather small beginnings, ... [with today's] widespread philosophical disillusion and scepticism concerning the cognitive status of science, despite the enormous strides it has taken since those early days."[1]

In this essay I wish to pursue the question of what is the cachet of 'scientific thinking' as compared to other modes of theory formation and ways of handling theories. I will also ask why modern science has developed in Europe, and only there, and submit the thesis that the rise of modern science has been made possible by particular institutional arrangements that developed in Europe, and only there. These institutional arrangements achieved the taming of political power, and that 'taming of the State' is a pre-democratic achievement.

1. THE DOMINANT ROLE OF SCIENCE IN OUR INTELLECTUAL CLIMATE

The intellectual climate of the civilized world is shaped by science in the sense that *scientific knowledge is the most respected sort of knowledge*. It is respected because it is objective in the sense of being intersubjectively testable so that the scientific merit of a theory can be appraised without taking into account how that theory was generated and what effects it may have on those who believe it, work with it, etc. For the broad mass of the populace the prestige of science is due to the efficacy of the *technology based on science*. Their lives have been changed by that technology, a development that has accelerated during the last six decades. For the thinking sections of the population the prestige of science resides also in the way it is progressing.

Fred D'Agostino & I. C. Jarvie, Freedom and Rationality: Essays in Honor of John Watkins, 163–181.

The world picture of the educated public has been shaped by the various scientific discoveries. We speak of a '*scientific world-picture*'. However, if taken literally, this expression would be a *contradictio in adjecto*. Creating a world-picture is not the task of science. A world-picture is created by philosophical cosmology and philosophical anthropology. They attempt to gather the relevant contributions of the various scientific disciplines and to mould them into a more or less *unified* picture.[2] In recent years, cosmology in combination with quantum theory and synergetics have strongly influenced our world picture while evolutionary biology and genetics have helped us to improve our image of man.

To understand the current situation of science it is necessary to know something about how this situation came about. The development of science forms part of a larger development: the 'European Miracle' and later the 'Rise of the West'. That phenomenon in turn constitutes a special case of a much broader development of cultural evolution. Before having a look at cultural evolution let us glance at the development of man's cognitive capacities.

2. THE EVOLUTION OF MAN'S COGNITIVE SENSORY APPARATUS - THE ROLE OF 'PRIMARY THEORIES' IN THE ADAPTATION TO THE MESOCOSMOS

Man's biological and intellectual make-up, which enables him to cope with the physical environment, is the result of biological and cultural evolution combined. Biology can show us the development of life from the level of bacteria about four billion years ago, for instance, the development of animal vision starting from the humble beginnings of photocontrolled movement of organisms that just had invented photosynthesis as a source of energy supply (Wächtershäuser 1987, 'Light and Life...'). According to Popper's "active Darwinism" organisms shape their senses in the course of their foraging activity (cf., e.g., Popper 1987, 'Natural selection...'; Bartley 1987, 'Philosophy of biology...'; Radnitzky 1987, 'Erkenntnistheoretische Probleme...'). In short it can teach us about the evolution of cognitive sensory apparatuses of various species (Radnitzky and Bartley, eds., 1987, *Evolutionary Epistemology...*). Man's cognitive sensory apparatus, sense organs and brain, are the result of a long evolution and are adapted to what has been called the 'mesocosmos', the world of medium dimensions with respect to the size of objects, to distances, velocities, etc. (cf., e.g., Vollmer

1987, 'Unity of science...').

It is suggestive to conceive the cognitive sensory apparatus in analogy with a computer as consisting of both hardware – the biological structures – and system software: the assumptions, expectations, hypotheses, theories, independently of whether they exist in the form of tacit knowledge or in the form of linguistically formulated hypotheses. Even in the simplest act of observing, of perceiving something as something, we transcend immediate experience by making hypotheses about what the thing perceived might be. We make these hypotheses on the basis of previous experience, and in any act of perceiving we at the same time also test these hypotheses. These hypotheses are structured by certain subconsciously held assumptions, expectations, or formulated theories – I use the term 'theory' here in the wide sense – that are ontogenetically *a priori*. (This, of course, does not imply anything about their validity; being genetically *a priori* does not entail being *a priori* valid.) They are the phylogenetic inheritance gained in the evolution of the species. I propose to subsume them under the umbrella expression of 'primary theories'. I follow here the anthropologist Robin Horton (1982).

Brain and mind have developed in a give-and-take relationship, i.e., what we have called the hardware of the cognitive sensory apparatus and its system software, the primary theories, have developed in such an interplay (cf., e.g., Sir John Eccles's work). Man has developed his brain by his dealing with the products of his mental activity, with the expectations, hypotheses, theories – by forming them and by testing them, and sometimes by replacing less good theories by better ones, i.e. by theories that provide a better 'fit' to the relevant aspects of reality (see also Jarvie 1972, Chapter 6). Examples of such theories are the Euclidean perception of space,[3] the one-dimensional concept of time, the tendency to interpret a sequence of events as causally connected, the push-and-pull conception of causality, and so forth. The usefulness of the sharp distinction between space and time, which is due to the properties of our sensory cognitive apparatus, illustrates how well that apparatus is adapted to life in the mesocosmos – the separation is useful because in daily life we deal only with velocities that are very small compared to the velocity of light (cf., e.g., Kanitscheider 1988, *Weltbild...*, pp. 84, 189). These synthetic *a priori* (ontogenetically *a priori*) theories constitute *the system software of man's cognitive sensory apparatus*. Hence, *the primary theories are common to all cultures, are human universals*. They guide man's dealings with the physical environment and hence have survival value.

3. FROM LIFE IN THE FACE-TO-FACE GROUP TO THE LARGE, ANONYMOUS SOCIETY: THE EVOLUTION OF TWO SEPARATE NORM SYSTEMS

Man belongs to those species that could survive only by forming groups. Man has lived in small face-to-face groups of gatherers, and later mainly as hunters, for about one to two million years. Because of the continuity between life in the primordial human horde and the chimpanzee horde, statements about when the transition occurred remain somewhat arbitrary and may well involve even a longer period (cf., e.g., Radnitzky 1987, 'Rise of Civilization...', 55, 62; Waal 1982, *Chimpanzee Politics*). The face-to-face groups were made possible by the evolution of a rule system that governed the behavior of the group members so that reliable expectations of reactions could be formed. I propose to call that rule system the '*solidarity norm system*' (cf., e.g., Radnitzky 1987, 'Rise of Civilization...', 65). It has evolved in, and has made possible, *life in small groups* where people can touch each other and be touched by their fellows, and where the group has a common aim. The 'solidarity norm system', adapted to life in the face-to-face group, is in important respects the *counterpart in social life to the primary theories*, (which, it will be remembered, are adapted to life in the mesocosmos and are relevant for survival). Indeed, in social life the counterpart to the mesocosmos is the face-to-face group. The *solidarity norm system* is deeply engrained in our emotional life. Mankind has lived in small face-to-face groups for about one to two million years, and to some extent the solidarity norm system is the endowment from our non-human primate ancestors – which extends the relevant period to about four million years (Waal 1982, *Chimpanzee Politics*; Radnitzky 1987, 'Rise of Civilization...', p. 55). During that period, because of the primitive state of technology, the planet's carrying capacity has been only between one to ten million people, and the total population of the planet as well as the medium group size remained fairly constant.

With increasing trade relations and the emergence of agriculture – a very late development, only about ten thousand years ago – the planet could sustain larger and larger populations, and the form of life changed from the small group to the large, *anonymous society*. Clans and larger tribes appeared and eventually nations and high cultures like China, India, and the Islamic empires. The anonymous society is characterized by the fact that now people serve the needs of others whom they don't know and profit from the efforts of others who do not know about them. They and likewise these others intend to improve their own lot and by doing so they simultaneously serve the welfare of the society, without intending to do so. A familiar quotation from Adam Smith epitomizes the situation best: "It is not from the benevolence of the butcher, the brewer or the baker that we expect our dinner, but from their regard to their own interests." Friedrich von Hayek introduced the term 'catallaxy' for that phenomenon; in certain contexts it means not only trading, but turning a foreigner into a friend, accepting him as someone with whom you peacefully do business. A central feature concerns the intra-group relationship: the primordial face-to-face group needs a common aim, but in the anonymous society, in order peacefully to cooperate, people no longer need any common aim – apart from the aim to preserve the institutional framework which makes it possible for them to cooperate peacefully, and which secures their economic freedoms and civil liberties. (If these are secured, the so-called 'political freedom' is of little significance in the daily life of the ordinary person.)

The transition from the face-to-face group to the large anonymous society has been made possible by the *evolution of a second norm system.* The thesis of the emergence of two moral systems and of their existence beside each other in modern industrialized society as well as the contention that, if in the public-political arena the abstract norm system were replaced by the solidarity system, economic welfare and freedom would be destroyed, from the central part of Hayek's theory of cultural evolution and of his social philosophy.[4] The rules of that norm system are abstract in the sense that they abstract from the particular time and place and apply equally to all members of the society. I propose to call this norm system the '*honesty system*' since '*pacta sunt servanda*' plays a central role in it. It enables the individual to transcend the realm of personal acquaintance. For instance, personal exchange is replaced by interpersonal markets.

The 'solidarity' system, the moral system that *successfully stabilizes the face-to-face group cannot fulfil that function for the anonymous society* – that is the central theme in Hayek's theory of cultural evolution. Of course, the 'solidarity' norm system has retained its applicability to all face-to-face groups, and, since many small groups – like the family and circles of friends – are interspersed in the anonymous society, the solidarity system applies to them, while the relatively free society, i.e., that special case of the anonymous society which is wealth-creating and which respects private rights can only be preserved if its public-political sphere is governed by the abstract 'honesty norm system'. In this sense the 'honesty norm system' with its abstract rules constitutes a social capital the importance of which can scarcely be overrated.

The anonymous society governed by the norm system of abstract rules opens the *possibility* of a free society, i.e. a society based on the market order, where private rights are respected. Such a society can develop institutional arrangements that make possible the rise of science. However, whether or not this possibility is realized depends upon whether appropriate initial conditions occur, which in turn depends on geographical constraints and historical accidents.

4. THE EMERGENCE OF 'SECONDARY THEORIES'

The more lively man's intellectual life became, the greater became the *demand for an explanation of phenomena which cannot be explained with the help of primary theories*. Such a demand is bound to arise even in daily life, since primary theories have a very limited explanatory power if only because of the limited concept of causality. In addition, perennial questions arise, like the question of the origin of the world, the origin of life, of man, or of a particular group. The demand for an explanation of phenomena of daily life not explainable in terms of primary theories and hence not controllable as well as the demand for an answer to questions that cannot be answered with the help of primary theories, can only be met by constructing a new sort of theories. Hence, the primordial group develops myths of origin and these myths guide rituals, which can be viewed as technologies designed to influence supernatural powers.[5] All high cultures have developed spiritualistic-personalized theories, which attempt to transcend the experience of the mesocosmos. Again following Robin Horton, I propose to use the umbrella word 'secondary theories' to designate that heterogeneous set of theories. Departing from Horton, I propose to define *secondary theory* as the class of theories that attempt to explain phenomena which cannot be explained by theories that are tied to our *sense organs*, independently of whether they involve supernatural entities or are empirically testable theories. Religious doctrines form an important subset of the class of secondary theories. They not only attempt to explain things but have social functions like providing coherence and supporting an order. Secondary theories owe nothing to biological evolution but are exclusively a product of cultural evolution. Hence, *they vary widely from culture to culture*. Since, by contrast to primary theories, they do not have survival value, *they can afford* to contain, as they often do, even patently absurd assertions or to postulate bizarre entities like gods with incompatible characteristics, or develop Cargo Cults (cf., e.g., Jarvie 1964).

From a *historical point of view*, it is appropriate to treat secondary theories as *one single* ideal type, and to draw attention to an interesting special case of secondary theory: to scientific theories. In cultural history the European development constitutes the exception, and one key feature of the 'European Miracle' is the accelerating development of science from the beginning of the 17th century.

From the *systematic point of view* it is recommendable to distinguish *two* types of secondary theories: spiritualistic-personalized theories and scientific theories, naturalistic-impersonal theories. In section (5), when institutional arrangements are considered, the historical aspect will be taken into account.

To clarify these two ideal types of secondary theories I shall compare them to each other, that means I shall first ask what they have in common and then ask in what respects they differ. By listing what the two subsets of secondary theories have in common we also sharpen the distinction between primary and secondary theories.

What do spiritualistic-metaphysical secondary theories and scientific secondary theories have in common? (1) All secondary theories, by definition, are designed with a view to meeting the *same sort of demand*: explaining what is unexplainable in terms of primary theories. (2) Both types attempt to meet that demand by means of a theory formation that involves *unfamiliar, 'hidden' entities*. (3) *Both* types of secondary theories have the *same input*, the same source of ideas for theory formation: experience structured by means of primary theories. The sources for analogies, for root metaphors, are but few. They will be taken from that sector of experience that is felt to show more regularity than any other realm. Thus, in stable primitive societies (like African tribes or Eskimo), where – because of the high information costs – there is almost continuous surveillance, no privacy, and behavior is strictly norm-governed, it will be the social realm that is used as a source of root metaphors. When social life becomes less stable, which in Europe is the case from the 13th century onward, the realm that provides the most reliable expectations is the arena of manual technologies, the realm of artifacts. Instead of sociomorph-anthropomorph models now *technomorph models* are the source of root metaphors. (The classic on this change is Topitsch 1958, *Ursprung und Ende...*; 1979, *Erkenntnis und Illusion*; cf. also Horton 1982, 'Tradition and modernity...', 238 and 253.) Technomorph models also inspired the mechanistic world picture, the metaphysics underlying early science. (4) *Both* types of secondary theory *transcend the realm of every day experience*, if only because they talk about entities not directly accessible to our cognitive sensory apparatus, which is limited to experience of the mesocosmos. The spiritualistic type of secondary theory postulates supernatural entities, all sorts of gods, demons, *genii loci*, and sometimes develops religious doctrines as well as demonologies. The scientific theories make assertions about the cosmos and about the micro-world. In our century we have witnessed three great breakthroughs that transcend our experience of the mesocosmos and that have changed our world-picture: with respect to cosmic dimensions, the theory of relativity; with respect to the micro-world, quantum theory; and with respect to systems far from equilibrium, chaotic systems and order arising out of chaos, synergetics. (5) However, *both* types of secondary theories *remain linked to the realm of everyday experience* to which man's cognitive sensory apparatus is adapted. They remain linked to it through the input (point (2) above), and even the spiritualistic-personalized theories, at least to some degree, submit to the test of experience, make some claim to a certain efficacy in explaining, predicting and hence also in controlling events. For instance, some of them claim that by means of certain rituals the gods may be induced to interfere in human affairs in certain predictable ways. The choice example is a rain dance. Scientific theories have, by definition, to submit to the crucial test of experience.

The question of what is the cachet of scientific thinking can be answered by investigating *in what respects personalized secondary theories and scientific secondary theories differ*.[6] (1) Each type of theory flourishes in a *different kind of intellectual climate*: the spiritualistic secondary theories in a climate of cognitive traditionalism (Horton 1982, 'Tradition and modernity...', 238); the scientific theories in the context of criticistic thinking, in which the fallibility of human knowledge is explicitly acknowledged. Hence, each type is inspired by a *different ideal of knowledge*. Cognitive traditionalism upholds an end-state conception of ideal knowledge: genuine knowledge should be proven to be true. It tends to fuse and confuse truth and certainty. The epistemological climate that is propitious for science favors a process conception of knowledge: the process of research is open-ended, and scientific progress is possible in spite of the principal fallibility of all human knowledge (cf., e.g. Radnitzky 1987, *Wittgenstein...*). (2) *The ways in which the two types of secondary theories are legitimized are completely different*. Typically, a spiritualistic-personalized secondary theory has recourse to an infallible source, and genuine knowledge is revealed knowledge that is handed down in accordance with the tradition. This type of legitimizing secondary theories has impregnated Western philosophy and shaped the latent epistemology that underlies ordinary language; ordinary language is impregnated with justificationist philosophy. (3) A particular way of legitimizing a theory is closely linked to a particular *mode of theorizing*. It is this respect that constitutes *the decisive difference* between the two types of secondary theories. Typically, *spiritualistic* secondary theories play an important role as stabilizers and legitimizers of social order, and hence the mode of theorizing is '*consensual*' (Horton 1982 'Tradition and modernity...', 239). An assertion shows its worth by being coherent with the tradition, at least with respect to principles. Hence, criticism from outside will be met by immunization strategies. *The scientific enterprise* is characterized by the *competitive* mode of theorizing (cf., e.g. Jarvie 1972, 145). Ideally, free *intertheoretic competition* reigns and rational theory preference primarily takes into account the differential problem-solving power of the competing theories. An assessment of this capability has to be based on the past record of the competitors in surviving falsification attempts. Prediction functions as a means of testing a theory. No attempts are made to justify a theory. A scientific theory cannot be 'justified'. No more than a species that survived so far is thereby 'justified', is a scientific theory justified by its explanatory and predictive success up to date. In each case, what matters is whether or not the theory, or the species, has shown itself to be *viable* (Bartley 1987, 'Philosophy of biology...').

The prototype of scientific theory, of course, is natural-science theory. In a considerable part of social science and the humanities the commitment of the researcher to extra-cognitive values has the consequence that the theories generated often have more affinity to the spiritualistic-personalized secondary theory than to the pure type of scientific theory (Horton 1982, 'Tradition and modernity...', 248). If intertheoretical competition functions properly, theory change is an objective process, and hence the better theory drives out the less good one (cf., e.g., Radnitzky 1987, '"Economic" approach...'). However, there is a big *difference* between *natural science* and the *sciences humaines*. In the latter, empirical testing is often not directly applicable – and sometimes it is even avoided. For instance, in certain historical narratives interpretation may be more important than explanation and, hence, there is a risk that moralizing will intrude. In the social sciences, peer review often functions poorly. Antony Flew (1985, *Thinking About...*, Chapter 1) gives some striking examples of the indifference of the scientific community to the exposure of social-scientific theories that systematically confuse description and wishful thinking.

The competitive mode of theorizing implies the requirement of intersubjective testability, which in turn entails that *scientific theories have certain important limitations* which spiritualistic secondary theories don't have. They cannot guide genuine, i.e. non-instrumental value judgements or norms, and hence they alone cannot settle any moral issues, and they have nothing to say on 'existential themes' like the 'meaning' of life, of death, of happiness, and so forth.

5. THE RISE OF SCIENTIFIC THINKING AS ONE OF THE EARMARKS OF THE 'EUROPEAN MIRACLE'

So far as we know, the competitive mode of thinking began with the Pre-Socratics. There, probably for the first time, criticism was encouraged, including criticism of the secondary theories which hitherto had constituted revealed knowledge handed down from generation to generation as something sacrosanct. Referring to Hayek I mentioned that the evolution of the large, anonymous society opened up the possibility of a development towards a free society – a possibility, not more. The breakthrough of scientific thinking was achieved in Europe and it is an integral part of the 'European Miracle' (I borrow this term from the title of E.L. Jones's book of 1981). The competitive mode, which is the growing ground of scientific thinking, can only thrive in a particular institutional environment. Let us ask why and how institutional arrangements that create wealth – be it economic wealth or new knowledge – evolved in Europe and only there.

So far as we know, all advanced civilizations have been despotisms, and the mass of the people have always lived in extreme poverty. A longitudinal performance comparison of economies and nations reveals that the 'Rise of the West' is the exception: a brilliant and perhaps fragile accident. About five hundred years ago, the high cultures of the Far East, in particular China, were superior to European civilization (cf. Needham 1969, *Grand Titration*, 190).[7] However, Europe embarked on an unparalleled development, which accelerated for the last two hundred years, so that at the turn of the century Europe literally ruled the world. Within the long-term cultural evolution a special case had appeared.

When we wish to explain '*How (and when and why) the West grew rich*', the '*West*' is defined as those areas where in the last two hundred years prosperity has touched the lives of more than the upper tenth of the population. (Cf., e.g., Rosenberg and Birdzell 1986, *How the West...*; Weede 1987, 'Eurosclerosis...'; 1988, 'Sonderweg...'; Engerman 1988, 'How the

West...'.) The phenomenon started in Europe and *has moved westward.* The itinerary is roughly this: in the 14–15th century, the North-Italian commercial cities, post-1500 Spain and Portugal, by 1600 Holland, post-1750 England, about 1900 the US, and today Japan, Hong Kong, Singapore, ROC, ROK (see, e.g., Rabushka 1987, *New China...*; 1988, 'Economic, civil...'). '*Rich*' in our explanandum is taken in the wide sense of 'welfare' comprising both financial and economic wealth *and* freedom. Disregarding the difficulty of concepts of national income accounting, it is clearly visible that the 'Rise of the West' brought a great advance in the welfare of human beings, as indicated by the means available to the great majority of individuals *themselves* to choose and shape the quality of the lives they lead (Engerman 1988, 'How the West...'). *Throughout history those economies grew fastest that were freest – and the same holds for scientific progress.*

The Rise of the West rests on three pillars: in the economy, the *market order*; in the political sphere, the *taming of the State*; in intellectual life, *autonomous science.* (1) The market order or *competitive capitalism* is the key factor. It presupposes and rests on private rights, i.e., that property rights are respected by the rules at least to such an extent that their predatory appetites are kept in check. (2) The existence of private rights (Hirshleifer 1980, 'Privacy...') in turn entails and presupposes the existence of some sort of *legal framework* protecting such rights. (3) Autonomous *science as a model of a particular cognitive mode* – the *competitive* mode – achieved its breakthrough only with Galileo and Newton. (Of course it had innumerable forerunners.) That breakthrough had been made possible by the evolution of institutional arrangements that foster wealth creation. *Wealth-creating institutions* have certain general characteristics that are independent of whether they produce economic and financial wealth or human capital – 'human capital' is here taken as comprising both epistemic resources, knowledge, and the producers of such epistemic resources or of economic wealth.

The *link* between *science* and *economic* wealth appeared only after the late 19th century. Science itself flourished best where its autonomy and freedom was fostered by the decentralization of invention and innovation and the ability of those who succeeded to capture their rewards – in contrast with the "ignominy memorialized in the eponym 'Lysenkoism'" (Rosenberg and Birdzell 1986, *How the West...*, 255).

Why did suitable political conditions develop in Europe rather than at other places? Like all historical processes the 'Rise of the West' is a singularity and a contingent process. We can explain the principle by pointing out certain possibilities which have opened up. Which of the possible developments has been realized, depends essentially on the *initial conditions*; these in turn depend on the crossing points of causal chains, which contain a substantial *chance* element.

In the case at hand, among the initial conditions natural constraints have played a key role. By contrast to the conditions which Asian and Islamic rulers were confronted with, European geography – soils, geology, climates, etc. – *varies* from place to place and core-areas are comparatively small. For European rulers it was far more difficult to project military power from the core-area to the periphery. Hence, the European states faced neighbors with roughly equal military capabilities, and the international game of power was characterized by a permanent and fierce *competition* (cf. Weede 1987, 'Eurosclerosis...'; 1987, 'Ideas...'; 1988, 'Sonderweg...'). Competition in economic markets provided the need for an autonomy of technology and later of science. *Competition* among holders of political power within a nation, a phenomenon common to Europe, boosted capitalism and economic growth. Historical accidents reinforced that development: one important factor was the power struggle between the State and the Church. Competition between states, together with practicable opportunities to exit, facilitated the spread of innovations throughout Europe.

Besides *competition between states in the form of trade, competition in the form of power politics and wars* was another factor that boosted economic development. Military technology has always been a pacemaker for civil technology (cf., e.g. Albert 1968, *Kritische Vernunft...*, 14; Andreski 1968, *Military Organization...*;' 1969, *Comparative Sociology...*). The role which the competition between states in the form of war played for European development can hardly be overrated.

The rulers, ruling over a limited territory, recognized that they could get a larger income from a steady stream of protection rents than from robbing merchants. Given these constraints for European rulers it was rational to keep in check their predatory and cleptocratic appetites and to concede relatively safe property rights (Weede 1987, 'Eurosclerosis...', 4; 1987, 'Ideas...', 2). Property rights were in turn the growing ground for a moral system of individualism and personal responsibility. Conceding property rights to individuals meant *limited* government; the protective purpose of the state became more and more pronounced: a *strong* (protective) state but not indulging in a wide range of activities. Thus limited government could evolve in an age of feudalism and absolutism. *Limited government is a pre-democratic achievement.*

In sum, sustained economic growth has been based on trading across an area large enough to be important but divided among a number of rival nation-states. Each of these states was large enough to protect property in a territory large enough to be important, but not so large as to become an empire with inefficiencies analogous to those of the monopolistic firm. The possibilities for capital mobility and the risk of losing valuable human capital operated as a deterrent to political confiscation or confiscatory taxes (Engerman 1988, 'How the West...').

6. THE CONSEQUENCES OF INSTITUTIONAL ARRANGEMENTS FOR SCIENTIFIC PROGRESS AND FOR INNOVATIONS

Science-based technology has become important only in the 20th century, and its importance has been accelerating for the last 60 years. To develop technologies certain institutional arrangements are required. This requirement has so far proved an insurmountable barrier to the development of *science-based high technologies* in totalitarian states. Thus, the USSR has been obliged to resort to technological espionage for practically all of its military technology (Mekler 1986, 'Soviet science...').

The body of scientific knowledge evolves like a spontaneous order, i.e., it is the result of the action of many individuals and groups, which neither individually nor collectively intend to bring about that particular state of the body of knowledge. The system is self-correcting at least in those disciplines where theory formation is strictly controlled by empirical testing. If the body of scientific knowledge evolves as a spontaneous order and the researcher (as an ideal type) acts as a discovery-maximizing entrepreneur, then *those institutional frames will best facilitate the improvement of the human capital that best approximate the market order. Competition* functions both as a discovery mechanism and as a selection process. Organization of science and of innovation will profit from *decentralization*. In science, the intellectual counterpart to free competition in economic markets is unlimited criticism of competing problem solution; the intellectual counterpart to protectionist measures in the economic system is the tendency towards cartelization within academe. In each case the improvement of wealth – be it physical and financial wealth or human capital involving knowledge – will be impeded and, if cartelization becomes virulent, progress will stop.

7. THE FUTURE IS OPEN

Evolution has no aim. Cultural evolution produced two striking developments: an increase in power, i.e. in man's opportunity set (Stone Age man possessed the same natural resources as modern man, the difference lies in the knowledge of how to use them in order to better his lot) and an increase in individual freedom. The secret of the 'European Miracle' has been the evolution of limited government. The phenomenon of the westward movement of the 'Rise of the West' can only be explained if we keep in mind that the rise of the West has been made possible by the evolution of freedom of the economic sphere from political influence as well as from religious restrictions that lead to the security of property rights. In Europe, the constraints constituted by geography, in combination with historical accidents, led to a *limited scale of political decision making and multiple sources of decision making*. By attending to changes in the political conditions we can also explain the development from the 'Economic Miracle' of Europe to 'Eurosclerosis', to the corporatist state.[8] The cancerous growth of government and the built-in escalation of welfare-state costs is likely to determine the future of Western Europe. This, however, is a problem, which all Western democracies face, even if in this respect some of them are better off than others (Buchanan 1986, 'Our times...', 35).

It is an open question whether the relatively free society, which can support autonomous sciences and is supported by it, which grew out of the 'European Miracle' and which constitutes a unique and fragile exception in human history, will be an episode or an enduring achievement. Much will depend on whether it will be possible to educate the educable sections of the population and above all the future decision makers so that they understand the functioning of modern society and economy. This is a cognitive and also an educational task. The *comparative institutions approach* outlines the consequences of various institutional arrangements: the ways institutions work out for people living under them, what opportunities various systems offer, what sort of life is possible under them. It will then be up to the individuals to choose betweeen giving individual freedom priority in the social and public sphere or to accept some form of slavery under a totalitarian system, including unlimited democracy in the sense of the dictatorship of the majority as a special case of totalitarianism. Thus, a position taking on value issues is indispensable. Hayekians posit the value of individual freedom. In my opinion, the contractarian approach to Constitution and State conceals the value issues. *Values are traded-off all the time.* Sometimes, people sell themselves into slavery if they are paid for it – as we witness in connection with the modern welfare state. *However, there is no trade-off between freedom on the one hand and economic success and scientific progress on the other hand.* The two are inseparable: economic growth has come from economic freedom and competition, and *scientific progress has come from a free market of ideas and intertheoretical competition.*[9]

University of Trier

NOTES

[1] Watkins, J.W.N.: 1987, 'A New View of Scientific Rationality', p. 64.
[2] This presupposes a certain 'unity' among the various scientific disciplines. See, e.g., (Radnitzky, Ed. 1987 and 1988, *Centripetal Forces...*).
[3] Purely visual space is not Euclidean (Heelan 1985, *Space Perception...*). Our Euclidean space perception is the result of a long process of adaptation to the mesocosmos in which various sense modalities have cooperated.
[4] Cf. e.g., Hayek 1973/76/79, *Law, Legislation...*; Hayek in Chapter 12 of Radnitzky, 1987, *Centripetal Forces...*; for an overview and reference to the literature see, e.g., Radnitzky, 1987, 'Rise of Civilization...'.
[5] The typical ritual is a pre-language communication system concerned with fairly constant themes. That system survives in modern societies, but it is also found in subhuman primate hordes. It may constitute a link between primary theories and spiritualistic secondary theories. Ritual may expand into pseudo-sentences like invocations and prayers. (Aijmer, 1987, 20), which may function like technologies.
[6] The problem of *demarcating science from non-science* has received consistent attention in the philosophy of science, and there is a vast literature on the explication of the concept of *empirical significance.* (A survey and appraisal is given, e.g., in Radnitzky 1968/73, *Contemporary Schools...*, 112–145; Radnitzky 1988, 'Wissenschaftlichkeit'.) Demarcating synthetic statements against analytic statements and hence empirical science from formal science is relatively unproblematic. The interesting problem is, within the class of synthetic statements, to demarcate empirical statements from the rest. For the present purpose, the following proposal may suffice. (1) A statement which *possesses empirical content eo ipso* runs the risk of being contradicted by reality, of being falsified. *Falsification* is a *methodological* concept. A theoretical system *TS* is falsified by falsifying premise *FP* iff *FP* is less problematic than *TS* and *FP* contradicts *TS*. A falsificatory argument is taken seriously only if the following is the case: in the standard model, if the conjunction of the theories, the auxiliary hypotheses and the statements of the initial conditions implies the prediction *P*, and the statement that describes the observed effect *P** entails that not-*P*, then *P** must describe a reproducible effect, i.e., be intersubjectively testable. While falsification is a methodological concept, *falsifiability* is a *logical* concept: it states the logical relationship between classes of sentence, i.e., *TS* is falsifiable (in the language *L*) iff *FP* (in the above argument) can be formulated in *L*. Being falsifiable in this sense is a

necessary and sufficient condition for qualifying as a statement with empirical content. (Usually, we call it "scientific" only if it also possesses a certain degree of systematicity.) For science as an ensemble of theories and as an activity, Karl Popper proposes the following characteristic: "...'Science' is differentiated from the older myths not by being something distinct ... but by a second-order tradition ..." (1963, *Conjectures...*, 127), that means if it is *placed in and handled in a non-justificationist context*, a context in which nothing is dogmatized. (Cf. Bartley, 1962/1984, *Retreat...*, and Part II of Radnitzky and Bartley, eds., 1987, *Evolutionary Epistemology...* .)

[7] How then does Needham explain that China, nonetheless, did not achieve economic growth? According to Needham it was institutional arrangements that "inhibited the rise of modern capitalism and of modern science" (1969, *Grand Titration*, 197). In other words, the great leap forward did not happen in China because China never had room for anything like the commercial cities of late medieval Europe, which offered economic freedoms. A supreme, because involuntary, compliment to capitalism coming from a lifelong Communist.

[8] A striking example is the wage structure in West Germany: it is determined by a highly centralized collective bargaining procedure; corporatism in the labor market is a political taboo. It has led to inflexibility and large unemployment and has grossly impaired the incentive system. The causes for the decline lie on the supply side: rigidities in the labor market, an excess of regulations for potentially dynamic sectors, subsidies and high taxes, a corporatist political climate. Other European countries appear to suffer to a similar extent, with the possible exception of Great Britain and Italy.

[9] In the words of Russell Kirk: "The struggle will be decided in the minds of the rising generation – and within that generation, substantially by the minority who have the gift of reason." (Quoted from Regnery 1979, *Dissident Publisher...*, 147).

REFERENCES

Agassi, J. and Jarvie, I.C. (Eds.): 1987, *Rationality: The Critical View*. Dordrecht: Martinus Nijhoff.

Aijmer, G. (Ed.): 1987, *Symbolic Textures*. Göteborg: Acta Universitatis Gothoburgensis.

Albert, H.: 1968, *Traktat über Kritische Vernunft*. Tübingen: J.C.B. Mohr; 3. verb. Aufl. 1975.

Albert, H.: 1986, *Freiheit und Ordnung*. Tübingen: Mohr.

Andersson, G.(Ed.): 1984, *Rationality in Science and Politics*. Dordrecht: Reidel.

Anderson, M. (Ed.): 1986, *The Unfinished Agenda. Essays on the Political Economy of Government Policy in Honor of Arthur Seldon*. London: The Institute of Economic Affairs.

Andreski, St.: 1968, *Military Organization and Society*. Stanford: University Press.

Andreski, St.: 1969, *The Uses of Comparative Sociology*. Berkeley/Los Angeles, CA.

Bartley, W.W., III: 1962, *The Retreat to Commitment*. New York, NY: Alfred A. Knopf; 2nd ed. 1984 with 100 pp. *Appendices*, La Salle, IL: Open Court.

Bartley, W.W., III: 1987, 'Philosophy of biology *versus* philosophy of physics' in (Radnitzky and Bartley, Eds., 1987), pp. 7–46.

Block, W. (Ed.): 1989, *Freedom, Democracy, and Economic Welfare*. Vol. II, Vancouver: The Fraser Institute.

Brunner, K.: 1985, 'The poverty of nations', *The Cato Journal* **5**; 37–50.

Brunner, K.: 1987a, 'Economic inequality and the quest for social justice', *The Cato Journal* **7**; 153–158.

Brunner, K.: 1987b, 'The perception of man and the concept of society: Two approaches to understanding society', *Economic Inquiry* **25**, 367–388.

Brunner, K. and Meckling, W.: 1977, 'The perception of man and the conception of government', *Journal of Money, Credit and Banking* **9**, 70–85.

Buchanan, J.M.: 1986, 'Our times: past, present and future', in Anderson (Ed.), 1986, pp. 29–38.

Butler, E. and Pirie, M. (Eds.): 1987, *Hayek on the Fabric of Human Society*. London: The Adam Smith Institute.

Eibl-Eibesfeldt, I.: 1976, *Grundriß der vergleichenden Verhaltensforschung*. München: Piper, 5. Aufl. 1978.

Eibl-Eibesfeldt, I.: 1986, *Biologie des menschlichen Verhaltens*. München: Piper.

Eibl-Eibesfeldt, I.: 1988, *Der Mensch das Riskierte Wesen*. München: Piper.

Engerman, St. 1988. 'Reflections of HOW (and when and why) THE WEST GREW RICH', MS for the 15th Interlaken Seminar on Analysis and Ideology, May 1988.

Flew, A.: 1985, *Thinking About Social Thinking*. Oxford: Blackwell.

Hayek, F.v.: 1973, 1976, 1979, *Law, Legislation and Liberty*. 3 Vols. London: Routledge and Kegan Paul.

Hayek, F.v.: 1978, *New Studies in Philosophy, Politics and Economics*. London: Routledge and Kegan Paul.

Hayek, F.v.: 1987, 'The rules of morality are not the conclusions of our reason' in (Radnitzky, Ed., 1987), pp. 227–236.

Hayek, F.v.: 1988, *The Fatal Conceit: The Errors of Socialism. (The Collected Works of F.A. Hayek,* Vol. I; General Editor: W.W. Bartley, III), London: Routledge.

Heelan, P.: 1985, *Space Perception and the Philosophy of Science*. Berkeley, CA: University of California Press.

Hirshleifer, J.: 1980, 'Privacy: Its origin, function and future', *The Journal of Legal Studies* **9**, 649–664.

Hollis, M. and Lukes, S. (Eds.): 1982, *Rationality and Relativism*. Oxford: Blackwell.

Horton, R.: 1982, 'Tradition and modernity revisited', in (Hollis and Lukes, 1982), pp. 201–260.

Jarvie, I.C.: 1964, *The Revolution in Anthropology*. London: Routledge & Kegan Paul.

Jarvie, I.C.: 1972, *Concepts and Society*. London: Routledge & Kegan Paul.

Jones, E.L.: 1981, *The European Miracle: Environments: Economics, and Geopolitics in the History of Europe and Asia*. (2nd Edition 1987). Cambridge: Cambridge University Press.

Kanitscheider, B.: 1988, *Das Weltbild Einsteins*. München: Beck.

Meckler, Y.: 1986, 'Soviet Science: How good is it?', *Discovery*, August 1986, 36–59.

Needham, J.: 1969, 'Science and society in East and West', in Needham, J. (1969), *The Grand Titration*. London: Allen & Unwin.

North, D. and Thomas, R.: 1972, *The Rise of the Western World: A New Economic History*. Cambridge: Cambridge University Press.

Olson, M.: 1982, *The Rise and Decline of Nations: Economic Growth, Stagflation and Social Rigidities*. New Haven, CT: Yale University Press.

Pitt, J.C. and Pera, M. (Eds.): 1987, *Rational Changes in Science*. Dordrecht: Reidel.

Popper, K.: 1963, *Conjectures and Refutations*. London: Routledge & Kegan Paul; 7th rev. ed. 1978.

Popper, K.: 1972, *Objective Knowledge. An Evolutionary Approach*. London: Oxford University Press (5th rev. ed. 1979).

Popper, K.: 1987, 'Natural selection and the emergence of mind', in (Radnitzky and Bartley, Eds., 1987), pp. 139–155.

Popper, K. and Eccles, J.: 1977, *The Self and Its Brain. An Argument for Interactionism*. Berlin: Springer International.

Rabushka, A.: 1987, *The New China: Comparative Economic Development in Mainland China, Taiwan, and Hong Kong*. Boulder: Westview, and San Francisco: Pacific Research Institute.

Rabushka, A.: 1988, 'Economic, civil and political freedoms: The cases of Singapore and Hong Kong', in (Walker, 1988), pp. 151–183 (followed by Discussion, pp. 184–213).

Radnitzky, G.: 1978, 'The boundaries of science and technology' in *The Search for Absolute Values in a Changing World (Proceedings of the VIth International Conference on the Unity of the Sciences*, San Francisco, Nov. 1977), New York, NY: International Cultural Foundation Press 1978, pp. II:1007–1036.

Radnitzky, G.: 1984, 'Scienza, technologia e azione politica', *Biblioteca della Libertà* (Torino) **20**, 75–119 (Luglio-Settembre).

Radnitzky, G.: 1986, *L'epistemologia di Popper e la ricerca scientifica*. (Edizione italiana a cura di Angelo M. Petroni). Roma: Edizioni Borla.

Radnitzky, G.: 1987, 'Erkenntnistheoretische Probleme im Lichte von Evolutionstheorie und ökonomie: Die Entwicklung von Erkenntnisapparaten und epistemischen Ressourcen', in (Riedl und Wuketits, 1987), pp. 115–132.

Radnitzky, G.: 1987, 'An economic theory of the rise of civilization and its policy implications: Hayek's account generalized', *Ordo. Jarhbuch für die Ordnung von Wirtschaft und Gesellschaft* **38**, 47–90.

Radnitzky, G.: 1987, 'The 'economic' approach to the philosophy of science', *The British Journal for the Philosophy of Science* **38**, 159–179.

Radnitzky, G.: 1987, *Entre Wittgenstein et Popper. Détours vers la Découverte: Le Vrai, le Faux, l'Hypothèse*. Paris: Vrin.

Radnitzky, G.: 1988, 'Wissenschaftlichkeit' in (Seiffert und Radnitzky, Hg., 1988).

Radnitzky, G. (Ed.): 1987/88, *Centripetal Forces in the Sciences*, 2 Vols., New York, NY: Paragon House Publishers.

Radnitzky, G. (Ed.): 1989, *Universal Economics: Assessment of the Achievements of the Economic Approach*. New York, NY: Paragon House Publishers, *forthcoming*.

Radnitzky, G. and Bartley, W.W., III (Eds.), 1987, *Evolutionary Epistemology, Theory of Rationality, and the Sociology of Knowledge*. La Salle, IL: Open Court.

Radnitzky, G. and Bernholz, P. (Eds.): 1987, *Economic Imperialism: The Economic Method Applied Outside the Field of Economics*. New York, NY: Paragon House Publishers.

Regnery, H.: 1979, *Memoirs of a Dissident Publisher*. New York and London: Harcourt Brace Jovanovich.
Ricossa, S. e di Robilant, E. (Eds.), 1985, *Libertà, Giustizia e persons nella Società Technologica*. Milano: A Giuffrà Editore.
Riedl, R. und Wuketits, F. (Hgd.): 1987, *Die Evolutionäre Erkenntnistheorie*. Berlin/Hamburg: Paul Parey.
Robilant, E. di (Ed.): 1987, *Libertà e società technologica avanzata*. Milano: Longanesi.
Rosenberg, N.: 1982, *Inside the Black Box: Technology and Economics*. Cambridge: Cambridge University Press.
Rosenberg, N. and Birdzell, L.: 1986, *How the West Grew Rich: The Economic Transformation of the Industrial World*. New York, N.Y.: Basic Books.
Seiffert, H. und Radnitzky, G. (Hg.): 1988, *Handlexicon zur Wissenschaftstheorie*. München: Ehrenwirth Verlag.
Smith, A.: 1776, *An Inquiry into the Nature and Causes of the Wealth of Nations*. (repr. 1937. New York, NY: Modern Library).
Topitsch, E.: 1958, *Vom Ursprung und Ende der Metaphysik*. Wien: Springer.
Topitsch, E.: 1979, *Erkenntnis und Illusion*. Hamburg: Hoffmann & Campe.
Vollmer, G.: 1987, 'The unity of science in an evolutionary perspective' in (Radnitzky, Ed. 1987), Chapter 10.
Waal, F. de: 1982, *Chimpanzee Politics*. London: J. Cape.
Wächtershäuser, G.: 1987, 'Light and life: On the nutritional origins of sensory perception' in (Radnitzky and Bartley, Eds., 1987), pp. 121–138.
Walker, M.A. (Ed.): 1988, *Freedom, Democracy and Economic Welfare*. Vancouver, BC: The Fraser Institute.
Watkins, J.W.N.: 1984, *Science and Scepticism*. Princeton: Princeton University Press, and London: Hutchinson.
Watkins, J.W.N.: 1987, 'Epistemology and politics', in (Agassi and Jarvie, Eds., 1987), pp. 151–167. (Originally published in *Proceedings of the Aristotelian Society*, 1957, **8**, 79–102.
Watkins, J.W.N.: 1987, 'A new view of scientific rationality', in (Pitt and Pera, Eds., 1987), pp. 63–82.
Weede, E.: 1985, 'Warum bleiben arme Leute arm? Rent-seeking und Dependenz als Erklärungsgrundlage für die Armut in der Dritten Welt', *Politische Vierteljahresschrift* **26**, 270–286.
Weede, E.: 1987, 'From 'the rise of the West' to 'Eurosclerosis': Are there lessons for the Asian-Pacific region?', *Asian Culture* **15**, 1-14.
Weede, E.: 1988, 'Der Sonderweg des Westens', *Zeitschrift für Soziologie* **17**, 172–186.

PART II

STUDIES IN THE HISTORY AND METHODOLOGY OF THE NATURAL SCIENCES

PAUL FEYERABEND

ANTILOGIKĒ

Plato's dialogue *Theaetetus* examines the nature of knowledge. It starts with the thesis that knowledge is perception. Socrates articulates the view, quotes authorities (Homer, Heraclitus and Protagoras among them) and describes how, following these authorities, perception must be defined: neither the object nor the perception can exist by itself: act, object and perception must form an indivisible block. The unperceived world contains motions, some swift, some slow, but without definite properties. The motions interact and produce the block with its well defined ingredients. This is a striking anticipation of some features of quantum mechanics (Einstein-Podolsky-Rosen correlations), for here, too, it makes no sense to speak of objects and their properties except as parts of a particular interaction. Finally, Socrates raises various difficulties. For example, he says that if knowledge really were the same as perception then we would cease to know when we close our eyes – but we still know because we remember what we have seen. Or he says that closing one eye we both know and don't know. Or he says that perception is certain, it is possessed by everybody, and relative to them so that teachers, such as Protagoras turn out to be superfluous. 'We must say, then', Socrates concludes his diatribe (which contains further counter examples) 'that each of the two [viz. knowledge and perception] are different from each other'. 'Obviously', mumbles Theaetetus, his fall guy.

Many readers are liable to agree with Theaetetus. Socrates thinks otherwise. 'But what are we doing?' he asks. 'Like a bad cock we forget our topic and start crowing long before we have been victorious.' The counter examples, he implies, are not decisive. He also explains why: 'all we did was to base our objections on the similarity [or dissimilarity] of words': the *word* 'knowledge' as introduced in the thesis (that knowledge is perception) leads to statements that clash with statements containing the same *word*, but used in accordance with commonsense, or after the manner of some noted philosopher. What we have is not an argument but a clash between two sets of statements, one popular and therefore taken for granted, the other unusual, but containing the same word and therefore apparently subjected to the rules of the first set. This, says Socrates, is *antilogikē*, word bashing; it has nothing to do with the search for wisdom.

Fred D'Agostino & I. C. Jarvie, Freedom and Rationality: Essays in Honor of John Watkins, 185–189.

Let me repeat the features Socrates finds objectionable. We start from a situation that is characterised by beliefs about a state of affairs *X*. Some beliefs are explicitly formulated, by philosophers, or by commonsense, others guide our speech in a less obvious way (rules of language). Some beliefs have strong empirical backing, others are more hypothetical. I call the totality of beliefs concerning *X* the field of *X* (at the time in question). Now a new thesis about *X* is introduced, say, '*X* is *A*'. How shall we examine it?

One way of examining the thesis is to compare it with its field and to reject it if it does not fit. For Socrates this is mere word bashing. What we have to do, according to him, is to adapt the field to the belief and see what happens.

For example, the objection from memory uses the assumption (which is part of the field of 'knowledge') that a perception and the corresponding memory both refer to a fact which becomes known via the perception or the memory. But the thesis that knowledge is perception denies perception- and memory-independent facts and thus should not be criticised by assuming their existence. The objection from teaching assumes that a teacher has privileged information about the world, that the pupil does not have this information and that the teacher imparts it to the pupil. According to Protagoras the pupil is in command all the time: to start with, it seems to the pupil that he is in a bad state, wisdomwise, and, according to Protagoras, he therefore really is in a bad state. It also seems to the pupil that a teacher may improve his situation and so he hires a teacher. The teacher arrives and produces words. He does so because it seems to him, the teacher, that words are excellent instruments for changing a state that seems bad to the pupil into a state that seems good to him. After a few lessons it seems to the pupil that he is indeed in a better state which means, according to Protagoras, that he has really improved... and so on. Lots of objections may be raised to this elaborate and strange-sounding account of teaching – but the original objection viz., that Protagoras is inconsistent in declaring that knowledge is perception and yet posing as a teacher – is not among them.

Socrates' advice is in direct conflict with certain versions of empiricism. Radical empiricists identify the field of a particular assumption with the known (empirical) facts and laws. Trying to change the field by way of the assumption means engaging in *ad hoc* moves. *Ad hoc* moves are to be avoided hence word bashing has to replace the search for wisdom (to use Socrates' terms).

Similar remarks apply to empiricist definitions of the content of a statement. Some empiricists separate meaning and empirical content, others identify the two; both assume that the scientific character and the fate of a statement are decided by its empirical content. Thus the field of a statement is made a measure either of its eminence, or of its truth – a typical example of word bashing.

Has Socrates perhaps made a mistake? Is word-bashing perhaps preferable to a search for wisdom? I don't think it is very profitable to discuss this question in the abstract. But there exists plenty of material that may help us to arrive at a decision. The material does not force us into certain procedures – but it makes word-bashing very unpopular indeed.

For example, take Lactantius' argument that the earth cannot be spherical as this would make the antipodes fall 'down'. The field of the idea that the earth is spherical here contains the assumption that the world is anisotropic and that things left by themselves move in a preferred direction, i.e. either 'up' (fire, mist), or 'down' (water, earth). Socrates advises against Lactantius and suggests redefining 'up' and 'down' in accordance with the hypothesis ('up' now means 'away from the center', 'down' 'towards the center'); in other words, he suggests introducing a central-symmetrical universe and to see how far we get. Today the spherical earth is a triviality and the move is obvious. (It was obvious already at the time of Lactantius). But as long as the matter is under investigation adapting basic notions to a new view means trying to save hypotheses by *ad hoc* maneuvers.

This becomes very clear in the case of Copernicus. Copernicus (Book i, Chapters 7 to 10 of his *De Revolutionibus*) enumerates the dynamical difficulties of the motion of the earth. He introduces the hypothesis that what is part of the earth also moves with it, no matter what other motion it has received. Expressed differently, he introduces a dynamics specially designed to fit the motion of the earth – a paradigm case of an *ad hoc* adaptation. It is a strange sort of dynamics (for contemporaries averse to occult qualities); it assumes that a distant body can as it were 'feel' where the earth is and react accordingly. Galileo vigorously attacked an analogous assumption in the case of the moon (tides). Yet the move was entirely in the spirit of Socrates and the resulting cosmology an advance according to Newton.

The argument of Einstein, Podolsky and Rosen (EPR for short) shows the reverse side of the coin. The thesis proposed is that a definition of the properties of elementary particles must include the experimental situation (or, in the case of cosmic processes, the mediating interaction). The thesis is criticized by confronting it with the conjunction of a complicated calculation (detested by Einstein, who ascribed it to Podolsky) and a 'principle of reality' according to which distant operations cannot affect the state of a system. But this is just the point at issue. A piece of wood in Australia changes its length when length is defined by a rubber unit-meter in Vienna and I go to Vienna, enter the unit-meter institute and stretch the meter. Have quantum mechanical magnitudes similar properties? Bohr thought so and defined them accordingly. His views may be open to objections, but EPR is not one of them. It is a clever and deceptive instance of word bashing (this observation is independent of the developments in connection with locality).

The proper procedure in such matters was outlined, long ago, by Galileo. Grassi had measured the distance of comets and had found them to be beyond the lunar sphere. Galileo objected (among other things) that triangulation works only if the nature of the things triangulated is already settled: it makes no sense to triangulate a rainbow. Ontological discussion precedes measurement (we might call this Galileo's requirement). EPR takes the requirement into account: it uses a 'principle of reality' to interpret the formal features of the case. The trouble is that the principle used assumes what is to be shown. Arguments that try to derive classical particle states from the Born interpretation lack even this already very tenuous connection with Galileo's requirement. They overlook the fact that statistics give us numbers, not the nature of the things counted; these can be particles with well defined positions and momenta, transitions from indefinite states to states with well defined positions or momenta – and many other things.

Antilogikē plays a large role in discussions of the mind-body problem. I have assembled some of the 'arguments' in my essay 'Materialism and the Mind-Body Problem', first published in the *Review of Metaphysics*, vol. 17 (1963), republished with comments as Chapter 10 of Vol. i of my *Philosophical Papers*, Cambridge 1981. I concluded my paper with the following remarks:

> A common feature of all the arguments [against materialism] is this: they try to criticize a theory *before* this theory has been developed in sufficient detail to be able to show its power. And they make established modes of thinking and of expression the basis of criticism. I have pointed out that the only way of discovering the faults of established modes of thinking is by resolutely trying out a different approach...

by which I meant the approach suggested by Socrates. I did not know his advice then, but I was very happy when I discovered that I had such a distinguished predecessor.

Socrates' procedure affects even the simplest cases. Take that familiar illustration of the force of counter examples: all ravens are black (I shall call the statement *R*). It seems we know exactly what the statement means and we also know exactly when we are going to give up: we can define its content and its potential falsifiers. But now assume that we find a raven that has been painted white. Is this a counter example? 'Obviously' not. Why not? Because when asserting *R* we mean 'intrinsic' color: in *R* 'black' means intrinsic black and the potential falsifiers have to be intrinsically white. But what are 'intrinsic' colors? Are they colors that are genetically determined or should we also regard the effects of longlasting illness as intrinsic? And what if we discover that the genetic makeup of a species can in special cases, occurring frequently in ravens, be changed by external means? Do we want to include ravens that were bleached by industrial fumes? Ravens that became white as a result of profound shock? And how can we decide the question? Not by an *ad hoc* definition based on a preferred field but by further research. Starting with our examination we have no idea at all what *R* really 'means' in terms of content and of falsifiers. (In the same way a Socratic researcher examining 'knowledge is perception' has not a fixed idea of the nature of perception – it may include memory and dreams, it may not.) The statement (and, for that matter, all statements claiming to express knowledge) is not a well defined semantic entity; it is an instrument for assembling a content which constantly changes and whose development is inextricably intertwined with the accidents of history.

There is only one area where word bashing seems legitimate, and this is the area of human behavior: desiring to lead a certain kind of life we can, and should, push aside all considerations that conflict with it. Aristotle's attitude towards Parmenides is an example. He criticizes him in two ways: 'scientifically', and by reference to the conditions of citizenship. The scientific criticism reveals hidden assumptions and lacunae; it belongs to the tradition encouraged by Socrates and Plato. But Aristotle also considered the practical effects of Parmenides' doctrine – it would create havoc. Now the question is: do we want havoc, or do we want to continue living as civilised (Athenian) human beings? Obviously the latter. Hence Parmenides' results are to be removed no matter how compelling the argument that produced them; the decision in favor of a certain form of life can eliminate even the most sublime piece of theorizing. This sounds like word bashing but is not; for what is opposed to the words of the theoretician are not other words, but our very existence.

University of California, Berkeley

J.N. HATTIANGADI

KUHN STUDIES

As a graduate student it was with great pleasure that I learned that John Watkins had decided to thank me publicly for helping him with a paper on Kuhn's view.[1] The help, such as I could give, was in Popper's seminar, twenty-five years ago. Watkins himself, and several others, contributed much more to the seminar than I did. (The seminar was run on the principle – to repeat J.O. Wisdom's quip – "thou shalt not speak whilst I interrupt"). Watkins was not just kind but generous in thanking me and I appreciated it very much then, as a mere graduate student. But over the years, I find that I appreciate it even more, especially as I compare it with practice common among philosophers. It is ironic that by the time Watkins' paper appeared (something resembling his view had by then already become the official reading of Kuhn's views among philosophers of science) my own views about Kuhn's book had changed.

Largely through Kuhn's own efforts, I came to see that I had indeed misunderstood Kuhn.[2] At one time, it seemed to me that Kuhn's claim that he was not understood was a philosopher's *amour propre*, who would rather be misunderstood than mistaken. But the mistake and the misunderstanding were mine. It seems to me now that a certain aspect of Kuhn's views, correctly seen, has something important to teach us about scepticism – a worry which has surfaced recently in Watkins's writings. I will remark briefly on how experiments might be helpful to the sort of quest Watkins is engaged in when he investigates scepticism. In most of this paper I shall merely renew a twenty five year old dialogue about Kuhn's book, *The Structure of Scientific Revolutions*.[3]

*

Kuhn's description of the normal scientist is at great variance with the descriptions found in the philosophy of science preceding the publication of *The Structure of Scientific Revolutions*. This is not to deny that precursors exist, but merely to point out that it is only after Kuhn's captivating picture was painted that we began to look for something similar in the nooks and crannies of the writings of earlier authors.[4]

Fred D'Agostino & I. C. Jarvie, Freedom and Rationality: Essays in Honor of John Watkins, 191–205.

If I were to select one aspect of Kuhn's picture today that serves as a point of departure from earlier philosophers of science, I would choose this feature, namely, that Kuhn portrays the normal scientist not as a thinker or as an author of doctrine – though a scientist is that, certainly – but as an actor of a certain kind, a technician who performs with consummate skill.

Kuhn's portrayal of the scientist as a skilful technician drew attention away from the traditional preoccupation of philosophers of science with what scientists say, to what they do and how they do it. Kuhn notices that the tasks taken up by scientists are highly complex, esoteric tasks very often calling for just the kind of skill that only normal scientists can bring to bear on them.

The coincidence is not, of course, accidental, for both the skills of the normal scientist and the tasks which require these skills have a common origin. This is what Kuhn calls a 'paradigm'. It was once customary among critical philosophers of science to think of Kuhn's 'paradigms' as a confused way of portraying theories. I remember in 1965 trying to clarify Kuhn's conception of a 'paradigm' by translating it into the perfectly clear expression 'a dominant theory'.[5] But we see in retrospect that however unclear Kuhn's notion of a paradigm was, it did not clarify matters to describe it as a 'dominant theory'. For to do so missed an important element of Kuhn's thought – a Wittgensteinian element perhaps – which I would now describe as entirely central to Kuhn's enterprise in *The Structure of Scientific Revolutions.*

The skills that a scientist brings to bear on his subject are not natural, or innate. They are learned skills. And implicit in Kuhn's description of normal science is the thesis that all learning is by example. A paradigm, then, is a skilful achievement from which others learn by admiringly repeating the technique employed. There is, of course, more than this to what Kuhn meant by 'normal science'. But I do want to emphasize that he did not intend less, and this much seems largely to have been missed by most contributors to that book of commentary which brought fame to both Kuhn and its senior editor, Lakatos: *Criticism and the Growth of Knowledge.*[6]

Paying attention to technique, we find that the source of technique is not a book, or a doctrine, but a tradition, a practice. How frequently does Kuhn describe the scientist as a practitioner, and how often does the philosopher of science miss this description? Having missed this, it is difficult to appreciate why science can only be described by a sociology, and not just a philosophy. For the esoteric tasks of the practitioner and the consummate skill are both acquired from the ongoing social enterprise of science.

Science is socially organized. The individual scientist depends on the other practitioners to recognize what is good and bad, what is routine and what is remarkable, what is acceptable and what is bizarre, in ongoing research in the field. It is this constant interaction between practitioners that leads to the developing skills of the novitiate. When describing science thus, Kuhn does not need to talk about truth, or logic, or facts. It is enough to say that scientists talk of such things while they do what they do with such finesse.

A little remarked feature of Kuhn's conception of normal science is that it is much like a Hobbesian theory of society. Extraordinary science, those brief periods between normal scientific periods, are described as thoroughly unpleasant for professional scientists. Lacking a homogeneous tradition to fall back upon, practitioners lose faith in the manner of practice. In order to preserve intellectual peace, as it were, scientists flock to a new sovereign paradigm, a new way of doing things that promises a measure of professional harmony. A paradigm, like a sovereign, is absolute in the sense that the practitioner adopts it uncritically, in much the same way that a speaker of English adopts the English language – not as a thesis which may be evaluated, but just as a tool for communication. So, too, the practitioner adopts a certain way of doing things. This is not a dogma, although Kuhn did once call it that. It is not a dogma but just good sense, because to do otherwise would leave scientists unsure of the criteria for success and failure, which in turn would lead them eventually to a slow professional death. A scientific or intellectual state without paradigm leads to professional careers which are nasty, brutish and short.

If we overlook this Hobbesian element in Kuhn's description of scientific practice, his description of science is remarkably close to Wittgenstein's description of ordinary language as it is practiced by an ordinary user, where different language games are practiced by imitating different linguistic paradigms. But the Hobbesian element in Kuhn's description is unavoidable, for unlike language, there are revolutions in science, and here lies a fundamental difference between science and ordinary language.

This Hobbesian underpinning of scientific traditions has an astonishing consequence: it leads Kuhn to dismiss the entire prehistory of science as relatively unproductive intellectual wrangling. For if there is not a common set of accepted techniques and tasks, there can be nothing but a Babel of tongues. Schools of thought flourish in pre-paradigm science. These schools waste their time and effort attacking the foundations of each other's thought. Each defends its own principles, and none therefore ever gets around to winkling out those extraordinary secrets of nature that scientists succeed in uncovering with their finely developed skills. To argue about fundamentals is not to get on with the job. One cannot find out about the world if one is going to pause at every moment to question oneself. So it becomes necessary to abandon one's critical stance, to adopt a definite way of doing things, to learn to do it well if not better than one's predecessors.

Preparadigm thought, we recognize clearly, is philosophy, or at least something philosophical, where normal science is technological. Kuhn's description of the rise of modern science is therefore profoundly correct in the respect that it describes a history of a relatively unproductive, philosohical debate which was abandoned for a suprisingly fecund technological endeavour. There is much truth to this description of what has happened in four centuries, more or less.

Philosophers of science have not been overwhelmed by his challenge. A whole army of them has tried to give accounts of science as a rational enterprise – that is to say as a philosophical enterprise – the most famous of these being accounts by Watkins, Popper and Lakatos, and following him Laudan and Sarkar, and the least known of these being some of my own efforts in the past.[7] But I believe now that such efforts are doomed to failure, because in fact, science has become technological. It has become in part big science, some of it having joined hands with what Eisenhower called the military–industrial complex. Only a philosopher would want to say that this ought not to be so, because it makes no difference to the fact that it is so. Kuhn describes science in such a way that we can see how it has become possible that we have big science.

The difficulty of looking for a philosophical rationality in scientific technology is much more pronounced than even Lakatos and Laudan and others have realised. They have addressed the difficulty that "a test is a test of the scientist and not of the theory", and also that "if a theory is refuted by the facts, then all theories stand refuted at all times". Kuhn does describe anomalies as refutations that scientists learn to live with, and perhaps even to love. So much effort has gone into showing how one can wriggle rationally out of falsification, that people have failed to see that on Kuhn's account, confirmation can lead to the overthrow of a theory, which is more astounding still, and quite outwits the brilliance of Lakatos. Let me illustrate.

Normal science consists of, among other things, a continuing rearticulation of the paradigm. Imagine two Newtonian scientists, both atomists, one of whom applies Newton's atomism to study elastic and inelastic collisions, and another who uses it to study chemistry. The rearticulations in the two cases are slightly different, but let us say that the difference is not one that causes concern. Two normal scientific traditions which are both 'Newtonian' in origin – the physical atom and the chemical atom – will develop side by side. After a century of rearticulation, however, they diverge quite significantly. But as long as chemists and physicists practice different techniques on different tasks, each can conform to his own 'Newtonian' theory quite happily, in blissful ignorance of any real or philosophical difference between the two.

Then one day Alessandro Volta discovers a curious electrical current flow from a hydro-chemical arrangement. Nicholson and Carlisle send this current back into water only to find that it decomposes water into its chemical constituents. All of a sudden, a series of confirmations of whatever scientists were doing when they followed their paradigms has led to a new hybrid field, electro-chemistry.

But which atoms does the electro-chemist adopt? Suddenly there are two paradigms, and without a single refutation, we find ourselves in the midst of a philosophical wrangle about fundamentals: we are in crisis. For with two atoms come two incompatible sets of standards and two styles of research, which is one too many.

A series of anomalies or failures is consistent with stable paradigm-based research because retooling is costly; at the same time a single unlucky success of our paradigm, or a confirmation, may lead to a crisis and the overthrow of a paradigm. How can a philosopher of science handle this?

I do not believe that modern science can be described in the rational terms that I have preferred over the years. Science is a social institution of great complexity, and we are foolish to try to find a rational basis to all of it. Would we expect a rational basis to royalty, or to kinship or to banking?

But a small question still remains – the last hope for a desperate philosopher of science. Is there at least something in science, or in the acquisition of knowledge, that remains philosophical? Is it *all* reducible to technique? Is there anything of value that can be retained of our nostalgic picture of Galileo or Newton, Darwin or Mendel, Einstein or Bohr, as *thinkers*? Is there any such thing as the growth of knowledge, as opposed to the proliferation of technique and skill?

If we are to find it anywhere, we will have to look at what used to be called pure science. But is there any such thing as pure science, and does it have the same structure as Kuhn ascribes to technological science? When we ask our questions in this modest way we begin to see a little room for rational debate.

It is not at all surprising that when we look upon science in the domain of skill and technique we gain great insight into it. Any activity of sufficient difficulty requires skill for its performance. The long and arduous training and rigorous testing of successful candidates entering the scientific professions attests to the great skill that is characteristic of its practitioners. Clearly, philosophers of science who forget about skills overlook a fundamental feature of science.

But there is another aspect to this neglect. For skill and technique are not unique to the scientific enterprise. Skills are an important part of all human activity of whatever degree of difficulty. From a biological perspective, all organisms who possess sensory motor systems and are capable of variable responses to the environment possess skills of varying degrees of complexity. So consummate is the skill of even simple organisms that they seem uncannily adapted to their environment, a fact which once formed the basis for a belief in God. This 'argument from design' has been replaced in more recent time by a belief in natural selection. Skills are ubiquitous.

Paul Feyerabend made fun of Kuhn's description of science – both normal and extraordinary – by showing in his essay 'Consolations for the Specialist' that even bank robbery exhibits the very patterns of normal and extraordinary periods that Kuhn describes as scientific.[8] Young practitioners go to reform school, or gaol, to train to be professionals. At any given time there is a set of techniques which is paradigmatic and widely practiced. For a long time, in normal robbery, practitioners may case a joint, use tools and meticulous methods to open safes with great stealth. Then there are revolutions. Dillinger came along and showed how a bank could be robbed not by stealth and careful planning but by daringly hitting it and running before the law could respond.

Kuhn responded to this amusing observation by saying that of course much besides science exphibits the same pattern, but that does not take away from the fact that *science does*. But we see in retrospect the weakness of this response. For while the scientist as a skillful technician is most interesting, studying science in the domain of skill misses just that aspect which distinguishes it from all our other consummate skills which form the very basis of our survival and existence, and even some which do not. All our activities are skill-based activities, to the extent that they are at all successful. In science – and in intellectual matters in general – the skills improve by virtue of our reflection.

It is of course not only in science that reflection helps improve skills. Dillinger and Gretzky can both improve upon their skills by thinking about their respective tasks from time to time. But where Gretzky wields a hockey stick on skates, and Dillinger a firearm on wheels, the scientist wields statements in a journal. No matter how much we are impressed by the lab coat and the instruments of the physicists or the biologists, they are primarily producers of copy. In studying a science, if we are to understand what happens in the lab, we have to read what is written. This is not necessary for Dillinger, or for Gretzky, who can be appreciated without intellectual recourse.

Intellectual activity is skilful activity, of course. But at the same time it is centrally an exploration of a thesis, a study of what is the case. Kuhn's portrait of the scientist might not be an inaccurate portrait, as far as it goes, but it misses something which we cannot afford to miss: at the very heart of science, where theoretical activity takes place, and where pure science is done, we find schools of thought, and debate.

When Kuhn describes normal science he describes it as a homogeneous whole. All scientists are joined together in common enterprise. When Kuhn visited the Institute for Behavioral Studies in Stanford before his book was written, he was struck by the fact that social scientists constantly question the validity of each others' work. In physics, and in the mature sciences generally, he thought this would be abnormal. Physical scientists listen to each others' work with respect, and without skepticism. This observation, he tells us, was seminal.[9]

Kuhn is an acute observer, and his remarks about the difference between mature science and social science or the humanities are true, by and large. But his Hobbesian argument that it is *because* natural scientists do not wrangle that they succeed, is not sound. (In the following I will sketch an alternative view.) If we look at any period of science which may be described as normal, we will not fail to see that it coexists side-by-side with theoretical disputes in another part of that science. Quantum mechanicists were doing fine normal science, thank you, even as Bohr and Einstein were debating on fundamentals. Kuhn's thesis that wrangling undermines the whole of a subject paints too uncomplicated a picture of the scientific institution. Consequently, it still remains to investigate why the science of Galileo and Newton was successful, while philosophy prior to that was not. And this is the little bit of work that Kuhn leaves for the philosopher of science yet to do. Our only consolation is that if this task was worth the attention of Immanuel Kant and Bertrand Russell, we can afford to play with it a little.

If we turn to the reason for the technological success of modern science, we find another clue in Kuhn's own ideas on the matter, which I will elaborate below. We need not accept his view that it is necessary for all scientists to abandon the debates of pure science, or the philosophical end of science, in order for some to practice technical skills. In the rich pluralistic world of science there is room for many types of practitioners.

Why has natural science been a technological success?

The classical answer to this question was that (a) science found the truth about nature; (b) that truth gives us power over nature; and (c) that what distinguishes science from the scholasticism of earlier times is its organon, namely the experimental method. Whereas logic was the sterile method of medieval philosophy, it was argued, experimental science, with induction, was hailed as the new organon for the discovery of Nature in its true light.

This conception of science had considerable influence in the centuries following the writings of Francis Bacon, but has since come under careful scrutiny, and has been largely rejected. Whether an experimental method and induction could ever establish any general truths, as Bacon had hoped, has been questioned by Berkeley and Hume, and following them by almost all subsequent philosophers of science.

The power that we have attained over certain natural conditions would seem to suggest that if we have not the truth about the physical world, then we must have at least a very close imitation of the truth. But logic tells us that this is not a persuasive argument. Every statement, and even every false statement, has true consequences. What distinguishes a system of true statements is simply that it has only true consequences. If all our physical, chemical and biological hypotheses were to prove to be a tissue of lies, it would still not detract from any true consequences that they may have. If all our scientific hypotheses should in time be replaced, they would not take away our technological competence.

The fact that we developed a wonderful science of mechanical and electrical engineering presupposing the truth of Newtonian laws was no less a fact after we learned to amend those laws at very high and very low energy levels. Relativistic and quantum variations of classical physics do not make a great deal of difference to the successful techniques of the past. But where they do, we can improve upon old techniques. Here, again, I believe that Kuhn has given us a valuable insight into the matter but not the one he relies on. The Kuhnian normal scientist is a practitioner. He translates abstract ideas into concrete laboratory practice, among other things. The unit of success is to be found in technique.

Classical philosophy, and even philosophy of science in more recent times, has missed this significant fact about science because we have been overwhelmed by the New Way of Ideas. We have understood the Experimental Method as a method of observation, as Locke or Kant in their different philosophies would interpret it. It seems to me that a most important aspect of the experimental method was missing from all this, until Kuhn illustrated so clearly that the experimental method is not just a means for us to perceive nature in an unfamiliar pose, as Bacon sometimes suggests, but that it is a practice, a successful exercise of skilful technique.

The empiricist tradition from Locke works with a model of human beings as perceivers, as passive receptors of images. Our knowledge is said to come ultimately from our senses, and it remains to show how the intricate, precise and immensely general laws of science can be derived from the fleeting, subjective and confusing perceptions of the scientist at work in the laboratory. But this empiricist tradition is misleading, because the scientist does not gain her knowledge of the laboratory techniques merely by observing, however acutely. She understands only by practicing her skills. The scientist does not necessarily test the abstract theories of physics or biology against the facts of 'experience'. More often the test is a performance, a well honed skill, or several of them, being used to produce a result with great predictability.

This criticism of modern empiricism is not itself new, because it is contemporary with Locke's *Essay Concerning Human Understanding*. Giambattista Vico made the central point of his critique of the new science and the new way of ideas that we can truly understand what we create, not what we merely observe. This theme was taken up subsequently by Romantics, who used it to decry science and the Enlightenment as false understanding, as compared to the understanding involved in the creations of the novelist or of the artist. But this interpretation tries to treat all of science – even its theories – as if they arise just out of the existence of technique. And this, I think, is also not accurate.

If we identify the empirical basis of science with experimental practice, I think we can begin to see that Vico did have something that we can feel confident that we understand. What Vico could not have known is that from the perspective of modern biology it makes much better sense to understand the knowledge possessed by an organism in terms of its overall adaptation to the environment than by looking to the content of its perceptions. For one thing most organisms do not possess perceptual systems. All plants and all small life forms, which make up almost all the living forms, both in the number of species and in biomass, lack them. Those organisms which do possess perceptual systems use these systems as an aid in responding to the environment only because they are part of their general adaptation. Moreover, the perceptual systems pick out features of the environment not with the mark of omniscience, but simply to direct its organism to react successfully in all matters which are of importance to its survival. Where the organism functions successfully, there it knows its environment well enough to satisfy its needs.

The human organism is no different. When we use a laboratory space to experiment, we learn techniques to adapt to it, to make it do what we want. It becomes our evolutionary stomping ground, or a new extension of it. We use our perceptual systems there with the consummate skill of any organism which navigates successfully in its natural surroundings. When the abstract theories of physics or biology need to be tested, we can rely on experiments not because they are empriical, but because they are well honed skills being practiced. We know the world in the only way that nature has fashioned us to know it, before we invented the socially constructed theories of science – namely by using our skills, which include, among others, our perceptual skills.

The technological success of modern science, then, is not a product of the abandonment of philosophical debate as Kuhn's picture might suggest, but the use of experimental technique to direct and improve upon this debate. When Bacon and Descartes predicted that their new science would benefit humanity, they intended this to be the fruit of the attainment of general truth. What seems to have transpired is that the experimental method which was merely to be an aid in the attainment of truth has spawned a technical mastery over nature entirely as a by-product. Whereas in the seventeenth and early eighteenth centuries the experimental scientists of the Royal Society were sometimes popularly lampooned as bumbling but innocuous philosophers lacking in common sense, the public image of the experimental scientist today is anything but that. This, too, is the mark of success.

In three hundred years science and technology have moved closer together in several areas. But in this time, some sciences have lent themselves to technological exploration more readily than others. As a broad generalization I believe nobody will challenge my claim that the natural sciences have had greater technological success associated with them than the social sciences.

Since this was, in a certain sense, the starting point of Kuhn's odyssey, let us consider this question at last. Why is it that in the natural sciences the proportion of technological success, and therefore the appearance of 'normal science' is greater than in the social sciences? Why can the social scientist not use the experimental method as successfully as the natural scientist?

If I am accurate in my assessment, there is no reason, in principle, why the social sciences cannot be as technologically successful as the natural sciences provided that they employ the same diligent experimental methods. The social sciences remain more philosophical and less technological only because there are certain important and unavoidable limits to experimentation, certain ethical constraints regarding the extent to which one human should act upon or 'use' another. In the intermediate case of bio-technology, I think the limits and the moral dilemmas are poignant and well known today. Experimental testing on live animals is an issue which is fraught with controversy. In the case of our proper studies, namely that of human beings, we are still less able to escape ethical considerations. The bulk of the experiments we perform of the social kind, when morally pursued, are thought experiments. The greatest of these thought experiments are the novels and plays that we value so much. They take experimental social situations and explore them with mock seriousness, because we cannot explore with equal freedom real human relationships.[10]

Thus, it seems to me that Kuhn's observation of the difference between the attitude of the normal natural scientists to each other and that of the normal social scientists to each other, is an astute one. It reflects the relative dominance of technology and philosophy in these two realms. But if I am right, there is a better and richer explanation for the difference between the two, and of the relative success of natural science, than Kuhn's Hobbesian claim that we must escape philosophical debate to adopt a single sovereign paradigm in order to reap technological rewards. This explanation has to do with experiments as the foundation of science. I offer this to John Watkins as an aid in his attempt to solve the problem of scepticism which has beset British empiricism.

The 'problem of scepticism' can be analysed in different ways and factored into different subproblems. My own preference is to look at the history of the issue to find out what might solve it. Looked at thus we can see that there are in fact two problems here which are closely related but which are usefully separated. There is, first of all Berkeley's critique of Newtonian science, which was so successful that we have wondered ever since how it is possible that we can have physical (mathematical) knowledge. We may call this Berkeley's scepticism, which is restricted to scientific knowledge. Berkeley saw no difficulty with common sense knowledge, or with our putative knowledge of spirits or with our putative knowledge of God. Hume, however, suggested a still more fundamental difficulty, namely, that if Berkeley's arguments undermine science then similar arguments undermine common sense knowledge, as well as other kinds of knowledge that Berkeley failed to challenge.

The form of argument which Berkeley employed, and which was in turn deployed by Hume against Berkeley, arise out of Locke's New Way of Ideas. Locke suggested that our understanding of the material world comes by way of things in the world making impressions on our senses. These impressions are the source of our knowledge, according to Locke. The things in the world itself are material mathematical objects (with 'primary qualities') which produce in our senses the qualities that we perceive in bodies.

Berkeley argues that if we can only perceive objects as ideas, and if it is our only source of knowledge, then we cannot cogently conclude that the objects are material and mathematical, as Locke and Newton claim they are. Thus Newton's physics, which relies on an interpretation of the world in terms of absolute space and time, on matter, and on gravity – none of which are fully representable in our senses (except indirectly) – must be fiction, albeit a useful fiction. Hume was a materialist, a Newtonian, who argued that if Berkeley's arguments against materialism are sound, then it also follows that all of our other knowledge, which Berkeley left unchallenged, is impossible. Hume concluded from this not that we do not know, but that the demand that we must found our knowledge of the material world on reason (as Berkeley and before him Locke assumed) cannot be fulfilled.

Scepticism is therefore a dual demand. Following Hume we want to know how we can know anything at all. Following Berkeley we want to know how we can know anything mathematical about the material world. The former question is closely connected in Hume's argument to action: Hume says that one cannot be a sceptic for very long before one has to act; and in acting we presuppose knowledge which we possess naturally, without good reason, as animals also do. Berkeley's argument, on the other hand, is closely related to the interpretation of mathematics. If mathematics is subjective or empty, then we may regard our knowledge of the physical world as simply a mathematical arrangement of common sense knowledge. It is clear, though that we cannot get to Berkeley's problem without meeting Hume's difficulty.

If we try to solve both problems at once we run the risk of introducing Reason where it manifestly does not exist. A bee which searches for honey, or a cheetah for a gazelle, is incapable of formulating reasons, at least so far as we know. But they are not inhibited in their action as a result. In this case a solution to Hume's problem of induction for the bee or the cheetah cannot be based on Reason. Berkeley's critique of mathematical physics cannot be met without considering how we can, as reasonable and rational human beings, assent to Newton's and subsequent claims about the physical world. These claims are, broadly speaking, that we live in a mathematical world whose mathematical structure goes beyond what we can perceive, even though all our knowledge of the world seems to be given to us in our perceptions of the world. Although we have replaced Newton's conception of mathematical matter by a new statistical conception of matter in which every stable molecule, or larger crystal, is a temporary thing, held together by London forces, the world still appears from the physical perspective to be irredeemably mathematical and quite remote from the sensible world that the scholastics had observed and studied.

When we locate the source of laboratory knowledge not in *percepts* but in *procedure*, we are finding in science a continuity with the knowledge of organisms. Organisms understand the world not by perceiving the world but by acting upon it successfully. When an organism acts appropriately in an environment whose global analysis neither the organism nor we can be said to know, the organism acts in response to a simple cue in the environment. When it acts successfully, its pattern of action presupposes the 'knowledge' (if this word may be used here) that the cue is a sign of the environment in which the action is successful. Now British Empiricism is based on the claim that the knowledge of the world comes to the organism in the cue. This is manifestly false, because if the cue were not simple it would be useless. To take an extreme case, if the organism reacted not to a simplified cue but to the environment as a whole all the time, then it would be successful only if it were omniscient. If an organism acts successfully in a *complex* environment by responding to a *simple* cue, then it follows that whatever allows it to anticipate nature is contained not in the cue but in the predilection to act that way in the presence of that cue.

I have described this as the knowledge 'when *S* obtains *E* obtains', where *S* is the cue and *E* the environment of the type in question.[11] If we ignore the question whether this is appropriately called 'knowledge' (another word will do just as well), there is, it seems to me no defence of British Empiricism and the New Way of Ideas in the light of modern biology.

If our knowledge is given to us in our predeliction for skillful action, and not in the cue (percepts being a special case of cues) then we can maintain both the mathematical picture of the world given by physicists and also that the experimental method gives us knowledge of the mathematical world. There is nothing impossible here, unless one insists that the knowledge comes to us not in our well-honed skills but in percepts. (I do not say 'perceptions' because these are actions, and so involve skills, and hence bring knowledge with them.)

We demand repeatability of experiments not only because they are then intersubjective in most cases, which is Popper's insight, but also because we, as organisms, can rely best on those things that we can do routinely. It is the experimental *routine* that makes it a source of knowledge and not that it leads to *observations*. In fact, this hint of the problem of scepticism has led me to embrace a Pythagorean or realist philosophy of mathematics, in the sense in which Kepler, Galileo, Descartes, Huyghens and Newton were geometrical realists (of course the geometry is different now!), and in the sense in which no one since the days of Berkeley and Leibniz has dared to be. Even Russell, who abandoned Kant in this subjectivist point, ended up by embracing Leibniz's subjectivist theory of mathematics (as analytic), and Russell's philosophy of mathematics has been the tail that has wagged the dog in the twentieth century.

York University (Canada)

NOTES

[1] J.W.N. Watkins, 'Against Normal Science', *Criticism and the Growth of Knowledge*, (Cambridge: C.U.P., 1975), p. 25. The seminars were held in October, 1963 in the series of fortnightly seminars that Popper chaired.

[2] *Truth, Acceptance and Agreement: A discussion of Professor Kuhn's Theories of Science*, was the title of my M.A. thesis (1965), one copy of which was copiously annotated by Kuhn with his pencilled comments, attesting to my misunderstanding, on almost every page. This copy was widely circulated among friends; one of them, presumably wondering why the copy was so defaced, kindly erased almost all the comments. While Kuhn's comments are erased, their effect remains, as I report in the text which follows.

[3] *The Structure of Scientific Revolutions*, (Chicago, Ill.: U. of Chicago Press, 1962).

[4] Most mentioned as precursors are L. Fleck and M. Polanyi.

[5] *Truth, Acceptance and Agreement, op. cit.*, p. 12. I remember a conversation with Popper during the seminars (cf. n. 1) in which we discussed what might be a better word than the word 'paradigm' to describe what Kuhn had in mind, and settled upon 'dominant theory'. I confess that though I had read Kuhn several times over, and quite carefully, it never occurred to me to point out that its nontheoretical component was quite as important as I think I should have realised.

[6] See n. 1.

[7] 'The Structure of Problems', *Philosophy of the Social Sciences*, 1978–79, 'A Methodology without Methodological Rules', *Boston Studies in Philosophy of Science*, (Dordrecht, Holland: Reidel, 1982) are not only little known, by and large, but are very poorly regarded among many who do know them. Upon reading the first in 1971, John Watkins kindly advised me to give up the subject of the philosophy of science, and to try my hand at something a little less difficult. On this point, I think, (immodestly, of course) that he missed something. At least as a curiosity, my views are more interesting than that suggests: mine remains the only naturalistic epistemol-

ogy which has worked out methodological standards as a product of the scientists' response to the logic of their (intellectual) situation.

[8] 'Consolations for the Specialist', p. 200, *op. cit.*

[9] T.S. Kuhn, *The Essential Tension*, (Chicago: U. of Chicago Press, 1977).

[10] Edward Davenport, 'Literature as Thought Experiment', *Philosophy of the Social Sciences*, 13, 1983.

[11] Jagdish N. Hattiangadi, *How is Language Possible? An Account of the Evolution of Language and Knowledge*. (La Salle, Illinois: Open Court, 1987), Chapter 8, 'Empiricism from a Biological Perspective'.

W.W. BARTLEY, III

UNFATHOMED KNOWLEDGE IN A BOTTLE[1]

I wanted to give John Watkins a birthday surprise, so I decided to say a few unexpected things about the unexpectable.

So far as I know, Watkins and I are the only persons to have made any use of one of the most useful and profound ideas in all of Popper's philosophy. I like to put the idea in this way: our best *existing* knowledge is *unfathomed and unfathomable.*[2] Which of course means that *we do not know what we are saying or* (since of course we act partly in terms of what we know) *what we are doing.*

What am I saying? Should I reply, "How should I know?"

*

One might suppose that I am simply expressing fallibilism in an unusual way. But the idea I have just voiced goes beyond ordinary fallibilism.[3] Ordinary fallibilism asserts that certainty cannot be achieved, and that anything that we may say, even anything which may have been incorporated into science, may be wrong. Ordinary fallibilism is also often suffused with the sense that much of what we have already discovered only touches the surface, and that many new things remain to be learnt about ourselves and the universe in which we live, things going far beyond our existing knowledge.

I am a fallibilist in this ordinary sense. But the idea that I am now proposing transcends this. Fallibilism acknowledges that our existing knowledge may be wrong, and encourages us to search beyond it; but I am also saying that we do not, and cannot, understand our existing knowledge. *We do not understand what we already have: knowledge is a product not fully known to its producers.* And the better our knowledge gets, the less we understand it. Even the creators of theories do not and cannot understand them. Which means that they do not understand their own theories even when the theories themselves, as may be, are correct. All of this needs to be explained, and I shall proceed to do so, to the limited extent that I understand what I am now asserting.[4]

Fred D'Agostino & I. C. Jarvie, Freedom and Rationality: Essays in Honor of John Watkins, 207–223.
Published by Kluwer Academic Publishers. Printed in the Netherlands.

*

To understand a theory, to understand what it asserts, requires, among other things, understanding its logical implications, understanding its content and its context, what historical problem situations it speaks to, what problems it can solve, and how it interconnects logically with other theories.

There are two closely-related ways to characterise the content of a theory logically: *logical* content and *informative* content, and these, taken together, go far towards explaining the idea of unfathomable knowledge.

The *logical content* of a theory is its Tarskian consequence class, that is, all the nontautological consequences that can be derived from it.[5] It is well known that the logical content of any theory must be infinite.[6] The infinity of the logical content of any theory is, in and of itself, a trivial matter.

The *informative content* of a theory, which is also infinite, is *the set of statements that are incompatible with the theory*. This formulation sometimes surprises people who are unfamiliar with the ideas of *The Logic of Scientific Discovery*, but it becomes easier to understand when one recalls the slogan that "the more a theory forbids, the more it says". One discovers more about what a theory says – one becomes better informed about it – by finding out something about what it prohibits. (Statements that do not prohibit anything, for example, "This normal die will, when tossed on a flat surface, turn up a natural number less than 7", or "It will either rain this afternoon or it won't", have negligible informative content.)

If one wants to gain information about the world, and to test and refine it through criticism, one will then prefer theories of high informative content, and be on the watch for criticisms of such theories. In particular, one will try to identify potential falsifiers: i.e., those experimental results that are forbidden by the theories and which would, if realised, lead to their overthrow.

Although the logical and informative contents of theories are distinct, there are some interesting connections between them. For example, the elements of the informative content of a theory stand in a one-to-one correspondence to the elements of the logical content of a theory. To every element in either of the sets there is to be found, in the other sets, its negation. Hence the logical and informative content of a theory increase and decrease together.

If the infinity of the logical content is a trivial matter, the infinity of the informative content is far from so.

Popper takes the relationship between Newton's and Einstein's theories of gravitation to illustarte this point. Call Newton's theory *N* and Einstein's *E*.

Any statement or theory that is incompatible with *N* will belong to the informative content of *N*, and similarly, any statement or theory that is incompatible with *E* will belong to the informative content of *E*. Since Newton's and Einstein's theories are incompatible with one another, each belongs to the informative content of the other.

But this means that the infinity of informative content is anything but trivial, for Einstein's theory is not simply incompatible with Newton's, it is historically connected with it in the important sense of having superseded it. This illustrates that *any* new theory that supersedes a reigning theory after, say, a crucial experiment decides against the latter, has to belong to the informative content of the superseded theory. That is, any existing theory includes in its informative content any theory that will eventually supersede it.

With this result in hand, one may return to logical content and find a similar situation. Since *E* belongs to the informative content of *N*, *not-E* belongs to the logical content of *N*. That is, *not-E* is entailed by *N*. Amplifying the closing sentence of the preceding paragraph, the situation may be summarised as follows: any existing theory includes in its informative content any theory that will eventually supersede it, and in its logical content the denial of any theory that will eventually supersede it.

*

By now, I hope that my initial claims, i.e., that our knowledge is unfathomed and unfathomable, and that we do not know what we are asserting or what we are doing, are becoming a little clearer. As Popper sums it up: "There is an infinity of unforeseeable nontrivial statements belonging to the informative content of any theory, and an exactly corresponding infinity of statements belonging to its logical content. We can therefore never know or understand all the implications of any theory, or its full significance."[7]

Thus, to emphasize the point, there is a radical sense in which the content of existing theories is inaccessible in principle even to their creators and other experts. Neither Newton nor other physicists prior to Einstein knew a crucial part of the informative content of the Newtonian theory. To them, the experiments that eventually overthrew Newton's theory were not only not yet imagined; they were unimaginable. I do not mean that there was anything defective about Newton's own imagination: the point is that, lacking Einsteinian theory, certain experiments were unimaginable to anyone.

Einstein's theory postulated *new situations and possibilities* that could not have been framed in terms of Newtonian theory. Einstein's theory instructed physicists to look in certain places, places that, lacking Einstein's theory, they could not have dreamt of, which would, when investigated, give results in conflict with Newtonian theory.

*

I want to consider an objection to what I have claimed, or at least to the form in which I have put it, but first I need to sketch some ramifications of what has already been said.

As hinted earlier, my argument makes the meaning and the significance of a theory – in the (admittedly rather vague) sense of its relevance and of the common understanding of it at any particular time – into a partly historical matter, as well as partly a matter of logic. It also helps to clarify what the difference between the objective content of a theory and the vague notion of the common understanding of a theory might amount to. The significance of a theory "depends on what has been discovered, *at a certain time*, in the light of the prevailing problem-situation, about the theory's content; it [that is, its significance] is, as it were, a projection of this historical problem situation upon the logical content of the theory".[8] I like to call this projection the "accessed slice" of the objective content of a theory.[9]

This point has been important to Popper in his effort to convince language analysts (and positivists preoccupied with the "explication" of "concepts" in the "language of science") that a theory's meaning has comparatively little to do with the particular words and terms used in it. In fact the meaning and content of a theory are not identical; the meaning, relevance, and common understanding (and thus also the *economic value*; see below) of a theory shift as we uncover more of, or gain access to a larger slice of, its objective logical and informative content.[10]

*

To turn to a possible objection: it might be countered that it is misleading to describe the history of scientific thought in such a way. That is, if Newton did not intend his theory to conflict with a theory that would be put forward only some two hundred fifty years later, and which he himself could not possibly have imagined, it is misleading and even quixotic to describe the later theory as part of the informative content, and its denial as part the logical content, of Newton's theory.

This is a reasonable objection; and although I shall reject it, it must be taken seriously. The objection is based upon a misunderstanding. For the issue here is not the intention of the author of a theory (indeed, it might be said that that is the whole point), but rather the objective content of the theoretical system itself. In many quite familiar examples these fail to coincide, and indeed often have hardly anything to do with one another. Anyone familiar with the history of mathematics will realise that mathematics generates problems wholly independent of the intentions of its creators. The early discovery of the distinction between odd and even numbers, and such questions as whether there exists a highest prime number, or highest twin primes, are unintended consequences of the creation of the series of natural numbers. Nor, when Russell and Hilbert began their work, did anyone intend or anticipate the discoveries of Gödel and Church about incompleteness and decidability. Finally, the feeling of oddness in our characterisation stems from a kind of time warp: with a case such as Newton and Einstein, where the theories are separated by some 250 years, the result may seem peculiar. But the peculiarity vanishes when we consider successive and incompatible theories close in time, or the unpredictable, yet soon to be discovered, content of current theories.

*

I mentioned earlier that our argument is extremely useful, yet hardly ever used. How can it be used?

Our argument immediately topples virtually all serious forms of determinism, and was the lynchpin in Popper's own refutation of determinism, both in *The Poverty of Historicism* and in *The Open Universe*.[11] Thus, confining our example to historical determinism, the course of human history is strongly influenced by the growth of human knowledge. Yet it can be proved logically that we cannot predict, by rational or scientific methods, the future growth of our scientific knowledge (i.e., we cannot predict, from present knowledge, something incompatible with it). Thus, to the extent that the growth of human knowledge influences the course of human history, we cannot predict the course of human history.

I have also found our argument useful in refuting Marx's theory of alienation, in refuting much of the psychology and sociology of knowledge, and in refuting virtually all forms of expressionism.[12] I shall briefly mention only the former here.

Marx complains how, in economic exchange, the fruits of the worker's labour are wrested away from him. His product *ceases to be within his own control and "stands opposed to him as an autonomous power...which he no longer experiences as his own"*. In a similar vein, in *The German Ideology*, Marx bemoans the *"consolidation of our own product into an objective power over us that outgrows our control, crosses our expectations, and nullifies our calculations"*.

Yet our argument shows that there is no way to avoid this, at least when the products are knowledge products (as almost all products in some sense are). For *objective knowledge is autonomous in just the sense that Marx had in mind.* It does outgrow our control, cross our expectations, and nullify our calculations – for just the reasons cited. Morever, this is inherent in the situation and cannot be remedied by any sort of social revolution or reform: our autonomous knowledge products escape our control whether we like it or not, whether we surrender them to other human beings or manage to keep them to ourselves. It may indeed happen, as Marx supposed, that our knowledge products are used by another against us. But they may also be used for us. Another individual – a capitalist or entrepreneur – may take an idea or tool that we have made, discover in it something that we had not noticed, and develop it in a way that aids us. This benefit to us may also occur contrary to the intentions of the person who obtained the idea or object from us. There is no need to assume either benevolence or malevolence, or to deny either exploitation or altruism. They are all present, and all play a relatively minor role. *The important thing is that the recipients of products cannot control them either*. These products continue to be autonomous: no matter who temporarily owns or "controls" them, they do continue to "outgrow our control, cross our expectations, and nullify our calculations". One function of a market is to get rid of purchases that do not turn out (at least as far as one can tell) as expected.

Thus it is wrong to characterise the process of exchange mainly in terms of exploitation. We need one another, and a competitive exchange process, to help objectify and probe our ideas and other products – and thereby to discover their potentialities.[13]

*

An argument that so readily topples both determinism and Marx's view of alienation is obviously useful.

However, I should like to turn now to some quite new and also very useful examples – ones in which the "time warp" mentioned earlier is absent.

Pills, and other drugs, are of course manufactured in terms of theories; and I entitled this paper 'Unfathomed Knowledge in a Bottle' in recognition of this fact, and in special recognition of the devastating AIDS epidemic, and of the treatments that have been discovered for it. Knowledge in this area is advancing so rapidly that the information I present in the next few paragraphs[14] may be obsolete by the time this essay is printed – but that will only go to support the point.[15]

The most widely known treatment for AIDS approved by the American Food and Drug Administration is a drug called AZT or Retrovir,[16] which hinders the AIDS virus from replicating. It is not a new drug, but was synthesised early in the 1960s, when it was hoped, wrongly, that it might be useful in the treatment of certain sorts of cancer.[17] Although over half of AIDS patients appear to be able to take AZT for extended periods without serious side-effects, the drug is highly toxic and of limited usefulness. It is also perhaps the most expensive drug ever marketed.

As I write this, another widely used treatment for AIDS, in California and elsewhere in the United States, is dextran sulphate.[18] Dextran sulphate was developed in the 1950s as an anticoagulant, and for more than twenty years has been sold across the counter in Japan, without prescription, as a treatment for arteriosclerosis. Some seventeen companies in Japan manufacture it. When used in recommended dosages it has no serious side effects, and two decades of use in Japan testify to its safety. Its application to AIDS is a beautiful example of what F.A. Hayek calls "dispersed knowledge".[19] Although there are few cases of AIDS in Japan, two Japanese medical researchers discovered that dextran sulphate stops the AIDS virus from replicating in vitro and that, moreover, its effectiveness is enhanced when it is used together with the antiviral drug acyclovir and with small, non-toxic, doses of AZT. The three drugs work together synergistically.

Dextran sulphate is not yet approved by the FDA for medical use in the United States, although it is currently being tested. But since mid-1987 hundreds of doctors throughout California have been recommending it to their patients, not only to those with AIDS and Aids-Related Complex (ARC), but also, prohylactically, to many persons who test positive in AIDS antibody tests. The reason for the latter course is that statistics seem to indicate that, if untreated, most of those exposed to the virus would eventually develop AIDS, even though the incubation period may in many cases be a decade or more. Thus thousands of Americans are now taking daily doses of 2.5 to 3 grams of dextran sulphate, which is smuggled into the United States either directly from Japan or by way of Mexico and Canada.

The parallel to our earlier example of Newtonian and Einsteinian theory should be obvious. But here the time warp is much diminished. When AZT and dextran sulphate were synthesised over two decades ago, nothing in the existing knowledge about them, or about immunology, suggested their current uses. It was unpredictable in principle, in conflict with all immunological theory of the time, that they should be used in treatment of a disease caused by a retrovirus that was itself completely unknown and which possibly did not even exist at the time. Indeed, retroviruses themselves were not known until early in the present century, and their manner of working has been understood for less than two decades.[20] It was part of the objective informative content of immunological knowledge of the time, and part of the background knowledge used in the production of these drugs, that such diseases did not and could not exist.

Thus, while it was part of the objective (yet unfathomed) content, and objective (yet unfathomed) potentialities, of AZT and dextran sulphate (both of which *did* exist), that they could be used in treatment of this disease, AIDS, that did not yet exist, there was no way to gain access to this 'slice' of their content. What is crucial about an item of objective knowledge – a book, or a pill, for instance – is its *potential* for being understood, or being utilized in some way that has not yet been imagined, a potential that may exist without ever being realized.

Our time warp should not only be much diminished, but should now begin completely to disappear. For there is nothing 'quixotic' or 'misleading' about this example. Rather, drug developers, manufacturers and distributors are well aware of such possible unforeseeable effects and uses of drugs. In the development, testing, patenting, and licensing of their products, they both hope for such results (if beneficial) and fear them (if harmful). When buying rights to and licensing such knowledge, one is speculating; and one takes precautions accordingly, using all the resources of the law. Moreover, existing patent law already largely takes account of (although it does not use our language or epistemological and logical analysis) the objective content of such discovered but unfathomed products.[21]

*

This suggests a line of useful research: I have been claiming that Popper's neglected idea is useful. Here is a research programme – one soaked in a rich and largely untapped mixture of logic, epistemology, biotechnology, economics, and law – that someone might pursue. Exchange of property ordinarily depends on very clear demarcation of what is being exchanged: for example, specific bank deposits or numbered securities, or surveyable pieces of real estate identified by surveyor's reports and district land records, and backed by title companies. But what about the exchange of knowledge products, which are in principle unsurveyable? To what extent, and how adequately, does current patent law provide for, and provide protection against, the unsurveyability, the unfathomability, of the knowledge that is regularly exchanged? To what extent does this unsurveyability provide incentives – and deterrents – to its development, and its exchange? To answer this question provides an education in epistemology, and also shows how irrelevant what is ordinarily called epistemology, in the philosophical tradition, has to do with such important practical questions.

Another fruitful direction for research might be to ask whether there need be any conflict between the objective theory of knowledge used here and the subjective theory of value advocated by the Austrian school, and indeed by most economists. One often hears in conversation that there is indeed a conflict here: I have been asked, for instance, how one who recommends an essay by Popper entitled 'Epistemology without a Knowing Subject' can also advocate subjectivism in the Austrian sense. How indeed do the two approaches relate?

The objective theories of economic value, such as the classical theory of intrinsic value or the labour theory of value, as held, for example, by Adam Smith, David Ricardo and Marx, are clearly untenable. But the objective theory of knowledge is not a theory of economic value. Demand for an item of knowledge, to the extent that it is in the marketplace, will be based on dispersed subjective understandings or estimates or preferences for its accessed slice, and on subjective speculations about its potentialities. It will not, and cannot, be based on its objective – unfathomable and autonomous – logical and informative content. This is not available or accessible for anyone to value.

If a theory is supposed to have no unknown or hidden potentialities – if it has become 'common knowledge' – its economic values may diminish.[22] Yet if something unexpected is then discovered in the theory, the situation may radically reverse itself. As the logical and informative content of an item of knowledge is probed, as its accessed slice alters, demand for it, and thus its economic value, may then change.[23] Far from conflicting, the subjective theory of value and the objective theory of knowledge go hand in hand.

On the other hand, the neoclassical equilibrium economics that dominates many professional departments of economics is incompatible with the theory of objective knowledge. Such economics tends to focus on situations where all relevant knowledge is already known to all concerned parties. Or it may even assume that all parties know in advance what they are ignorant of, or have *deliberately chosen* whatever existing ignorance they enjoy following a cost-benefit analysis of the cost of remedying it.[24]

*

After having very briefly considered economic value, I would like to inject yet another different element into our discussion: morality. For in stressing that we do not know what we are asserting, we should not forget that we also do not know what we are doing.

John Watkins is, I believe, the only person to have attempted to base ethical theory on Popper's philosophy of science. His account, based on Popper's passing recommendation[25] that the utilitarian principle of maximising happiness might better be put in the negative form of "Minimise suffering!", he called "negative utilitarianism".[26] Negative utilitarianism suggests that moral postulates, principles or standards be treated hypothetically, in a way parallel to scientific theories; and that feelings of moral revulsion (as against cruelty and suffering, for instance) be treated as analogues to basic observation statements. As basic statements may falsify scientific theories, so might feelings of moral revulsion (as in the face of suffering) rebut moral injunctions.

Apply this perspective now to one of the conditions that contributed initially to the rapid spread of AIDS: sexual promiscuity. Sexual promiscuity is frowned on by most, if not all, forms of religion,[27] and also by much conventional non-religious morality. An exception, in religion, is to be found in those forms of Buddhism that, seeking freedom for their adherents, object not to promiscuity as such but to *attachment* to promiscuity, to enslavement to sex.[28] In the twentieth century these prohibitions have weakened: penicillin became readily available in the early 1940s, providing quick and easy cures for the then-known venereal diseases;[29] and the birth control pill and other means of contraception diminished the risk of unwanted pregnancy. In the three decades following, sexual promiscuity seems to have increased considerably.

Quite rational and prudent persons came to the conclusion, based on reasoning that conforms with negative utilitarianism, that there was considerable pleasure to be gained, much pain and stress to be avoided, and little if any harm to be suffered, from engaging in sexual promiscuity.[30] One need not invoke the examples of Bertrand Russell or extreme advocates of 'free love'. Take the case of Oscar Ichazo, a Chilean 'guru' in the tradition of Gurdjieff, and student of religions and techniques of self-mastery and self-liberation. In the 1970s Ichazo moved to New York, where he established a training programme known as 'Arica' which achieved considerable attention and respect, and even became fashionable for a time.[31] Like many of the Asian teachers from whom he had learned, Ichazo taught, as a goal, spiritual liberation from vice. Reflecting however on the strain of ordinary living, and on the way in which many people sought relief from the pain of stress through indulgence in various vices – gluttony, drinking, drugs, sex, and so on – Ichazo suggested thinking of these vices as 'doors', as temporary escapes from the pressures of life, and casually remarked that, among them, the only 'door' that did no damage physically was sex. His position should not be misunderstood: Ichazo did not advocate promiscuous sex or free love; he advocated spiritual freedom from craving; but he saw promiscuity as the least, and least harmful, among the ordinary vices.[32] Since AIDS, of course, all this is changed: indeed the anecdote just told itself sounds like what Gurdjieff would have called a tale of the miraculous. But the example reminds one that, whatever moral attitude one strikes towards promiscuous persons, prudence is also a traditional virtue; and one cannot blame most of those who were promiscuous prior to mid-1982 (when it was discovered that AIDS was caused by a virus) for flagrant *imprudence* about their health. What was then known about sex and disease suggested that promiscuity carried minimal physical risk. Yet, at least according to many medical researchers, the majority of those who now carry the AIDS virus had already acquired it, without knowing it, prior to 1982. They did not know what they were doing.

Take another example which, if it has less to do obviously with morals, yet concerns responsible action while also being less charged with controversy. Asbestos was introduced into construction because of its fire-retardant qualities and thus to minimise the suffering and destruction due to fire. Only later was it discovered that its fibres would get into the air and cause lung cancer. It is still often unclear what one should do in particular cases about the asbestos that has already been installed: for to remove existing asbestos also puts its fibres into the air.

Such examples make an end to John Watkins's suffering-minimizing negative utilitarianism (and, I should think, to any other kind of utilitarianism)[33]. Even if one were to agree *ex hypothesi* to the dubious claim that whatever was most conducive to the least suffering was good, this would itself be a doctrine without utility in view of the hard logic of unfathomable knowledge: one can calculate neither what will be conducive to maximum pleasure nor what will be conducive to least suffering.

Not that negative utilitarianism was ever a promising account of moral evaluation: Watkins's analogies between science and morality are rough and unsatisfactory, most especially his analogies between decisions about basic statements, on the one hand, and decisions about unsatisfactory consequences of moral standards, on the other.[34] Nor was his notion that one could reject hypothetical moral postulates on the basis of negative feelings ever plausible. As George Bernard Shaw wrote in the preface to *Major Barbara*, "It is exceedingly difficult to make people realise that an evil is an evil".[35] On those controversial questions that Watkins hoped to settle with his doctrine, it is hardly clear that our negative feelings really do, as he suggests, tend to lead in the direction of agreement, or that, in case they did, that that would even be relevant in constructing an adequate account of ethics. Many people are remarkably, if regrettably, callous about the suffering of others – particularly where the violation of religiously endorsed taboos is concerned. If sexual morality leads to suffering, then, it is often still argued (as, for instance, against AIDS victims), this is only proper punishment for sinners. Nor is the ineffectualness of suffering as an argument for moral reform a new phenomenon. "In the Victorian age", as Turner reports,

> a reformer stood a better chance of success if he could present his reform in such a way as to show that the victims of injustice were in moral danger; and even today this is by no means the weakest card to play. What shocked the middle classes, who read the reports on conditions in the mines, a little more than a century ago, was not so much the system under which children crawled on all fours, dragging sleds behind them, or in which men ruptured themselves lifting loads on to their daughters' backs; it was the revelation that lightly-clad young women working in proximity to naked men at the coal face made no strenuous efforts to save their honour when molested, which was fairly often.[36]

Incidentally, since Watkins called on Popper to buttress negative utilitarianism, it ought to be mentioned in passing that Popper himself is not a utilitarian, and that he thinks of himself as a modified or critical Kantian in ethical matters.[37] His recommendation to "minimise avoidable suffering", which by itself is hardly objectionable, was made in passing, in connection with social policy, not as part of an attempt to formulate a criterion for moral evaluation.

*

The notion of unfathomed and unfathomable knowledge is useful and neglected. What have I done with it in this brief sketch? I have distinguished it from fallibilism, clarified the difference between the common understanding of a theory, its economic value, and its objective content, and rebutted determinism and Marx's theories of labour and alienation. I have introduced some new and readily understandable examples of unfathomable knowledge in a bottle – examples that have obvious relevance to biotechnology and current social and moral issues. I have sketechd two promising lines of research: one having to do with the adequacy of patent law for dealing, in exchange, with the unfathomable character of knowledge; the other having to do with the relationship between the objective theory of knowledge and the subjective theory of economic value. And I have knocked out negative utilitarianism, and perhaps most other forms of utilitarianism, into the bargain. This is perhaps not a bad score. *But what have I really said and done?*

Hoover Institution on War, Revolution and Peace, Stanford University

NOTES

1 The author expresses his gratitude to Sephen Kresge and Leif Wenar for helpful suggestions and criticism.

2 Popper used the idea in his lectures for many years, and has discussed it in many of his writings. A very clear and brief summary appears in his *Unended Quest* (La Salle: Open Court, 1976), Section 7. Watkins used the idea briefly in his *Hobbes's System of Ideas: A Study in the Political Significance of Philosophical Theories* (London: Hutchinson, 1965), pp. 22–23, although confining himself chiefly to that aspect of the idea that deals with its logical content; I have used it in a number of places including my 'Ein schwieriger Mensch', in Eckhard Nordhofen (Ed.), *Physiognomien: Philosophen des 20. Jahrhunderts in Portraits* (Königstein/Ts.: Athenäum, 1980), pp. 43–69; 'Knowledge is a Product Not Fully Known to Its Producer', in Kurt Leube

and Albert Zlabinger (Eds.), *The Political Economy of Freedom* (Munich: Philosophia Verlag, 1985), my 'Alienation Alienated: The Economics of Knowledge *versus* the Psychology and Sociology of Knowledge', in Gerard Radnitzky and W.W. Bartley, III (Eds.), *Evolutionary Epistemology, Rationality and the Sociology of Knowledge* (La Salle: Open Court, 1987), pp. 423–451; and in the 'Afterword, 1985' to my *Wittgenstein*, Second Edition, revised and enlarged (La Salle: Open Court, 1986; London: Hutchinson, 1987) as well as in my forthcoming book, *Unfathomed Knowledge, Infinite Ignorance*. See also my *The Retreat to Commitment*, Second Edition, revised and enlarged (La Salle: Open Court, 1984), Appendix 2, 'Logical Strength and Demarcation', pp. 185–209.

[3] Lord Quinton, who wrote the article on fallibilism for the *Fontana/Harper Dictionary of Modern Thought*, expresses fallibilism in a justificationist way that no Popperian could accept. So I do not want to insist on the word 'ordinary', and am willing to allow Lord Quinton the credit for being ordinary.

[4] I shall on the whole follow Popper's exposition in *Unended Quest* in the next few paragraphs, but I shall also add some ideas of my own, so that Popper should not be held responsible for my errors.

[5] Alfred Tarski, 'On the Concept of Logical Consequence', Chapter 16 of his *Logic, Semantics, Metamathematics* (Oxford: Oxford University Press, 1956), pp. 409–420.

[6] See Watkins, *Hobbes, op. cit.*, p. 23, n. 1, and Popper, *Unended Quest, op. cit.*, pp. 26–27 and n. 18, p. 199.

[7] *Unended Quest*, p. 28.

[8] Popper, *Unended Quest*.

[9] For discussions of objective content, see Popper, *Objective Knowledge* (London: Oxford University Press, 1972), as well as *Conjectures and Refutations* (London: Routledge, 1963).

[10] This would also have to be so since a new theory both explains *and corrects* its predecessor, as well as correcting the corroborating evidence for it. See 'The Aim of Science' in *Objective Knowledge* (London: Oxford University Press, 1972). See also *Realism and the Aim of Science*, Vol. I of *Postscript to the Logic of Scientific Discovery*, ed. W.W. Bartley, III (London: Hutchinson, 1983).

[11] K.R. Popper, *The Poverty of Historicism* (London: Routledge, 1957), especially the preface; K.R. Popper, *The Open Universe: An Argument for Indeterminism*, vol. II of *Postscript to the Logic of Scientific Discovery*, W.W. Bartley, III (Ed.) (London: Hutchinson, 1982), especially sections 20–24.

[12] See my 'Knowledge is a Product Not Fully Known to Its Producer', *op. cit.* and my 'Alienation Alienated: The Economics of Knowledge *versus* the Psychology and Sociology of Knowledge', *op. cit.* See also the Afterword to my *Wittgenstein*, Second Edition, Revised and Enlarged (La Salle: Open Court, 1985).

[13] See my discussion in 'Alienation Alienated', *op. cit.*

[14] I am writing in late May 1988.

[15] Even whether AIDS is caused by the AIDS virus (i.e. HIV) has been a matter of some controversy. See Peter Duesberg, 'Retroviruses as Carcinogens and Pathogens: Expectations and Reality', *Cancer Research*, March 1, 1987, as well as the report in *Discovery*, June 1988, pp. 62–68 sketching a preview of the report to be presented in Stockholm in June 1988 to the effect that although (as argued by Duesberg) T-cell infection, on its own, probably cannot cause AIDS, HIV does nonetheless wage a

crippling onslaught on a different class of immune-system cells: macrophages. Another mechanism that destroys uninfected T-cells in the formation of syncytia.

[16] Azidothymidine.

[17] When it was discovered not to be effective as hoped in treating cancer, its economic value fell. See the article on the drug, and on the work of Jerome P. Horwitz in developing synthetic AZT in the 1960s, in *The New York Times*, September 20, 1986, pp. 1 and, on p. 7, 'A Failure Led to Drug Against AIDS'. See also Erik Eckholm, 'Test Results Are Due on Drug to Fight Aids', *New York Times*, September 18, 1986, p. A22, and 'AIDS Drug is Raising Host of Thorny Issues', *New York Times*, September 18, 1986, p. 38. See also Marilyn Chase, 'Promise Is Seen in Tests of Drug for Treating AIDS', *Wall Street Journal*, March 14, 1986, p. 18, and 'Burroughs-Wellcome Test of AIDS Drug May Be Halted, Suggesting Initial Success', *Wall Street Journal*, September 17, 1986.

[18] Kozo Yamada, Humio Kuzuya *et al.*, 'Studies on Some Actions of Sulphated Polysaccharides on Arteriosclerosis (IV), Oral Administration of Dextran Sulphate', *Japanese Circulation Journal*, **25**, June 1961; US Patent No. 3,126,320, March 14, 1964; Yuki Tashiro and Shinichi Fukumoto, 'Therapeutic Results of Arteriosclerosis by Dextran Sulfate (MDS)', in *Japanese Circulation Journal*, **29**, March 1965; I. Hirono *et al.*, 'Carcinogenicity of Dextran Sulphate Sodium in Relation to Its Molecular Weight', *Cancer Letter* **18** (29), 1983; Ryuju Ueno and Sachiko Kuno, 'Dextran Sulphate, a Potent Anti-HIV Agent in Vitro Having Synergism with Zidovudine', *The Lancet*, June 13, 1987, p. 1379, reporting also a personal communication from H. Mitsuya and S. Broder, of the US National Cancer Institute; M. Berenbaum, 'Anti-HIV Synergy Between Dextran Sulfate and Zidovudine', *The Lancet*, August 22, 1987, p. 461; H. Mitsuya and Samuel Broder, 'Strategies for Antiviral Therapy in AIDS', *Nature* (1987) **325**; 773–778; Ryuji Ueno and Sachiko Kuno 'Anti-HIV Synergism Between Dextran Sulphate and Zidovudine', *The Lancet*, October 3, 1987; M. Ito, *et al.*, 'Inhibitory Effect of Dextran Sulphate and Heparin on the Replication of Human Immunodeficiency Virus (HIV) *in Vitro*', *Antiviral Research*, **7**(6), 361 (1987); H. Nakashima *et al.*, 'Purification and Characterization of an Avian Myeloblastosis and HIV Reverse Transcriptase Inhibitor, Sulphated Polysaccharides Extracted from Sea Algae', *Antimicrobial Agents and Chemotherapy* **31** (10); 1524, (1987). See also Marilyn Chase, 'Defusing a Bomb; Doctors and Patients Hope AZT Will Help to Stave Off AIDS', *Wall Street Journal*, April 28, 1988, pp. 1 and 19, in which dextran sulphate is discussed briefly.

[19] See F.A. von Hayek, *The Fatal Conceit*, W.W. Bartley, III (Ed.), being vol. I of *The Collected Works of F.A. Hayek* (London: Routledge, 1988). See also 'Economics and Knowledge' and 'The Use of Knowledge in Society' in *Individualism and Economic Order* (London: Routledge, 1948).

[20] But they were of course not then described and understood as they are today in terms of DNA, RNA, and reverse transcriptase, these aspects of genetic theory being at the time not yet discovered! Reverse transcriptase was discovered only in the late 60s, and David Baltimore and Howard Temin were awarded the Nobel Prize for it as recently as 1975. See Dani Bolognesi, *Human Retroviruses, Cancers, and AIDS: Approaches to Prevention and Therapy* (New York: Alan R. Liss, Inc., 1988).

[21] I am much indebted to Dr. Günter Wächterhäuser for instructing me in patent law. In order to obtain a patent, one must show that the knowledge that one seeks to patent

was not predictable from prior art; and one must state explicitly in one's patent everything that one believes to be derivable from it.

[22] One sign that this may be happening is that the key terms of the theory become matters of definition or convention, as happened to Newtonian theory in the late nineteenth century immediately prior to the discovery of Einstein's theory.

[23] Valuing can, however, affect the history and development of the accessed slice of a theory. It may influence one part or aspect of a theory to be investigated prior to another.

[24] See Esteban F. Thomsen, 'Knowledge, Discovery, and Prices', in *Humane Studies Review*, **5**, (1), 1–17 (Fall 1987).

[25] *The Open Society and Its Enemies* (London: Routledge & Kegan Paul, Ltd., 1945), Vol. I, Chap. 5, n. 6 (2); Vol. I, Chap. 9, n. 2.

[26] *Proceedings of the Aristotelian Society, Supplementary Volume*, 1963. See also Ninian Smart, *Mind*, October 1958; J.J.C. Smart, *An Outline of a System of Utilitarian Ethics* (Carlton: Melbourne University Press, 1961); H.B. Acton, 'Negative Utilitarianism', *Proceedings of the Aristotelian Society, Supplementary Volume*, 1963; and Richard Robinson, *An Atheist's Values* (Oxford: Oxford University Press, 1964), p. 18. See also Eduard von Hartmann, *Das sittliche Bewußtsein* (Berlin: Duncker, 1879), and Aurel Kolnai, 'The Thematic Primacy of Moral Evil', *Philosophical Quarterly*, January 1956.

[27] But see John Boswell, *Christianity, Social Tolerance, and Homosexuality: Gay People in Western Europe from the Beginning of the Christian Era to the Fourteenth Century* (Chicago: Chicago University Press, 1980).

[28] See my *The Retreat to Commitment, op. cit.*, Appendix I.

[29] A qualification is needed. It was suspected in the 1960s and became known in the 1970s that hepatitis was a venereal disease. Although most hepatitis passes fairly quickly without serious damage, a tiny percentage of hepatitis cases are fulminant, and usually fatal. A somewhat larger number of hepatitis cases develop into persistent or chronic hepatitis, which in turn sometimes leads to serious disease and death.

[30] Watkins's paper was, incidentally, published in 1963, when homosexuality was still illegal in Britain, and was widely being referred to as a 'victimless crime'. See *The Wolfenden Report: Report of the Committee on Homosexual Offenses and Prostitution*, Authorized American Edition, with Introduction by Karl Menninger, M.D. (New York: Lancer Books, 1964), and Edwin M. Schur, *Crimes Without Victims: Deviant Behavior and Public Policy: Abortion, Homosexuality, Drug Addiction* (Englewood Cliffs, N.J.: Prentice-Hall, Inc., 1965).

[31] See Winifred Rosen, 'Down the Up Staircase', *Harper's*, June 1973, pp. 28–36. Notes on Ichazo and even a cartoon about Arica also appeared in *The New Yorker*. See the discussion of, and dialogue with, Ichazo by John C. Lilly, *The Center of the Cyclone: An Autobiography of Inner Space* (New York: Bantam Books, 1973).

[32] See also Robert S. De Ropp, *The Master Game: Beyond the Drug Experience* (New York: Delta, 1968), pp. 154–157.

[33] Even where we are most certain about what we are saying and what we are doing, we cannot – because of the unfathomable content and unpredictable consequences that are silently interwoven into virtually every statement we utter, into all our reflections, and into every act we commit – know what we are saying or what we are doing. There is every point in taking into account, when acting, the best-examined information

available. But there is no point in trying to base morality on calculations, positive or negative, that are impossible in principle.

[34] Nor, as Watkins thinks, does one need decisions for basic statements. Moreover, it is doubtful whether criticism even in natural science is centered chiefly on observational test statements. See *The Retreat to Commitment, op. cit.*, Appendix 3.

[35] (New York: Dodd, Mead & Co., 1941).

[36] E.S. Turner, *Roads to Ruin: The Shocking History of Social Reform* (London: Penguin Books, 1966).

[37] *Unended Quest, op. cit.*, 82, 193; *The Open Society and Its Enemies* (London: Routledge & Kegan Paul, Ltd., 1945), pp. 107, 108, 139, 235, 238, 239, 256, 304, 324, 384–85; *Conjectures and Refutations* (London: Routledge & Kegan Paul, Ltd., 1963), pp. 181, 345, 361; *The Self and Its Brain* (New York: Springer, 1979), p. 168.

PETER MUNZ

WATKINS'S EVOLUTIONISM BETWEEN HUME AND KANT

Towards the end of his paper on the unity of Popper's thought (Watkins, 1974: 403), John Watkins addresses the old problem of the irreconcilability of commonsense realism and scientific realism. "Realists", he says, "are faced with a painful choice: either cling to commonsense realism (grass really is green, etc.) and interpret science in some nonrealist way (for instance, instrumentally) or interpret science realistically and adopt a nonrealist view of experience (grass is not really green, we mentally 'paint' it green'). ... *If we take science seriously*, we must conclude that, although my perception of the orange is indeed caused by external realities, these are about as different as they could be from my idea of an orange – colourless, tasteless, very small. The *orange* exists only in my mind."

On the following page Watkins proposes that this dilemma can be solved by evolutionism and states explicitly that this solution will not be 'hit' by the facts by which most classical theories of knowledge, Hume on one side and Kant on the other, are hit. "Evolutionism", he writes, "suggests that the information about the physical realities of our near-environment that our sense organs provide is generally rather reliable, otherwise we would not be here... We can be confident that they do not systematically misinform us about those features of the world realistic appraisals of which are sometimes a matter of life and death." Watkins here attaches himself to a now well established tradition of biological thinking. Thus, for example, Simpson wrote as far back as 1963 that it is no accident that we have a good perception of space, for we owe it to our ancestors who used to live in trees: "The monkey who had no realistic perception of the branch towards which he jumped, would soon have been a dead monkey and could thus not have become one of our ancestors" (Simpson, 1963: 93).

On the face of it, this argument could be taken to be a positivist's dream. Positivists had argued ever since Locke and certainly long before the term 'positivism' itself was introduced, that we acquire knowledge by sense observation and that only such knowledge as can be derived from or reduced to sense observations is valid knowledge. Without a theory of evolution which can place the reliability of sense experience on a biological basis, this doctrine of positivism had always seemed rather shaky. Why, one could

Fred D'Agostino & I. C. Jarvie, Freedom and Rationality: Essays in Honor of John Watkins, 225–246.

easily have asked, should the senses, of all things, be considered such a specially reliable source of knowledge, especially when such authoritative sources of knowledge as revelation or self-evidence, were available? Positivism had, however, recommended itself because it had paraded as 'progress', as the last, rational stage in the history of the human mind. This historicist consideration could in itself not be a philosophical justification of positivism, but it went a long way toward explaining why positivism, right down to the Vienna Circle and the early Wittgenstein, continued to have such a hold in spite of its dubious philosophical credentials (Munz, 1978a). One could easily argue that the evolutionary perspective provided the answer to the question why, of all things, we should rely on the senses. We should rely on the senses, because they have evolved through natural selection and if the senses sytematically misled us, we could not have evolved to be here, thinking about the matter.

Such an argument would indeed be a positivist's dream: we learn, positivism said, by accumulating sense-observations. Now that we know why sense-observations are the reliable source of knowledge, this doctrine about learning is vindicated. For it tells us that our senses are genetically selected to be instructed by our environment and therefore pick up correct information. It is true that even on this view, positivism is left with the unenviable task of having to explain how such veridical sense observations, which must, by the nature of the case be *particular* observations, can be transformed into verbal propositions and, what is even more, into verbal propositions of a *general* nature. There has never been a lack of philosophers who have believed that there is some kind of philosophical magic by which these philosophical rabbits can be produced out of a hat. As Watkins's argument stands, it could, therefore, be taken to provide comfort to positivists. Such comfort is clearly not part of Watkins's intention because the whole of the preceding argument is an argument in favour of Popper's falsificationism and, by implication, an argument against induction and against learning by picking up instructions from the environment. Moreover, Watkins says that classical theories of knowledge find themselves between the devil of Hume and the deep blue sea of Kant. This must be taken to mean that the Kantian way is as mistaken as the Humean way. As a closer examination of the contribution evolution has made will show, this is not the case. With evolution we are very close to Kant and very far from Hume.

In order to eliminate the ambiguity in the evolutionary argument presented by Watkins, we have to go right back and examine carefully the core of evolutionary epistemology. It will then appear that evolutionary epistemology is anything but a comfort to positivism.

The leading principle of evolutionary epistemology goes back, of course, to Darwin himself. "Plato", he wrote in 1837, "says in Phaedo that our *'imaginary ideas'* arise from the preexistence of the soul, are not derivable from experience. – Read monkeys for preexistence." On another occasion he remarked that "he who understands a baboon would do more toward metaphysics than Locke." The two remarks between them show that Darwin was completely even-handed because he appealed to baboons against both Locke and Plato, that is against the two diametrically opposed strands of philosophical thinking. This fertile contribution to philosophy fell on deaf ears and almost a hundred years later when all of Darwin's thinking had revolutionised just about every single science in the world, Ludwig Wittgenstein declared portentously in 4.1122 of his *Tractatus* that the Darwinian theory has nothing to do with philosophy.

Wittgenstein's remark was not just a casual aside, although he himself might have regarded it as one. Wittgenstein was a deep died in the wool Linnaean. By this I do not mean that he was a 'creationist' or did not accept the idea that man had evolved. I mean something more fundamental. Like a 20th century Linnaeus, Wittgenstein believed that all cultural and linguistic systems are absolute and emerge spontaneously and are not descended from one another, so that one cannot think of one as related to any other in an evolutionary way. Each one is an absolute and irreducible phenomenon. With this view, Wittgenstein, *Philosophical Investigations*, Part II, xi, p. 224e, explicitly adopted a position incompatible with evolution (Munz, 1987b, p. 72); for in evolution nothing emerges spontaneously and everything has a predecessor. Had such a belief simply echoed Linnaeus, Wittgenstein's philosophy would most probably have remained a curiosity. But it so happened that Wittgenstein with his theory of the spontaneous and non-evolutionary emergence of cultural language systems hitched his thinking to a lively Linnaean tradition in philosophy. At the very beginning of our century there had taken place the confluence of two ideas. One was derived from the tradition of German cultural history and found its best formulation in Ranke's epigram that all societies and their cultures are 'equidistant' from God (Munz, 1980 p. 141). The other found its expression in the epigram that "there are no fossils", coined by the Polish anthropologist Malinowski who was a major influence in British social science for several decades. It takes very little reflection to see that these programmatic epigrams between them really harden people against all thoughts of evolution. If every thought is equidistant from God, it cannot have evolved from earlier thoughts; and if there are no fossils, any existing culture must be a permanent, stable institution and cannot have evolved from an earlier culture. Given these powerful anti-evolutionary thrusts in both history and anthropology, we can see that Wittgenstein merely dotted the 'i' when he said that every language-game is a form of life which is a spontaneous occurrence and that none of this has anything to do with Darwin. Compared to this influential trend in philosophy and the social sciences, Bishop Wilberforce's misgivings about evolution were child's play.

I have dwelt on this matter at some length in order to indicate how deeply rooted the philosophical neglect of Darwin has been. It was not just that philosophers were ignorant of Darwin and therefore failed to appreciate the relevance of Darwin to philosophy. It was above all that philosophers were given to a way of thinking which effectively and efficiently precluded reaction to theories of evolution in both philosophy and social science. On top of this there was a general reaction to theories of evolution in human history because of their potentially unethical implications. Biologists are readily prepared to consider an amoeba a more 'primitive' form of life than a chimpanzee. But it is not so easy to hold that a tribe in New Guinea the members of which are now supposedly citizens of the Republic of Papua and New Guinea which is a member of the United Nations as well as of the Commonwealth, are a more 'primitive' form of life than the inhabitants of Manhattan. Added to this were the unfortunately simplistic versions of so-called social Darwinism. When I was a student, I recall that it was with some considerable relief that most of my teachers were able to refer me to Julian Huxley's *Synthesis* in which it is stated categorically that the study of culture has nothing to do with natural selection.

It is hardly surprising then, that philosophers have been so slow in taking up Darwin's suggestion that monkeys and baboons can solve Plato's as well as Locke's problems. In order to arrive at a correct estimate of the import of evolution in philosophy, let me try to flesh out the casual remarks Darwin had jotted down in his Notebooks.

When philosophers in ancient Greece started to ponder the fact that we have knowledge and questioned the idea that the knowledge we have is simply information handed down by traditional authorities, it soon struck them – and we must here start with Plato – that any knowledge we have of the world we are living in is not simply picked up by observation; but involves in one way or another the functioning of our own 'minds' or, as we would say today, of our nervous system. There are presuppositions, assumptions, cognitive structures which precede any particular information we get and which make the reception of such information possible. If this was readily granted, there came in the wake of this insight a formidable problem. If some of our knowledge is *a priori*; or if the forms in which all our knowledge is cast are *a priori*, how can we be sure that our knowledge is true knowledge about the world? If *a priori* categories or forms play a crucial part, should we not rather suppose that knowledge is not different from myth? At best, should we not conclude that all knowledge has to be interpreted symbolically or allegorically or metaphorically or in some other as yet unknown manner?

These and allied questions were debated endlessly without conclusive success. People espoused now one now another of the leading solutions but the fact that the debates continued and the schools of thought kept facing one another as if they were forms of revealed religion, ought to have indicated very clearly that somehow the problem is either unresolvable or that as yet something vital had been missed.

For a short period, philosophers who tried to argue that the correct account of knowledge was an account without *a priori* categories, made a fleeting impression, and some sort of mutual support system may have existed between Locke and Newton. "Mr Newton in his never enough to be admired book", Locke wrote in Book 4, vii, 3 of *An Essay Concerning Human Understanding* and stated that Newton, for one, never made use of *a priori* 'maxims'. Newton himself seems to have imagined that he had followed Locke's theory as to how one gains knowledge. First, he said, one reveals by analysis some simple results and then one generalises them by induction. At least this is how I.B. Cohen summarizes the scenario which Newton claimed to have followed (Cohen, 1980: 13).

This brief euphoria was destroyed by David Hume. He granted that Newton had been a good practitioner of Locke and had proceeded by "admitting no principles but such as were founded on experiment...but resolute to adopt every such principle." This praise of Newton in Ch. 71 of *The History of England* is well known, but Hume's scepticism about Newton with which he continues, is not so frequently quoted: "While Newton seemed to draw off the veil from some of the mysteries of nature, he showed at the same time the imperfections of the mechanical philosophy: and thereby restored her ultimate secrets to the obscurity in which they ever did and ever will remain." Hume, in other words held that owing to Newton's well-nigh Lockian procedure, no real advance in knowledge had been made.

Watkins agrees with Hume that no theoretical science can emerge from Hume's epistemological principles. He writes that "Humean empiricism with its key dogma 'that all our ideas are deriv'd from correspondent impressions' can hardly be squared with those highly theoretical, experience-transcending ideas that are so characteristic of modern science and cosmology." (Watkins, 1974: 406) But where Hume had resigned himself to the fact that "mechanical philosophy" must remain imperfect, Watkins, on the contrary, takes the fact that theoretical science exists as proof that Hume must have been mistaken.

There can be no reasonable doubt that Watkins is right and that Humean empiricism must be wrong. But it is peculiar that Watkins has to invoke the existence of theoretical science which, by definition transcends sense impressions, in order to rebut Hume's scepticism. Long before one thinks of theoretical science, there are very compelling reasons for believing Hume to have been mistaken. It was part of Hume's tacit assumption that sense-impressions are magically transformed into words and he might, had he been pressed, have believed that all sense-impressions are 'worded' or tagged with words – so that when we see an orange we can have a sense-impression of the orange and then transform that sense-impression into the word 'orange'. The assumption was tacit because in Hume's days nothing was known about neurology and neuroscience did not exist. Hence, had he been questioned, Hume would not have thought that any kind of magic is needed to produce the wording of sense-impressions. We must certainly forgive Hume. Not only is neuroscience very much a 20th century achievement but even avowed practitioners of neuroscience as well as neuroscience oriented philosophers who now call themselves 'neurophilosophers' have been able to appreciate the size of the problem. And for that matter the initial Humean belief that sense-impressions magically transform themselves without further ado into words was still shared in the early decades of our own century by the Vienna Circle who believed in 'protocol-statements' – that is, in the existence of observations that were, provided the observations were simple enough, automatically transformed into verbal expressions.

To be sure, Newton knew that there is a problem and often called red light 'rubrifick' (i.e. 'red-making') because it consist of nothing but rays which are disposed to propagate a movement in the sensorium in which they are sensations of those movements under the form of colour. Even though both Locke and Newton knew that one must make a distinction between wavelengths that impact upon the body and colours we are conscious of, they were not aware that there ought to be a second distinction – the distinction between the subjective awareness of colour and the attribution of a worded label to that awareness. We are most probably conditioned to identify a certain awareness by a certain word; but the rules governing the employment of the word are very different from the laws which govern the emergence of the subjective awareness. With the help of a word, we drag the subjective awareness into the public arena. But the coming together of a word and an awareness is like an alliance between two independent rule-systems and cannot be taken for granted. Our knowledge of when to deploy the word 'red' comes from the language-game we participate in; whereas the awareness we label by that word comes from the lengths of the light-waves we sense.

The belief that the second distinction is unnecessary and that, given the sensation, the word must emerge by magic is indeed deep-seated and will not die. It must be one of the tacit assumption behind the so called 'Received View' of the philosophy of science according to which scientific theories are axiomatic calculi which are given a partial observational interpretation by means of correspondence rules (Suppe, 1977: 3). There is nothing in any of these correspondence rules to govern the transformation of physical or chemical stimulation of the nervous sysem into articulated verbal expressions. That transformation is not mentioned and seems believed to be magical. If the Received View is innocent of neuroscience, the same cannot be said for Patricia Churchland. In her book entitled *Neurophilosophy* (Churchland, 1986: 455) she states that conscious and, presumably verbal expression of colour, taste and odor have as their "physical substrate a neuronal phase space whose axes are [in the case of colour] three streams of neurons carrying information about reflectance efficiency values at three wavelengths." Based on Land's results (Land, 1977), she is on firm neurological and optical ground. What is amazing and daring is the jump into what is after all a worded awareness and a verbal expression. She makes this jump with tacit ease as if it went without saying that once a neuronal pathway is discovered, the verbalisation follows automatically.

Or let us look at the way a well-known neurologist treats the matter. Jean-Pierre Changeux says, for example that "mental objects with a realistic component, like pictures...mobilise neurons in the right hemisphere, while those with a more verbal content – concepts – recruit neurons in the left hemisphere" (Changeux, 1985: 163). Changeux maintains that this is evidence that the barriers that separate the neural from the mental have been destroyed (Changeux, 1985: 168). He provides other examples to show what he means by 'destruction'. There are, he writes (Changeeux, 1985: 109) areas where cell bodies and processes contain a specific neurotransmitter – "*dopamine* ... In the hypothalamus and the brainstem, dopamine synapses can thus be regarded as "pleasure" or "hedonic" synapses 'where the cold information regarding the physical dimensions of a stimulus is translated into the warm experience of pleasure'"(Wise, 1980).

Changeux' view is seductive; but his rhetorical devices disguise the real difficulty. He says that neurons are "recruited" and that "cold information" is translated into "warm experience". There is no justification for these rhetorical leaps. Dopamine, undoubtedly, generates a certain feeling inside the body. But that feeling is opaque, inchoate and blind, because unlabelled. There is no reason why we should simply assume that the term 'pleasure' is the *correct* label. Indeed, as is well known, many of those feelings would respond equally to the term 'pain' – hence, for example, the psychological phenomenon known as 'sado-masochism'. Whatever mental awareness is neuronally generated, is an unworded or unlabelled awareness. The identification of such awareness by the ascription of a word such as 'pain' or 'pleasure' cannot go without saying, for neurons do not carry labels and their circuits do not flash verbal messages. Even though neuroscience has made progress in bridging the gap between neuronal and mental events, the belief that it can also bridge the gap between physical events such as neuronal circuits and mental representations, is therefore illusory. There is no easy bridge from neuronal substrates to verbally labelled mental images and concepts.

The euphoria of some contemporary neuroscience was due to important discoveries. It was one thing to discover that there are certain areas of the brain concerned with vision and that injury to a set of cortical regions, located in the dorsal (parietal) region of the cortex, was critical for surveillance, attention and arousal: its injury results in indifference and in the *"loss of a sense of caring about one's own person"* (Gardner, 1985: 267). But it is quite another to conclude from such evidence, that there is a one-to-one correspondence between neuronal events and the verbal designation of the inchoate feelings they produce. We really are confronted with three sets of events. There is, first, the neuronal circuits; second, the inchoate or blind feeling (emotion?) they generate; and third, the verbal identification of that opaque feeling as 'orange' or 'pain'. The first two events are, as the evidence now indicates, very intimately linked. But the third type of event is a hypothetical superimposition and cannot stand in a causal relationship to the two other types.

Consider, for example, the very striking contribution made by D. Marr to the theory of vision. Marr's investigations have shown that there is direct energy transfer from the object seen to the nervous system. He discovered the emergence in the nervous system of the 'primal sketch', which is a primitive but rich description of the intensity changes that are present in an image. "All subsequent analysis reads the primal sketch, not the data from which it is computed. The primal sketch therefore acts in a genuine sense as the interface at which visual analysis becomes a purely symbolic affair... Downward-flowing information will not affect the line-finding state...at all. Its most usual *modus operandi* is in choosing which processes are to be used to read the primal sketch..." (Marr, 1976: 516–7). There is "acquired knowledge which is stored and catalogued in our brains...The problem of searching for catalogued information...occurs only at this final stage of visual processing. Nobody had imagined that so much information about shapes could be extracted from the retinal images before a search of catalogued information would be necessary" (Rosenfield, 1984: 56). Marr's work documents in great detail the relationship between input and final awareness and demonstrates beyond all shadow of reasonable doubt that the input is no more than a primal sketch of the object seen and that what we call vision is the brain's computation and analysis of the data provided not by the object seen but by the primal sketch it causes. But not even this detailed account of the process of vision can bridge the last gap between neuronal computations performed on the primal sketch and the moment at which we say, e.g. 'chair', that is, the moment where we provide a word. Marr insists that the primal sketch corresponds very closely with the image we are conscious of, i.e. with the moment at which we say 'chair'. But close correspondence is not the same as identity and really can mean no more than that the primal sketch does not falsify the moment at which we say 'chair'. Somewhere during the process of downward-flow of information to the primal sketch there occurs a verbal superimposition which is not part of the neuronal computation. Or, if it is part of it, it uses information stored in the brain from a source other than the primal sketch.

It seems indeed that the most important contribution neuroscience has made so far to our old philosophical problem is that it has explained how energy is transferred from objects seen or smelt or touched to the nervous system and how this energy is processed inside that system, so that we can be confident that perception of something real does take place and that the events in our nervous system are related to something *real.* But at the same time neurological inquiry into nervous circuitry has also demonstrated that at no stage in this circuitry the events are or become demonstrably verbal. In all cases, they remain neuronal. We can get from outside objects to nervous events and from nervous events to feelings and awareness. But these feelings and awareness are and remain unworded. We are thus left, from neuroscience, with the assurance that the outside world makes a very direct and traceable impact upon our nervous system; that that impact is noticeable in terms of opaque and inchoate feelings; but that these feelings are indeed opaque and inchoate, because non-verbal. There is every reason to suppose that these neuronal circuits mobilise or activate neuronal circuits in which the appropriate words are encoded until they are uttered. But the point is that the source of the verbalising circuits and the source of the circuits which compute Marr's primal sketch, do not coincide. The first circuits come from a language-game we participate in; and the second, from the image or object in front of our eyes.

Once this is accepted, it follows that not even contemporary neuroscience can solve Hume's problem. Long before we reach the heights of theoretical science, we are confronted by the fact that there is no straight line which leads from impression to verbal expression. Even quite ordinary experience is fatally divorced from the verbal description of that experience, even though nowadays, with the assistance of neurology, we can understand how and why objects outside our body worm their way into our nervous system and generate there feelings and opaque emotions. These feelings, however, are as unlabelled as the neuronal circuits which produce them. All we can say – and this probably is an advance over Hume as well as over the Vienna Circle – is that the gap which we cannot bridge does not lie between the outer world and our body-mind, but between the circuitry of our nervous system and our language. Neuroscience has shifted the problem from one place to another, but has not solved it. In that sense, we are still with the difficulty which Hume had not been able to cope with.

A real step forward was taken by Kant. Plato had argued – and I think this is one of the main reasons why his arguments failed to command consensus – that the a priori element in our knowledge consisted of preexistent knowledge of every single *type* of object we find in this world. This is not a very likely story. Among other problems, it places all universals in our heads and raises the great question which nobody had been able to answer satisfactorily, as to how the particulars which Plato had located in the world, could be related to the universals which he had placed in our heads. Kant introduced real precision into this matter when he suggested that the a priori element in all knowledge consists of general categories and of forms of perception, not of particular types of objects. He argued that all human beings have certain innate forms of understanding and of perception and that these forms are common to all human beings. For this reason all human beings keep on making those observations which prove Newton's Mechanics correct. This way Kant had dispelled Hume's doubts about Newton, but he had to admit that if the world were 'observed' by a non-human being who did not possess those same categories from birth, the world might indeed turn out to be a very different place from what we think. With such an admission, Kant, like Hume, turned out, after all, to be a sceptic too. Not a sceptic about the world of nature as we understand the term 'nature'; but a sceptic about the world as it is really and ultimately in itself. Nature, Kant said, is only the world as it appears to human beings. The laws of nature we know, are the laws which the human mind with its specific cognitive structure prescribes to nature – not the laws, if any, which are *really* here.

Leaving aside the idealist consequences of this way of reasoning, let us examine the difficulties of the Kantian argument itself. As is well known, Kant arrived at his notion of the a priori element in our knowledge by transcendental deduction. He started from what we observe and went behind the scenes, so to speak, in order to deduce from the observation the conditions that are necessary for the observation to have been possible. This transcendental deduction raises either of two difficulties. Bertrand Russell believed it to be a logical fallacy (Russell, 1946: 740). If one holds that there are *a priori* categories and forms at work, and if there is no compelling reason why they should be the way they are, it must be possible that I should wake up tomorrow and find that in all human beings the categories have been switched during the night. If that should happen, all of us might, for example, fail to see the sun rise or notice that effects precede causes, and so forth. What is more, if the categories have been switched in all human beings during that same night, nobody would think that there was anything amiss. Russell concluded that he had to reject Kant because Kant's argument did not preclude the possibility that nature could unpredictably change its appearance over night. He admitted that this consideration was no proof that Kant was wrong; but he added that his own notion of the reality of nature would not permit him to entertain such a possibility. Russell had obviously spotted a fundamental difficulty in Kant's theory because he refused to accept the validity of Kant's transcendental deductions. Even if one is more sanguine than Russell about their validity, the problem would remain. For different observations would produce different transcendental deductions and so we would never know whether during the night a switch had occurred, because the next morning the world would simply oblige us to make different observations and these different observations would yield different transcendental deductions and thus everything would still appear to be fine.

Watkins raises a different objection, to Kant's strategy. He agrees with

Popper (Popper, 1963: 95) that Kant, in 'answering' Hume, went too far in the opposite direction: "by displaying contemporary (i.e. Newtonian) physics, not as a possible conjectural system, but as a *necessary* product of our organising categories, he made the problem no longer how Newton could make his discovery but how everybody else could have failed to make it" (Watkins, 1974: 406).

This objection is as cogent as Russell's objection, but easier to rebut. It is perfectly true that if these categories must be there *a priori*, one would have expected people to discover the laws of planetary motion centuries, even millennia before Newton. The reason why they did not has, however, nothing to do with Kant. Theories about planets as well as theories about everything else under the sun play a very contrary role in human history. For the most part, they are not regarded as knowledge at all; but are used in order to consolidate social bonds (Munz, 1988). One way of fomenting social bonds, though by no means the only way, is to get a group of people to subscribe to a catechism, that is, to a body of knowledge. That knowledge becomes the founding charter or catechism of a society and helps to promote solidarity, cooperation and division of labour (Munz, 1985: 74; Albert, 1987: 154; O'Hear, 1988: 87). Knowledge plays, therefore, a very important but non-cognitive role in human history. It is a bit like using paper money as wallpaper to line rooms rather than as a means of exchange. But the really cussed thing about this misuse of knowledge is that it is adaptive and advantageous only if the knowledge so used happens to be false knowledge. Thus one can build a society around the belief that the earth is flat or that water is solid, because there are only a few people who are thus identified as flat-earthers. One cannot build a society around a true belief, because everybody in their right mind would automatically be a member of that society. In order to promote solidarity and cooperation and division of labour it is essential that most human beings be *excluded.* A society of which all of mankind is part, would be a very inefficient and unsurveyable society. Only false knowledge can act as an exclusion principle.

Once we understand this nefarious but useful role which knowledge had always played in human history (Munz, 1988: 84) we can understand why, in spite of the universality of Kantian categories, people are mostly inhibited from making the Newtonian or, for that matter, any other discovery and why science is something of a 'sport' – a luxury which societies which must depend on catechisms for social bonding, cannot afford. Every society that is built around a piece of false knowledge must dogmatically protect that knowledge from criticism and from other scrutiny. Such a society could be called a form of cognitive mercantilism. Its survival depends on the survival of a given piece of false knowledge and in such a society nobody can afford to allow that false knowledge to be examined and questioned. Hence Newtons can emerge only under very special and, as a rule, very unusual social conditions, i.e. only when knowledge is not required for social bonding. This explains why science (= knowledge which withstands criticism) is not coextensive with the human race, even though Kantian *a priori* categories are or, at least, may be.

In the long run, therefore, Russell's difficulties with Kant are more fundamental than the Popper-Watkins difficulties. Russell's difficulties turn upon the absence of a necessary relation between the external world and the categories. We have seen that even if one grants, unlike Russell, that the transcendental deduction is correct, we would never know whether the world has changed during the night or not. If one wants to link the *a priori* categories to the external world, something more than a transcendental deduction is required. If nothing more is added, it remains indeed possible that the categories could be switched at any moment. One cannot quite see by what agency they could be switched; but then, in Kant's own argument, one cannot see by what agency they are uniformly present either. So Russell's imaginary agency which could switch them, may be no more imaginary than Kant's agency, which puts them into place.

As far as Kant's major works are concerned, the matter must rest there. But it so happens that in one of his minor writings, Kant himself admits the problem, gives it some thought and comes up with the possibility that the *a priori* categories are 'innate'. If they are innate (though Kant, in his day, without much knowledge of biology could not understand this), they must have evolved; and if evolved, they must have been placed there by natural selection. But Kant's thoughts on the matter open the way for a biologisation of Kant's critical philosophy and since the passage in question (Kant, 1790: 68 and 1958: 337–338) is not widely known, I will quote it in my own translation: "...there has to be in the subject a ground which makes it possible for our representations to originate the way they do and not in a different way and especially in a way in which they refer to objects which are not yet given. That ground is innate." It would be quite wrong to impute intimations of evolutionary epistemology to Kant; but this passage indicates that he himself suspected that the coincidence between our knowledge of objects and the objects themselves has some cogent ground which points in the direction of biology. It seems that he himself suspected that there was more to that coincidence than a transcendental deduction and that that coincidence had roots or 'grounds' beyond mere logic (Vollmer, 1981: 10). Needless to say, at the end of the 18th century there was no way in which Kant could suspect where the real explanation for that coincidence was to be found.

What with Russell's doubts and Kant's own hints, philosophers should have taken up Darwin and evolution. Since they did not, it is high time that we should do so now. Kant, so it seems, really made a hypothesis about the categories for which he himself could have no evidence. It speaks a lot for Kant that since the discovery of evolution by natural selection, his hypothesis should have found independent confirmation. In my view, this series of philosophical events is a typical example of discovery. It is very similar to the role of Darwin himself. When Darwin advanced his theory, he did not know how old the earth was and he had no inkling of the true mechanisms of genetic inheritance. In other words, Darwin had a grand idea which put all sorts of separate pieces of evidence together but left enormous question marks (Hull, 1973). And then, later, independent discoveries were made in both geology and genetics which turned out to provide precisely those links in Darwin's argument which had been missing.

Russell's misgivings about Kant were dispelled by a biologist. In the early years of the second War, when he was – by a strange coincidence – professor in Konigsberg, Konrad Lorenz published a seminal paper (Lorenz, 1941). He argued that Kant's scepticism about what the world is really like was unjustified because the cognitive structure which enables us to know the world had evolved through natural selection. The reason why our minds have this particular and no other cognitive structure, Lorenz pointed out, must be that we have evolved. The categories of Kant or something very similar, are not accidental. They are not created arbitrarily and could not be different. They are adaptive. Whatever they tell us, must be truths about the environment as it really is. The Kantian distinction between the world as it appears to us and the world as it really is in itself becomes unnecessary once we take evolution into account. If evolution is taken into account, Kant's Transcendental Idealism is transformed into Hypothetical Realism. Lorenz's utterly persuasive argument was very slow to catch on. This may have had something to do with the War. It was not until much later when he published his *Behind the Mirror* in 1973, that philosophers began to take notice. By that time, the ice had been broken by Karl Popper (1972) and by D.T. Campbell (1974 and 1960) and in 1975 there appeared the first edition of Gerhard Vollmer's *Evolutionäre Erkenntnistheorie*, unfortunately not yet translated into English.

The force of Lorenz's argument is not confined to a re-writing of Kant.

One could think of the human cognitive structure which precedes any particular knowledge of the world in more concrete terms. Take our senses. We are born with them and we know that the structure of the retina, the nerve endings on our skin or the organisation of the inner ear leave a marked imprint on our awareness of the world. But today, thanks to our understanding of evolution, we also know that every one of our senses is adaptive and would not have evolved by natural selection if it were systematically unable to spot what the world is really like. The mere fact that we have an eye proves that there is something out there to be seen by the sort of organ our eye is. If there were not, we would not have an eye. Our sense impressions must therefore be broadly trustworthy. Moreover, if our nervous system is such that it adds so called secondary qualities to the perceived primary qualities – if, e.g., we sense wavelengths as colours – there must be an adaptive advantage in our ability to do so. And this means that colour awareness enables us to make distinctions which are really there. This does not mean that 'colours' as such are really out there. But it does mean that colours correspond to distinctions which it is to our advantage to make; and that organisms unable to make them are disadvantaged when they are competing for limited resources with us. The secondary quality is therefore not just an arbitrary, decorative adjunct to a wavelength.

G.M. Edelman's most recent and first comprehensive theory of brain functioning seems to confirm this view by providing a neurological explanation of how the *a priori* elements are put into place. It explains how it comes about that the organism can be guided accurately by its brain through the world and why the brain's computations refer to actual configurations of events. At birth, Edelman says, the organism contains a superabundance of nerve endings, some of which are selected to survive atrophy and to connect with each other by the organism's interaction with the environment through its senses. The process of natural selection, in other words, continues after birth so that the brain is made to contain a model of the environment (Edelman, 1987; Changeux, 1985: 234, 242, 268). The fine tuning of the genetic programme, so to speak, is done again by a process of Darwinian selection. This is how the organism 'learns' from the environment. Obviously, for reasons of economy, these maps which correspond to the organism's sensory surfaces, must contain general, rather than particular information about the world. That is, they must constitute information on the general and regular features of space, colour, etc., which enables the organism to recognise, e.g., a dandelion; rather than information about a particular dandelion. Thus these maps correspond quite closely to Kantian categories and forms of perception.

The moment this is realised, we come up against two difficulties. The first is the fact that there is a world of difference between the way the world appears to, say, a frog and the way it appears to us. Our eyes are very different from the eyes of a frog. This, however, does not mean that the evolutionary argument breaks down. It merely shows that every organism lives in a specific world of its own. But since the several environments to which organisms are adapted are all part of the same large universal environment, it follows that no matter what the differences between the frog's *Umwelt* and the human *Umwelt* are, the two appearances must be compatible with each other. It is a very important part of the scientific enterprise to work out precisely what the terms of such compatibility are. The compatibility itself, cannot be rationally questioned.

The second difficulty is of a different kind. If our senses must be trustworthy, one might suppose that the philosophy of science of the positivists who maintain that the human mind is a *tabula rasa* waiting to be impressed upon by our senses, is the correct philosophy. This conclusion in favour of positivism is false. A philosophy of knowledge which takes evolution into account comes, on the contrary, to the opposite conclusion. Every organism is born or comes into being with a minimal cognitive structure. This is as true for a virus as it is for *homo sapiens*. The ability of an organism to learn by observation is extremely limited, though there is some variation in those degrees. The cognitive relationship between every organism and its environment is genetically programmed. The reason why this programme is in all cases moderately correct lies in the fact that every organism has evolved by natural selection. The correctness of its *a priori* knowledge is due to the fact that over vast spans of time all those organisms with a less correct a priori knowledge have been weeded out for they could not reproduce equally fast. No organism has acquired its a priori cognitive structure by observation or by being instructed by its environment to store this and no other particular information. In the early 18th century, when it was first surmised that there may be connections and affinities between animals and human beings over and above those indicated in Genesis, Lord Monboddo (1714–1799), a dedicated aposteriorist, believed that animals would eventually learn to be human beings if they could be made to live longer, that is, learn more. The real link between animals and human beings is of a completely different kind.

Every organism has acquired the particular cognitive structure it has through natural selection. The reason why it has such an *a priori* cognitive structure is precisely because it was, literally, not born yesterday but is descended from countless generations of naturally selected ancestors. This is not only cold comfort to positivists; but is as cogent a falsification of positivism as one could possibly think of. The theory of evolution is the only viable explanation of the presence of our *a priori* categories. The fact that our genetic programme is due to natural selection indicates that our *a priori* knowledge is more, not less, reliable than the bits and pieces we pick up *a posteriori*.

If we look at the acquisition of knowledge in this way, we can see that every organism is like a theory about its environment. I mean this quite literally. We orient ourselves by deducing, with the help of observed initial conditions, a prognosis from an *a priori* general theory. An organism embodies a certain cognitive structure. It takes in signals or stimuli from the environment and immediately interprets or decodes them in terms of its structure and then responds. This is exactly what happens when we test a theory. The cognitive structure is the general theory and is present *a priori*. The stimulus from the environment is the initial condition and natural selection has endowed every organism with at least some ability to pick up these stimuli correctly. The response is the prognosis or prediction which we deduce from the general theory with the help of the initial condition. To be precise, therefore, I would suggest we call an organism an embodied theory about its environment.

All organisms taken together are an immense body of information about parts or aspects of the world they live in. They constitute a lot of knowledge. This knowledge is a representation of those parts of the world and by looking at it, one can learn a lot about the world. Eyes teach us about light; ears, about sound waves; hoofs, about the steppe, and so on. This information is a representation because the relationship is not symmetrical. The steppe does not tell us about hoofs and light waves tell us nothing about eyes. The question then arises how the information about the steppe is transferred from the steppe into the hoof when the steppe itself contains no information about the hoof, i.e., when the relationship is asymmetrical. Obviously, the way the world presents itself to the organism is very different from the way the organism represents itself to the world. The organism cannot be thought to be using the way the world presents itself in order to decode or transform that presentation into a representation. Presentation and representation do not coincide. The information transfer must be thought of as a progress of random mutation and selective retention (Campbell, 1974, 1960), that is, as a long drawn out and very time consuming process.

As we ascend the evolutionary ladder from amoeba to *homo sapiens*, organisms become more complex embodied theories. And when we come to *homo sapiens* himself and herself we discover suddenly that we are at the crossroads. The human nervous system is so complex that it not only decodes sensations but decodes them in such a way that we become consciously aware of them. Now if one is aware of the stimuli, that is of the initial conditions, it would be reasonable to assume that one can use them as instructions and that one can, unlike pre-human organisms, actually learn from the environment by observation. Conscious intelligence, one is tempted to conclude, makes it at long last possible for presentation of the world to the organism and representation of the world by the organism, to coincide. One could argue that such a conclusion is the final delivery of an evolutionary theory of knowledge.

I am afraid that such a conclusion is unwarranted. The emergence of consciousness which raises the possibility of learning by instruction also, at once, makes genuine learning by instruction impossible. For the moment we become conscious of a stimulus, we give it a label and identify it linguistically. The system of verbal labels we have at our disposal, is however, a semantic convention which we owe not to our situation in a given natural environment (so that one might presume that natural selection has provided the correct labels) but to the fact that we are living in a social community with other human beings. For this reason every single observation we make is instantaneously laced into the particular social or cultural convention we are a member of. This does not necessarily prove that it is mistaken; but it does indicate that it loses the veridical quality which it possesses for pre-conscious organisms. (I am aware, of course, that languages themselves have evolved by natural selection. But the evolution here has taken place by natural selection in a social environment; and not by competition of language users in a natural environment. Hence it is impossible to treat established languages as mirrors of the environment in the way in which one can look upon all organisms as mirrors of their environment. This view does not necessarily contradict Chomsky's theory that the ability to acquire a language is part of our genetic programme. I am not here concerned with the linguistic potential as such; but with the specific semantic systems in use in different communities. It is those semantic systems, not our genetically conditioned linguistic potential, which coerce our observations into being something other than neutral reports of what there is.)

We must not conclude that the emergence of consciousness plays a purely negative and inhibiting role in the acquisition of knowledge, although it is true that there have been many philosophers like Schopenhauer and

Nietzsche who have worshipped the blind life-force and been of the opinion that the arrival of consciousness is a purely life-inhibiting factor. But the truth of the matter is that the arrival of consciousness merely disappoints the common sense and folk psychology expectation that consciousness enables us to learn by observation and to use induction to arrive at knowledge. In another direction it proves an enormous asset. Below the level of conscious organisms, knowledge has to be acquired by the pace of evolution itself. Every variation of every organism is a hypothesis put to the environment. The errors are gradually weeded out through the differential rates of reproduction. All this takes time and such error elimination is tied to the life and death cycle of organisms. As soon as a hypothesis can be put to the environment by conscious representation, that is, in the shape of symbols, this process of error elimination can be speeded up. A consciously formulated hypothesis can be eliminated, often in a matter of minutes (Popper, 1972: 261; Watkins, 1974: 406). Only in the rarest of cases does one have to wait for a string of generations to die out before an erroneous hypothesis is falsified. In short, with the emergence of consciousness, the situation is turned inside out. Where organisms were embodied theories, consciously proposed hypotheses can be regarded as disembodied organisms.

Every theory is very much like an organism which we ourselves have invented, so to speak. When we can specify an initial condition, we can deduce, with its help, from the general theory, a prognosis. If the prognosis turns out to be wrong, the theory should be dropped, that is, it cannot survive. One should also notice that the fuzziness which creeps in when one is seeking indiscriminately for observations to be used as initial conditions, is here eliminated. For the general theory gives precise specifications as to the initial conditions one is looking for. Thus we can continue and extend the process of evolution in our minds at an unprecedented speed.

As an aside I would speculate that the evolution of consciousness which makes disembodied organisms possible, is a step which speeds up evolution and comparable to that great, earlier step which also speeded it up – the evolution of sexual reproduction. The analogy is worth pondering. Consciousness not only makes it possible to have disembodied organisms as proposals – but consciousness also makes it possible for us to make mistakes in a way in which no animal can make mistakes. Thus it multiplies the occurrence of errors. Since adaptation and truth are the result of error elimination by natural selection, the occurrences of errors rather than the avoidance of error, is the optimal mechanism for selection to operate. In any case, natural selection is possible when and because errors are made. Thus we can spot a new meaning in the concept of rationality. Rationality, we can now say, consists in making as many mistakes as possible and in allowing selection by uninhibited criticism. It does not consist, as is so commonly supposed, in the avoidance of error and the employment of a 'correct' method.

When we look at the knowing organism and compare what comes in with what comes out, we see very clearly that the positivist's attempts which go back to Locke and Hume to prove that what comes out is in direct proportion to what comes in, have failed. This is particularly so for animals and only slightly less so for human beings. There is always meagre input – mostly no more than the picking up of an initial condition – and a torrential output. Obviously, the nervous system supplies the difference. Watkins suggested that "classical theories of knowledge are hit either by the fact that theoretical science exists at all or by the fact that it is far from coextensive with the human race." He sees himself somewhere in the middle between these two mistaken extremes of Hume and Kant. We now find that there is no such symmetry. Hume and post-Humean attempts have failed. But the conclusion we have reached is very close to Kant, so that we find ourselves not in the middle, but very much on the side of Kant or of a biologised Kant. Watkins' objections to Hume have been shown to have been not formidable enough. But the objection which Watkins raised against Kant can be explained away and the process of natural selection can explain something which Kant himself had not been able to explain well: why every organism is born with a more or less suitable genetic programme. Given that the programme is genetic and innate, Kant said that we *prescribe* laws to nature. In the biology oriented perspective which is largely derived from Popper, we say that we *propose* laws to nature. There is a striking similarity between the two expressions as to where the initiative comes from. But there is also a great difference between 'prescribing' and 'proposing'. Kant thought of 'prescribing' and considered that since no matter what is prescribed, there must be feed-back through the same channels, the feedback must at least *look* right. An elegant argument, but at considerable cost: for prescription and feedback between them constitute an effective screen over reality through which nobody can break. The idealism with which Kant explained Newton's laws may have been transcendental; but idealism it was all the same. Where, for Kant, was the real world of the planets?

Looking back upon our whole argument, we can see now that between the idea and the reality there had fallen the shadow of Darwin's genius which had diagnosed the philosopher's quandary correctly. The suggestion that the cognitive structures we have are innate and, if innate, must be due to natural selection, is not so much an overthrow of Kant, as a profoundly important amendment of Kant's assumption that the innate cognitive structures are there either by accident or by rational, i.e. transcendental deduction. Since the real environment itself rather than a mere process of transcendental reasoning let alone an accident, was responsible for putting them into place through natural selection, the information they mediate must be information about the *real* world. Idealism, even when qualified as transcendental, now becomes unnecessary and some kind of realism must take its place. Popperian proposals are thus confidently to be substituted for Kantian prescriptions and, with the help of Darwin, Kant's transcendental idealism is transformed into Popper's hypothetical realism.

Victoria University of Wellington

REFERENCES

Albert, Hans: 1987, *Kritik der reinen Erkenntnislehre*, Tübingen, J.C.B. Mohr.

Campbell, D.T.: 1960, 'Blind Variation and Selective Retention in Creative Thought as in Other Knowledge Processes', *Psychological Review*, **67**, 380–400.

Campbell, D.T.: 1974, 'Evolutionary Epistemology', in: Schilpp, P.A. (Ed.), 1974.

Changeux, Jean-Pierre: 1985, *Neuronal Man*, Engl. Transl. New York, Oxford University Press.

Churchland, Patricia: 1986, *Neurophilosophy*, Cambridge, Mass., Harvard University Press.

Cohen, I.B.: 1980, *The Newtonian Revolution*, Cambridge, Cambridge University Press.

Edelman, G.M.: 1987, *Neural Darwinism*, New York, Basic Books.

Gardner, Howard: 1985, *Frames of Mind*, London, Paladin Books.

Hull, D.L.: 1973, *Darwin and His Critics*, Cambridge, Mass., Harvard University Press.

Kant, Immanuel: 1790, *Ueber eine Entdeckung nach der alle neue Critik der reinen Vernunft durch eine ältere entbehrlich gemacht werden soll*, Königsberg, Friedrich Nicolovius. (*Werke*, Wiesbaden: Insel Verlag, 1958, Vol. III).

Land, E.H.: 1977, 'The Retinex Theory of Colour Vision', *Scientific American*, **237**, 108–128.

Lorenz, Konrad: 1941, 'Kants Lehre vom Apriorischen im Lichte gegenwärtiger Biologie, *Blätter für deutsche Philosophie*, **15**, 94–125.

Marr, David: 1976, 'Early Processing of Visual Information', *Philosophical Transactions of The Royal Society of London*, Series B, **275**, 483–524.

Munz, Peter: 1980, 'Finches, Fossils and Foscarini', *The New Zealand Journal of History*, **14**, 132–152.

Munz, Peter: 1985, *Our Knowledge of the Growth of Knowledge*, London: Routledge.

Munz, Peter: 1987a, 'The Unity of Science and the Dubious Credentials of Positivism', in: G. Radnitzky (Ed.), 1987.

Munz, Peter: 1987b, 'Bloor's Wittgenstein or the Fly in the Bottle', *Philosophy of the Social Sciences*, **17**, 67–96.

Munz, Peter: 1988, 'The Philosophical Lure of the Sociology of Knowledge', in: G. Radnitzky (Ed.) 1988.

O'Hear, Anthony: 1988, 'The Evolution of Knowledge', *Critical Review*, **2**, 78–91.

Popper, Karl R.: 1963, *Conjectures and Refutations*, London, Routledge.

Popper, Karl R.: 1972, *Objective Knowledge*, Oxford, Clarendon Press.

Radnitzky, Gerard (Ed.): 1987 and 1988, *Centripetal Forces in the Sciences*, 2 Vols., New York, Paragon House.

Rosenfield, Israel: 1984, 'Seeing Through the Brain', *New York Review of Books*, **XXXI** (15), October 11.

Russell, Bertrand: 1946, *A History of Western Philosophy*, London, Allen and Unwin.

Schilpp, P.A. (Ed.): 1974, *The Philosophy of Karl Popper*, La Salle, Open Court.

Simpson, G.G.: 1963, *This View of Life: The World of an Evolutionist*, New York: Harcourt Brace.

Suppe, Frederick: 1977, *The Structure of Scientific Theories*, Urbana: University of Illinois Press.

Vollmer, Gerhard: 1981, *Evolutionäre Erkenntnistheorie* (3rd rev. ed.), Stuttgart, S. Hirzel Verlag.

Watkins, J.W.N.: 1974, 'The Unity of Popper's Thought', in: P.A. Schilpp (Ed.), 1974.

Wise, R.A.: 1980, 'The Dopamine Synapse and the Notion of Pleasure Centres in the Brain', *Trends in Neurosciences*, **3**, 91–95.

FRED D'AGOSTINO

THE 'OPTIMUM' AIM FOR SCIENCE

In his book *Science and Scepticism*[1] John Watkins proposes to defeat the 'rationality-sceptic' within the limits defined by Humean scepticism about the possibility of knowledge of the world of external objects and of future events. Rationality-scepticism is according to Watkins (p. 59) that view – advocated perhaps and if so most prominently in recent discussions by Paul Feyerabend[2] – that "we never have any good reason to adopt a hypothesis about the external world" – or, perhaps more clearly, that we never have any good reason to adopt from amongst those available any particular hypothesis.

According to Watkins (pp. 123 & 279) it would be possible rationally to choose amongst competing scientific hypotheses (i) if we could "specify a nonarbitrary aim [for science] ... an aim to which all members of the republic of science could subscribe" (or perhaps to which they should rationally subscribe) and (ii) "if we could know, at least in a good many cases, which one of a set of competing hypotheses, given the present evidence, best satisfies" such an aim. Watkins claims (*ibid.*) that we would in this case "have the best possible reason to adopt" such an hypothesis. For it is the hypothesis which best satisfies the 'optimum aim' for scientific research; it is the hypothesis which is with respect to such an aim the best of those under consideration by us. Watkins has in mind, I think, an argument of roughly the following kind.

- Suppose that we have shown that the theory **T** is with respect to certain rules for theory-preference superior to the theory **T′**.

- Suppose also that superiority with respect to these rules is reflective of superiority with respect to the aim which is according to Watkins the 'optimum aim for science'.

- In this case, we could if the aim specified were indeed the 'optimum aim for science' conclude that the theory **T** was in some important and 'objective' sense – sufficient to defeat the rationality-sceptic – superior to the theory **T′**.

It will be clear even from this very brief account that the claim that there is

Fred D'Agostino & I. C. Jarvie, Freedom and Rationality: Essays in Honor of John Watkins, 247–256.

and that he has specified an 'optimum aim for science' is the key to Watkins's antisceptical methodological project. Suppose that we were able to show that the aim specified by Watkins as the optimum aim for science is not in fact the optimum aim for science, whatever that might be. In this case we could undermine his claim to have defeated within the limits of Humean scepticism the rationality-scepticism which he takes as the main target of his argument. And we could do so by pointing out that

- the fact that the theory **T** better than the theory **T′** satisfies some aim specified by Watkins as the optimum aim for science provides, if it is false that the aim specified by Watkins is the optimum aim for science, no ground strong enough to defeat the rationality-sceptic for preferring **T** to **T′**.

(Why prefer **T** to **T′** simply because the former better than the latter satisfies some aim specified by Watkins? It is only if Watkins has correctly characterized the optimum aim for science that satisfaction of the aim specified by Watkins would really matter.)

The success of the antisceptical enterprise of Watkins in other words depends on the strength of the case for taking the aim he specifies as 'the optimum aim for science' *as* the optimum aim for science. If we could show that the aim specified by Watkins is not in fact the optimum aim for science then we could altogether undermine his claim to have defeated within the limits of Humean scepticism the rationality-scepticism of Feyerabend and others.

Is the aim specified by Watkins in fact the optimum aim for science? In order to answer this question, it will be helpful to proceed as Watkins himself does. Watkins argues (Section 4.1) that there are five adequacy requirements for aims of science which must be met by any specification of the aim of science which is plausibly to be presented as the 'optimum aim for science'. These requirements are (p. 124) that "any proposed aim for science must: 1. be coherent; 2. be feasible; 3. serve as a guide in choices between rival theories or hypotheses; 4. be impartial; 5 involve the idea of truth." Watkins then (Section 4.2) examines with respect to these requirements a 'utopian' account of the aim for science which is, he says, implicit in the early modern epistemological theories of Bacon, Descartes, and others. He is able to show that the utopian aim does not satisfy the adequacy requirements which he has specified, and must therefore be modified. It is interesting to examine in some detail and in particular Watkins's analysis of the demand for ultimate explanations which is he says a component of this utopian aim for science. In this we will be concerned especially with those requirements of impartiality and of feasibility which Watkins himself stresses in this context.

Watkins suggests that the utopian aim included as a prominent feature the demand **UE** for explanations in terms of ultimate essences. He says, adopting the point of view of a scientist or an epistemologist of the seventeenth century (pp. 127–128):

> We believe that we have already made quite good progress in the discovery of underlying factors in terms of which empirical regularities can be explained. But these underlying factors themselves now stand in need of explanation in terms of still deeper factors. And our supreme demand is that our science will penetrate deeper and deeper until eventually it achieves ultimate explanations of all phenomena.

But Watkins rejects by reference to the idea of feasibility the demand **UE.** He says (p. 130) that "[t]he feasibility requirement obviously requires us, as a first step, to recast [the utopian aim] as the ... aim of progressing towards these far-off goals without necessarily attaining them", and therefore in this instance of progressing towards the goal of ultimate explanations without necessarily attaining this goal.

I believe that Watkins's discussion of this matter is somewhat less clear than it might be. He says (p. 124, emphasis added) that "an aim is infeasible if we *know* that it cannot be fulfilled". In conjunction with the claim that the demand **UE** is not a feasible demand, this therefore suggests that we know that:

> we cannot fulfil the goal **UE** of providing for phenomena explanations in terms of ultimate essences.

But how indeed could we know this? We could it seems to me know that it is impossible to explain various phenomena in terms of ultimate essences only if we could know that classical atomism is false – where classical atomism is, of course, that theory according to which as Watkins says (p. 131) "there are ultimate bits". But how could we know any such thing – at least at this stage of the argument, where we are still trying to establish some aim with respect to which theories of precisely this kind might eventually be evaluated?

It seems altogether clearer to say, as does Watkins himself at a later stage of the argument (see Section 4.21), that the demand **UE** must be rejected, not because it is infeasible, but, instead, because it could be known to be feasible only if we already had grounds for accepting one of a number of competing hypotheses about the fundamental structure of the world – or, in other words, because it presupposes for its feasibility one of a number of competing hypotheses and therefore fails to be impartial amongst them. (How could a methodological demand which in this way presupposes a metaphysical theory be neutral with respect to that theory and its rival(s)?)

The demand **UE** is I think appropriately rejected on the grounds that it is not impartial between classical atomism and a doctrine which might be called infinitism – the doctrine, advocated by Leibniz (see p. 131), according to which there are no 'ultimate bits', but, instead, an infinitely multilayered reality.

Anyway it is certainly on these grounds that Watkins himself rejects that 'progressivized' variant of the demand **UE** for ultimate explanations which he at the next stage formulates and which amounts to the demand **PUE** that (p. 130) we are "to progress towards ultimate explanations with theories that are ever deeper". (According to Watkins (pp. 130–131), the demand **PUE** must be rejected on the grounds that it "violates our impartiality requirement since it takes sides ... in an important cosmological issue" – namely, of course, that between atomism and infinitism. Progress towards ultimate explanations is feasible only if there are ultimate explanations towards which progress can be made, or, in other words, (again) only if atomism is true. (For if atomism is false there are no ultimate explanations and progress towards ultimate explanations is therefore impossible.) The demand **PUE** must therefore be rejected on the grounds that its feasibility presupposes the truth of one amongst a number of competing cosmological hypotheses and that it therefore fails the impartiality requirement.

Watkins at this point proceeds further to 'deflate' the still utopian demand associated with the idea of depth of explanation. He says (p. 113):

> So if [the demand associated with the idea of depth] is to be impartial between this metaphysical thesis [i.e. that of 'infinitism'] and the thesis of classical atomism that there are ultimate bits, we should [altogether] discard [rather than simply watering down] the ideal of ultimate explanations and retain that of ever deeper explanations.

Watkins in other words proposes substituting as a component of an 'optimum aim for science' for the demands either for ultimate explanations or for progress towards such explanations the demand **DE** for ever deeper explanations. And this aim is he clearly implies one which is, unlike the demands **UE** and **PUE**, both feasible and impartial in character, it is a demand which can rightly be considered a component of the aim specified by him as the optimum aim for science.

I believe that Watkins is wrong to include as a component of the aim specified by him as the optimum aim for science the demand **DE** for ever deeper explanations. I believe that this demand, like those for ultimate explanations and for progress towards ultimate explanations, is one which falls foul of the feasibility and impartiality requirements which Watkins himself imposes on the demands associated with the optimum aim for science.

Is the demand for ever deeper explanations one which is feasible? This demand is, of course, one which it will be possible always to satisfy only if it is true, as is implied by infinitism, that there is no level of explanations which is ultimate. For were it instead true, as is implied by atomism, that there is such a level, then it would be impossible, once this level had indeed been reached, that we provide explanations which are (genuinely, not speciously) deeper than those available at that level.

The demand **DE** for ever deeper explanations is therefore feasible only if the infinitistic theory is true. The demand **DE** must therefore be rejected on the grounds that its feasibility presupposes the truth of one amongst a number of competing cosmological hypotheses and that it therefore fails to satisfy the impartiality requirement imposed by Watkins himself on any demand rightly associated with the optimum aim for science.

Exactly those considerations which move Watkins to reject as demands legitimately associated with the optimum aim for science the ideas of ultimate explanations and of progress towards ultimate explanations require that we likewise reject the demand which he says can legitimately form a component of the optimum aim for science – that for ever deeper explanations. This aim too is 'partial' in the sense – here perhaps made clearer than it is by Watkins – that its feasibility presupposes the truth of one amongst a number of competing theories about the fundamental structure of the world.

The aim specified by Watkins as the optimum aim for science includes as a component the demand for ever deeper explanations. This demand does not satisfy the adequacy requirements specified by Watkins and any specification of an optimum aim for science which includes it therefore fails also to satisfy these requirements. The aim specified by Watkins as the optimum aim for science therefore cannot legitimately be taken to be the 'optimum aim for science'. But in this case, the fact (should such it be) that some theory **T** better than some other theory **T′** satisfies the aim specified by Watkins as the optimum aim for science provides us with no grounds (strong enough to defeat the rationality-sceptic) for preferring **T** to **T′**. For the rationality-sceptic can always in this case ask why superiority with respect to the aim specified by Watkins should constitute grounds for theory-preference. (Anyone can specify an aim and show with respect to that aim the superiority of one theory to another. The problem is not to specify but to justify the aim specified, and therefore to justify the preferences which it supports.) But in this case, Watkins has provided us with no grounds for theory-preference of a kind sufficiently strong to defeat the rationality-sceptic.

What are we to conclude in relation to the failure of Watkins's project (here allegedly demonstrated)?

(a) We might conclude that Watkins was wrong to impose as adequacy requirements on any proposed aim of science the requirements of impartiality and feasibility, which, as we claim here to have shown, rule out as inadequate the very aim which he himself specifies as the optimum aim of science. Perhaps the aim specified by Watkins could with respect to some set of adequacy requirements less demanding than that specified by him be shown to be the optimum aim of science.

This response to our demonstration is not I think an attractive one. Feasibility and impartiality are quite reasonable constraints on any specification of the so-called optimum aim of science, and ones which the rationality-sceptic can, I think, quite reasonably insist on, so long as we persist in pursuing in this way the question of the rationality of science. You can readily imagine, I think, with what relish the rationality-sceptic would greet any proposal that the aim of science be specified in a way which need be neither impartial nor feasible. (Any such specification would be said, I am sure, to have 'all the virtues of theft over honest toil' as Bertrand Russell has somewhere in another context complained.)

(b) We might conclude, on the other hand, that Watkins has failed not in his specification of the adequacy requirements for proposed aims of science, but, instead, in his characterization of the so-called optimum aim of science. Perhaps there is an optimum aim of science, though not of course the one characterized as such by Watkins, which does indeed satisfy the requirements specified by him for any such aim (or such other requirements as might be reasonable).

This response to our demonstration is, at least superficially, rather more attractive than that canvassed at (a) above. Perhaps it is possible so to specify as the optimum aim of science some aim which is – as Watkins's so-called aim is not – coherent, feasible, and impartial, which involves the idea of truth, and which provides a basis for theory-preference. But a little reflection shows, I think, that this idea is chimerical, indeed rather obviously so. How could it be possible, within the limits of impartiality, to specify an aim of science which provides a basis for theory-preference? Any standard which provides a basis for the formation of preferences must ipso facto be 'partial', not impartial, with respect to the theories which it enables us to rank; if it were indeed genuinely impartial it would provide no basis for the relative ranking of theories. The idea that there is some specification of the aim of science which is genuinely impartial with respect to competing hypotheses (especially of a metaphysical character), is, as Feyerabend has in effect pointed out, a nonsense: every specification of the so-called optimum aim of science must fail either the impartiality requirement or that which demands that it provide a basis for theory-preference; *there can be no impartial basis for theory-preference*. (Feyerabend put essentially this point with reference to the 'partiality' of methodological rules. He said (op. cit., p. 295) that "every methodological rule is associated with cosmological assumptions, so that using the rule we take it for granted that the assumptions are correct", and therefore are biased a priori in favor of those theories which are compatible with the truth of the assumptions which are taken for granted: we will ceteris paribus prefer such theories to theories which are not compatible with the truth of the relevant assumptions.)

Any specification of an optimum aim of science must therefore fail at least one of the adequacy requirements which Watkins has specified. Some specifications, like Watkins's, will fail either impartiality or feasibility requirements (or both). Others, like those contemplated in the preceding paragraphs, will fail either the impartiality requirement or that associated with the idea that a method should provide a basis for theory-preference.

(c) I think that the conclusion which is properly to be drawn from our demonstration is a rather more radical one than either of those so far considered. That it is not possible to specify an aim of science which satisfies plausible requirements on the adequacy of any such specification shows, I believe, that a *teleological* approach to the question of the rationality of science is misconceived and that we must instead adopt an approach which is non-teleological or *deontological* in character.

It will be helpful here to remember that Watkins is struggling to cope with the problem of the rationality of science. There are according to the rationality-sceptic no reasons why one amongst a number of competing theories should be preferred to the others; no actual preference can therefore rationally be grounded. Watkins proposes to meet the challenge posed by rationality-scepticism. He claims that it is possible in a non-arbitrary way to specify for science an optimum aim, and that it is possible to show that the realization of this aim is 'promoted' by choices (amongst competing theories) made in accordance with some methodological standard (roughly, degree of corroboration). But it is in this case possible to defeat the challenge posed by the rationality-sceptic. The methodological standard specified by Watkins provides a basis for the choice amongst competing theories. The basis it provides is a rational one in the sense that choices made in accordance with such a standard better promote realization of the non-arbitrary aim of science. (If we choose in accordance with the standard, we choose those theories which better realize the optimum aim of science. What could be more rational than that?)

Watkins's approach to the problem of the rationality of science is teleological in a fairly transparent sense. It is rational to employ such a standard of theory-preferability as he has specified because or insofar as employing such a standard enables us to realize the aim or goal of scientific enquiry (which he has specified). Watkins's approach is teleological *meta-methodologically*: the methodological standard specified by him is a reasonable one insofar as it enables us better or more reliably or more efficiently to realize the goal of scientific enquiry, the optimum aim of science.[3]

Now there are very good reasons for adopting a teleological approach to methodological issues, foremost amongst which is the absence, apparently, of any viable alternative. But the case of Watkins is only one amongst many in which such an approach cannot under critical scrutiny actually be sustained.[4] Perhaps *this* lesson should 'inductively' be drawn: a teleological approach is doomed to failure – we seem to be unable either to specify in any adequate way the goal of scientific enquiry or to recognize when this goal has and when it hasn't more completely been realized. But what are we in this case to do?

If we are unable to account within a teleological framework for the rationality of science, how indeed (if at all) are we to do so? This is the question we have to face, unless the rationality-sceptic is to prevail, in the face of the failure of Watkins's project amongst many others.

The rationality of science is in my view not to be found teleologically, in the adaptation of means (for the determination of theory-preferability) to ends (the optimum aim of science). It is to be found, instead, deontologically, in the conformity of scientific discourse to certain rules which are as it were 'instinct' in this enterprise.

To see, at least in a sketchy way, what I have in mind here, it is helpful to consider what I call the *circumstances of method*. These are the conditions with respect to which it alone 'matters' that there are two or more incompatible theories in some particular domain of scientific investigation. It will matter to the theorists concerned that others do not accept the theories they have themselves articulated only if these people are committed to four basic principles – realism, fallibilism, respect, and rationalism. According to realism, no two logically incompatible statements can both be true; according to fallibilism, it is always possible, except in the case of certain trivial first-person beliefs, that we be wrong in the beliefs we have formed; according to respect, it is always possible that those who disagree with us have reasons for doing so; according to rationalism, it is intolerable knowingly to hold to be true claims which there is better reason to believe are false.

It is only in case we are committed to these principles that it will matter to us that others have not formed the same beliefs as we have. It is therefore only if we are committed to these principles that we will look to method to provide the means whereby our disagreement with them might one way or another be resolved. These principles in effect define the circumstances of method. But they provide as well the basis for deriving, as 'instinct' in them (i.e. as already implicit in our commitment to them) certain rules for the *resolution* of disagreement. It is, for instance, 'instinct' in the circumstances of method that disputants can reject as reasons for abandoning their theory only those considerations which they have reasons to reject; they cannot simply ignore as impotent to affect their commitments the 'reasons' for abandoning them offered by their opponents. Such a rule of responsible argumentation is not justified teleologically. It is not justified on the grounds that conformity with it better enables us to realize some so-called goal or aim of scientific enquiry. Such a rule is justified, instead, deontologically; on the grounds that we have reason, in exactly those commitments with respect to which it alone makes sense to take seriously the fact of disagreement, to conduct our discussions in accordance with this rule. (If you're committed to believing what there's best reason to believe, and, at least defeasibly, to the view that those who disagree might have reasons for doing so, then you have reason not to reject except for good reasons those reasons for abandoning your beliefs which your opponents have presented.) As far as rationality is concerned, it will not matter if discussions conducted in accordance with such a rule more reliably realize some so-called aim of science; what matters, what makes conformity with such a rule rational, is that the rule is implicit in just those conditions or commitments which themselves define the boundaries and character of the epistemic enterprise.

Can the rationality-sceptic be defeated? Not perhaps as Watkins has defined the terms of the debate. The rationality of science cannot be shown to consist in the means-ends adaptation of the rules of scientific method to some alleged goal of scientific enquiry. But science may nevertheless be a rational enterprise, at least in the sense associated with deontological rationality: science is rational because or to the extent to which the disputes which arise within the scientific community are addressed within a framework of discursive rules which are themselves implicit in the so-called circumstances of method.

University of New England

NOTES

[1] London: Hutchinson, 1984.
[2] See Feyerabend, *Against Method* (London: New Left Books, 1975).
[3] I discuss within a broadly teleological framework various problems of meta-methodology in my paper "Meta-Methodology and Methodological Nihilism", *Methodology and Science*, **17** (1984), 143–161. See also the references cited there.
[4] For documentation of this claim in relation to truth-oriented approaches to teleological meta-methodologizing, see my paper 'Adjudication as an Epistemological Concept', *Synthèse*, forthcoming.

JOHN WORRALL

WHY BOTH POPPER AND WATKINS FAIL TO SOLVE THE PROBLEM OF INDUCTION

Accepted science not only does, but should, inform our technological practice. If someone wants to build a bridge that will stand up tomorrow or a plane that will fly tomorrow she should assume in particular that currently accepted low-level generalisations will continue to hold tomorrow. Someone who claimed (without evidence) that falling bodies will soon start to fall with an acceleration which increases as the cube of the time of fall would be regarded as downright irrational. Someone who encouraged passengers to fly on an aeroplane built on that supposition about future falling bodies would be regarded as criminally irresponsible. But we know, following Hume, that, since all the observational evidence we have for the generalisations accepted by science is of necessity evidence about the past, and since deductive logic is not content-increasing, we certainly cannot deductively infer that accepted generalisations will continue to hold in the future from any amount of evidence we may have. But what then is the basis for these very firm judgments about rationality and responsibility? This is, of course, the notorious 'pragmatic problem of induction'. John Watkins has recently joined the long list of philosophers who have attempted to solve the problem.

In this paper I shall argue that his crisp and challenging 'solution' entirely fails. However, he himself calls his attempted solution 'neo-Popperian' and sets it against the background of what he sees as Popper's own failure to solve the problem. I begin therefore by explaining exactly why I agree with John Watkins that Popper's solution fails. I feel rather apologetic about this: since I have nothing of real substance to add to the points already made against Popper's alleged solution in the 1930s by Reichenbach and Feigl and later by Ayer, Lakatos, Salmon, Grünbaum, Newton-Smith, and many others.

However, there are some philosophers – notably David Miller – who continue to believe that these criticisms miss their mark and that Popper's solution remains viable.[1] Since the point which seems so obvious to me is clearly not universally regarded as obvious, there may be some merit in trying once more to set out the argument as perspicuously as possible. Moreover, although Watkins accepts that Popper failed to solve the problem, it is not always clear that he accepts the full extent of that failure, since his

Fred D'Agostino & I. C. Jarvie, Freedom and Rationality: Essays in Honor of John Watkins, 257–296.

own treatment seems to be to inherit some of the faults of Popper's. My treatment of Popper's 'solution' and Miller's defence of it in Sections 1 and 2 of my paper set the scene, then, for my arguments against Watkins's 'solution' in Section 3.[2]

1. POPPER'S FAILURE TO SOLVE THE PROBLEM OF INDUCTION

Chapter 1 of Popper's *Objective Knowledge* begins with the following striking claim:

I think that I have solved a major philosophical problem: the problem of induction.[3]

And David Miller agrees – Popper's development of falsificationism means, Miller says, that "the problem of induction is at last well and truly solved".[4] However, as Popper complains, and as Miller documents, almost no other philosopher of science agrees that he has done any such thing. I'm afraid that on this issue the rest of the philosophical world is right, and in this section I try to explain why.

First of all, what *is* the problem of induction? Well, most philosophical problems tend to proliferate into whole sets of subtly different problems once you start to investigate them closely, and 'the' problem of induction is no exception. But the problem on which I shall concentrate (sometimes called the 'pragmatic problem of induction') can be posed by adapting slightly an example of Lakatos's.

Suppose you are admiring the view from the top of the Eiffel Tower and fall into conversation with a fellow view-admirer. You have both decided to return to ground level to continue your conversation over a cognac, when your new companion suddenly tells you that, rather than take the lift, he intends to leap over the balustrade and float gently to the ground. This fellow is, let's suppose, entirely sincere, not consciously or subconsciously suicidal, not into hallucinogenic drugs and has no hidden parachute under his T-shirt, no hidden super-powered motor in his pocket – he just genuinely believes that he will naturally float gently to the ground below. He turns out to be no intellectual slouch. He has taken A-level physics and knows all about Galileo's law (or, has he prefers to call it, 'alleged law) of free fall and he accepts that all the evidence so far in accords with this law – or at any rate with suitably weakened versions of it. For he also knows history of science and accepts that Galileo's law, as originally formulated, is strictly false. This is partly because of frictional effects of the air, and partly because, even in a vacuum, falling bodies would not be constantly accelerated, instead their acceleration depends on their distance from the centre of the earth (though this distance changes so little in 'normal' falls as to make Galileo's law good enough for practical purposes). He does not want to base his unorthodox proposal on points of detail such as this – he is happy to accept that there are weaker versions of Galileo's law, which so far as past evidence goes, are exceptionless. For example the generalisation that humans stepping out from heights comparable to that of the Eiffel Tower and not meeting any obstacle on the way arrive at the ground with sufficient velocity to prove horrifically fatal. And he accepts that what he proposes depends on a prediction which is quite inconsistent with this so far universally instantiated hypothesis. (Let me, for brevity, carry on referring to 'Galileo's law' – though with the understanding that this is to be understood in the above weakened sense.) Finally your new friend has also studied and accepted deductive logic; and indeed he insists that only deductively valid inferences infallibly transmit truth from premises to conclusion. However, he also insists – clearly correctly – that all the actually observed evidence is evidence *about the past*. Since deductively valid inference is non-content-increasing, it cannot then follow deductively from this evidence, no matter how much of it there may be, that Galileo's law, or alleged law, holds for all past *and future* falling bodies. In particular it cannot deductively follow from the evidence that this law will continue to hold for bodies that are released from a height in 5 minutes' time, which is when he intends to start his float. So far as deductive logic is concerned (the only sort that *guarantees* transmission of truth from premises to conclusion), all the evidence about what happened to bodies in the past is consistent with *any* assumption about what will happen to falling bodies in the future: it is consistent, for instance, *both* with the assumption that future freely falling bodies will, just like their predecessors, fall with roughly constant acceleration *and* with the assumption that they will quickly achieve and then maintain a gentle 'terminal' velocity' of, say, 1″ per second. This means, your new acquaintance points out, that your belief that he will come to a grisly end if he carries out his plan is no more entailed by the existing evidence than is his assumption that all will be well. Moreover, he points out, your own assumption that the safe plan is to take the lift is no better grounded than his own.[5] Your plan takes for granted various general claims, for example, about the tensile strength of steel cables of certain constitutions. These claims have admittedly been well-tested and have passed these tests – but all of them were of course performed in the past. So all *this* evidence is perfectly consistent, so far as deductive logic is concerned, with those steel cables in just 5 minutes' time suddenly exhibiting the tensile strength of, say, a normal cotton thread (not because of any 'fault' in this particular cable, but 'because' that's the way steel cables in general are going to be). If this were to happen then you'd better hope that he's right about the future of falling bodies. He makes no claim, therefore, that his plan is any *more* rational than yours, but only that his plan and your own *stand on a par* so far as reason is concerned. Indeed he had earlier flipped a penny intending to 'float' or take the lift depending on whether heads or tails came up; and heads had won. Wishing you luck in your venture, he prepares to climb over the balustrade.

What would you say to your new acquaintance? Well, one thing you *might* say is "Yes; on reflection you're right. There *is* no reason to prefer my proposal to yours. If we have good reasons for accepting anything at all, it is only for accepting the results of observation and the principles of deductive logic: anything else would call for a justification, and the justification for a justification and so on; so that any other claim that we relied on would ultimately have to count as ungrounded assumption. And your argument is precisely that nothing follows from established observation statements by deductive logic about what will happen to falling bodies in 5 minutes' time. My gut reaction is still that you will certainly come to a grisly end, but since I am a man of reason, I accept that there is no good reason to prefer my proposal to yours: go ahead you *may* be right."

You *might* say this, but you wouldn't. As Russell said, it might be impossible to fault the inductive sceptic's logic but it is downright impossible to take his claims really seriously.[6] Someone who was suffering from a bad case of over-philosophising might convince himself that he would let the chap on the Eiffel Tower go ahead, but put in the actual situation, then like me and you he would say that the chap is frankly 'bananas' and, if he persisted in trying to carry out his plan, would restrain him while sending someone else off to summon psychiatric assistance.

So that's precisely what you start to do. But then you have a better idea: since our floating friend is clearly a believer in argument and certainly *seems* otherwise sane enough, perhaps what the chap needs is not a psychiatrist but a *philosopher*. You vaguely recollect that the difficulties he is raising have to do with something called the problem of induction and you remember that there's a famous philosopher who actually claims to have *solved* this problem. So you put in an emergency call to Sir Karl himself, who agrees to come, accompanied by David Miller, who has interesting views on how to handle any criticism of Popper's solution that might come up.

Our friend, very keen to cross intellectual swords with the great man, agrees to postpone his float for a few hours. While waiting, however, you begin to get cold feet, knowing that some philosophers sometimes hold pretty dotty views and even sometimes have strange views on what counts as a 'solution' of a problem. Luckily someone has left some philosophical books lying around up there and one of them happens to be *Objective Knowledge*. So you check; and yes there it is: Popper definitely claims to solve the problem in the positive sense – he *does* positively recommend the lift as opposed to the air as the rationally preferable action if your aim is to get to the ground safely. Popper gives the following characterisation of the pragmatic problem:

Which theory should we prefer for practical action, from a rational point of view.[7]

And he gives the following unequivocal answer:

My answer...is...we should prefer as a basis for action the best-tested theory.[8]

Not, you are pleased to notice, '*will* as a matter of fact prefer' but *should* prefer "from a rational point of view".

Given this, and Sir Karl's reputation for being hard-headed about claims to *know* anything, you feel confident that you have called on exactly the right person – he will surely convince your new friend that, while his arguments might be cogent, his proposal to float down is irrational.

So Sir Karl and David Miller presumably choose the airplane based on the best corroborated theories of aerodynamics and choose to take the lift up the Eiffel Tower rather than, say, try to levitate: and, one way and another, arrive safely. However your new friend has spent the intervening time reading *Conjecture and Refutations, Objective Knowledge* and some of the other books that were lying around up here and himself starts off the debate.

FLOATER: I found what you had to say Sir Karl about the scientific method all very congenial; that we can never have positive reason to think a scientific theory true, that the best we can have is positive reason to think such a theory false. This seemed to be very much in line with my own view that, no matter how many freely falling bodies we have in the past observed to accelerate roughly uniformly, we cannot know that the next one will accelerate roughly uniformly, nor does all that positive evidence even make it more probable than not that the next falling body will roughly uniformly accelerate. You stress more than once that your methodology delivers only corroboration appraisals, that is, accounts of how well various theories and generalisations have stood up to empirical testing *so far*. You stress more than once that these corroboration appraisals have no synthetic forward-looking element. You stress more than once that the attempt to use the fact that a particular theory is best corroborated so far as an argument for believing it to be true or probably true, and hence for believing its predictions to be true or probably true is an inductivist error. All seemed to be confirming my view that it is no 'more rational' to take the lift than to take to the air. I was just about to use your arguments to quieten this tiresome chap who keeps calling me irrational, when I turned to Chapter 1 of *Objective Knowledge* and found to my astonishment that you assert that the only rational course of action is to act (in the future) on the best corroborated theory. This advice just seemed to be slapped down against the whole spirit of all your previous, powerful arguments and certainly without the least justification.

This fellow Wes Salmon, whose paper also happens to be lying around, seems to sum up very clearly the puzzlement that I (and apparently many others) feel:

> We begin by asking how science can possibly do without induction. We are told [by Popper] that the aim of science is to arrive at the best explanatory theories we can find. When we ask how to tell whether one theory is better than another, we are told it depends on their comparative ability to stand up to severe testing...When we ask whether this mode of evaluation does not contain some inductive aspect, we are assured that...since this evaluation is made entirely in terms of past performance, it escapes inductive contamination because it lacks predictive import. When we then ask how to select theories for purposes of rational prediction [and hence as a basis for rational action], we are told that we should prefer the theory which is 'best tested'..., even though we have been explicitly assured that testing...[has] no predictive import.[9]

As I understand it, Salmon (like your other critics) is ready – at any rate for the purposes of this argument – to concede that your deductivist account of theory-testing is entirely correct. Certainly no one any more believes (if indeed anyone ever *did* believe) that scientific generalisations are arrived at or justified by simple induction by enumeration. Everyone accepts that Galileo did *not* arrive at his 'law' by simply observing a whole series of freely falling bodies, noticing that they were all uniformly accelerated and then inductively generalising. The process by which Galileo's law became accepted as part of science is altogether more subtle – undoubtedly involving conjectures, refutations of alternatives, checks to see that other factors do not affect the observational outcome, to say nothing of a certain amount of idealisation. Let's assume that the judgement that Galileo's law (or, better, the weakened version mentioned above) is scientifically superior to any rival is as pure-white deductivist, as 'corroborationist' in your sense, as you like. Still the question arises whether this acceptance has any implications for *future* action. You say eventually that it *does*: but what else is your assertion that it is rational to work on the future predictions of the so far best corroborated theory than an inductive principle, connecting past performance with (likely) future performance?

So I ended up feeling lost: your philosophy in the end does contain an inductive principle, and, just like all the other inductivists, you can give no reason for this principle. You *tell* me that it is 'irrational' to try to float down, but *why*? Surely you don't expect me just to take this on trust? I concede of course that I may be in a minority of one – but isn't it the fate of all great innovators to be called 'irrational' by their stick-in-the-mud contemporaries? I also concede, of course, that it is easy for you to produce an account of rationality which will brand me as irrational by just making the bald *assertion* that it is rational always to act on the best corroborated theory. But I could equally well create a different theory of rationality which baldly asserted that it is rational to act on the worst-corroborated theory or fourth best corroborated theory – or, surely a better idea, since it seems downright rash *always* to act on either the best or the worst corroborated, I could create a theory of rationality which asserts that it is rational to act on the best corroborated theory on Mondays, Wednesdays and Fridays and rational to act on the worst corroborated theory the rest of the week, or, most rational of all, my own policy: play 'corroborationist' or 'countercorroborationist' on the flip of a fair coin. What makes your theory of rationality, your bald assertion of the inductive link between past and future, correct and any of my theories of rationality incorrect?

SIR KARL:[10] I see that, like everyone else, you have totally misunderstood my straightforward solution of the problem. Let me hand you over to David Miller to see if he can root out the sources of your misunderstanding.

DAVID MILLER: Well, first of all, you are correct that Popper's position yields the positive advice to prefer the predictions of the best-corroborated hypothesis that we have. And this means, if you like, acting *as though* the best-corroborated theory were true. For Popper, however, this is not because we have any reasons for supposing that it is true, but because there are no reasons for supposing that it is *not* true. 'The hypothesis that has best survived the critical debate is the one that we have least reason to think false.'[11] Now to quote a paper of mine:

> This answer manifestly involves no recourse to any principle of induction. There is no suggestion that the proposal that best survives criticism is one that we have any reason to expect to be successful. In fact we will not have any reason to expect it to be successful; but this hardly means that it will not be successful.[12]

FLOATER: Now I'm really lost: if anything is manifest here it's surely precisely that this answer *does* involve recourse to a principle of induction. But let's proceed one step at a time, the first question I want to raise is this: why exactly has the critical debate so far given me least reason to think Galileo's law false? After all, I accept of course that all the so far observed falling bodies have instantiated it, but this means, since all those observations were made in the past, that they equally well instantiate this hypothesis: that all falling bodies up to 5 minutes hence will fall with roughly constant acceleration and all future bodies achieve the gentle terminal velocity I mentioned. It seems clear to me that I have just as little reason – *good* reason in yours and Popper's sense – to regard this alternative as false as I have for regarding Galileo's law as false: both equally fail to be empirically refuted so far.

DAVID MILLER: Ah, I see you have been reading that gruesome stuff by Nelson Goodman. You really ought to learn to use your time better, especially since you seem to have so little of it left. Gruesome hypotheses are banned from consideration.[13]

FLOATER: I assume that you would agree that if you need to ban such hypotheses from consideration in order to solve the problem, then you would need to cite good, *non-inductive* reasons for the ban – and I can't see any cogent Popperian reason to rule them out. But let's leave this for the moment and go along with your ban.[14] This will force me to change the details of my position slightly, but the essential point remains the same. The universal theory whose prediction I now propose to act on is simply the 'natural' generalisation that '*All* freely falling bodies accelerate up to 1″ per second and then retain that velocity'. Now I accept of course that I have plenty of reason to regard this as false – as I said right from the start I accept observation results and deductive logic, and my theory has of course been multiply refuted. And given that I'm accepting your gruesome ban, I accept that the only articulated generalisation already in this area which there is no reason so far to think false is Galileo's.

Still, as you and Popper have often emphasized, *all* the universal theories that we have so far articulated – perhaps even all the universal theories that we *can* articulate in our language – may well turn out to be false. Certainly as far as observation and deductive logic go we have no reason to think that Galileo's law won't be falsified in exactly the *next* instance. So why exactly doesn't the advice you give – namely to act as though it *won't* be the case that Galileo's law will be falsified in the next instance, given that it has failed to be falsified in all the cases so far – why exactly doesn't that advice 'involve recourse to a principle of induction'?

Let's, in order to make things simple, assume that the total duration of the universe is finite – that the theory that everything, including time itself, started with the big bang and will end with a big crunch – is correct. In that case, there will only ever be finitely many bodies released above but close to the earth's surface and the generalisations we are considering will all reduce to enormous finite conjunctions. Let's suppose, just to pluck a few figures out of the air, that there have so far been 10^{25} falling bodies and that there will be 10^{250} further falling bodies before the big crunch. And suppose finally that my initial gruesome hypothesis is in fact true – as I'm sure you will concede is logically possible despite your ban. This means that the universal theory I am now proposing is overwhelmingly *less false* overall (in a perfectly definite sense now that we are dealing with only finitely many instances) than Galileo's theory which admittedly looks less false *so far*. Surely then, in advocating the action based on the prediction of Galileo's theory as rational, you are taking it that it is rational in circumstances like these to suppose that the sample we have so far is representative of the whole population – a classic inductive assumption. Of course everyone agrees with me and you that the prediction *may*, logically speaking, be wrong – but everyone apart from me seems to think it is rational to act on that prediction nonetheless. Your way of describing the present evidential situation in this negative way, of least reason to think certain theories false, may be preferable to the descriptions adopted by other philosophers, but when it comes to linking the past evidential situation to rational prediction, you all seem to be in the same boat. In taking it that it is reasonable to rely on the predictions of a theory we presently have least reason to think false, you are invoking an inductive principle, no less than if you had taken it that it is reasonable to rely on the predictions of the theory we have presently most reason to think true.

DAVID MILLER: What you say is 'really quite silly'[15] and the possibility that you point to of a change in the course of nature really quite 'boring'[16] – I am beginning to think that you don't exist and are nothing more than the product of a philosophically-overheated imagination. After all, Galileo's law is an accepted part of science, indeed part of background knowledge, and it gives conclusive reason to regard your prediction as false. No one can really hold the view you claim to. If you have a *serious* criticism of Galileo's law please state it. If not then *mere* 'doubt is not criticism'.[17]

FLOATER: I have a whole list of things to say about that. One is that you would surely find it difficult to say what a "change in the course of nature" *means* if you insist on language-independence, as you do in your interesting work on verisimilitude.[18] But my response to the main point is this.

The question, of course, is not whether there are many (or even any) people like me. The question is about *why* I am in such a minority. The question is *not* about whether the possibility I point to is 'silly', it is about what *grounds* you have for ruling it out. The question is *not* whether science *does* as a matter of fact accept and 'project' into the future certain theories that my opponents seem to think are the intuitively right ones to project. Of course it does this – that's why I'm such a revolutionary figure in refusing to go along with what is, so far as I can tell, a universally adopted but entirely groundless practice. Of course, if you are allowed to appeal to the fact that science just does contain many generalisations on whose predictions everyone else relies, and if you are allowed to count *that* as a solution of the problem then the problem was "solved" long before Sir Karl by David Hume himself. Hume never for a moment thought that he was bringing induction into serious practical doubt or that the logical possibilities he pointed to were other than 'silly' from the practical point of view – we all do of course use induction. But Hume – surely more accurately – saw this as *constituting* the problem, *not* as the solution of it.

You say that I am not raising a serious criticism of Galileo's law, but only a 'boring' doubt. But what if science had accepted the gruesome hypothesis that I think true? In that case someone who suggested that the accepted theory *might* be false because the next falling body might fall with roughly constant acceleration would be raising a 'mere doubt' not a serious criticism. You might say that this itself is not a serious possibility. But *why* not? Surely the answer is because science is just unthinkingly inductive in the usual sense. But then in assuming science or any part of it, such as so-called background knowledge, you are taking induction on board. You can't presuppose the very point at issue and then claim that you have solved the problem! The history of philosophy is littered with the corpses of just such attempted solutions.

DAVID MILLER: O.K., let me try a different argument. Your claim that Popper's position and that of the self-confessed 'inductivists' are in the same boat here is quite wrong. The inductivists *do* invoke a principle of induction because they take the past evidential record as justifying the prediction of the so far best confirmed theory – that is, as lending that prediction inductive certainty or, at any rate, greater rational credibility than any conflicting prediction. Thus an inductivist would tell you that you are entitled to a greater degree of rational confidence in your eventual safety, if you decide to take the lift, than if you take to the air. Popper's position, on the contrary, involves no inductive principle because he does *not* assert that you will be entitled to any more rational confidence in your eventual safety if you abandon your present plan and you decide to take the lift. According to Popper, if you are rational then you will not *assume* that you can have greater confidence in the predictions of the so far best corroborated theory; it's just that if you are rational you will always act on those predictions and therefore will always act *as though* you were entitled to such a greater degree of confidence.[19]

FLOATER: Let me get this straight – if you're saying what I think you're saying, then in the immortal words of Frankie Howerd 'never has my flabber *been* so gasted'. I'm no philosopher of course, but are you seriously claiming that you can issue the advice always to act *as if* you assumed that the future will (in these very specific respects) be like the past, that you can state that this is not a simple *description* of what everyone apart from me will in fact do but instead that it is the rational way to act, that you can issue this allegedly rational advice to act *as if* a certain assumption were true and yet still not invoke that assumption and therefore avoid any obligation to defend it if challenged?

Let me chance an analogy – a rather tasteless one, I'll admit, but one that seems to make the point, and after all my life is at stake here. Consider a white South African whose every action is *as if* he held that blacks are innately inferior, he prevents them from having any political power, he bars them from every place that whites like to frequent, and so on. However he asserts that the *assumption* that blacks are innately inferior is no part of his theory of the world and indeed, if asked, explicitly allows that he has no good reason to think so. Does he on this account escape the charge of being racist?

Why, returning to the induction problem, should it be rational to act *as though* the best corroborated theory were true, if it is not rational to have greater faith in the best corroborated theory's prediction? What you seem to be saying is that the difference between the 'inductivist' and the Popperian man of rational action is that, while both take the lift, the latter is as scared as hell while doing so!

SIR KARL: Let me step in here. Despite your claim to know the history of science, you must in fact be quite ignorant of it. No one who knew the history of that great intellectual adventure could possibly think that it is rational to have complete faith in the prediction of even the best best-corroborated theories – even the best-corroborated theories are eventually empirically refuted. Moreover, as I have shown, the whole enterprise of producing a probabilistic inductive logic which would justify rational *degrees* of confidence has run into sand.

INTERESTED BYSTANDER: I couldn't help overhearing your interesting debate. I wonder if I could say a few words at this point on behalf of the so-far silent majority. It seems to me, Sir Karl, that you are not quite appreciating how revolutionary a position our friend the floater adopts.

Everyone nowadays is, I take it, a fallibilist about scientific *theories*: by this I mean not a fallibilist *in principle* (this position seems to be dictated by logic alone) but a fallibilist *in practice* – the history of science clearly shows that even the most successful high level theories may eventually be rejected (even if they do standardly 'live on' as 'limiting cases').[20] Certainly a well-corroborated theory may make empirical predictions which are *false*. Suppose some 19th century man of science was planning an action where success, rather improbably, depended on the details of Mercury's observed orbit over the succeeding six months. No one, I take it, would say that his only rational course of action was to depend on the predictions of the best corroborated high-level theory available to him. The best corroborated theory was Newton's and, as is well known, it got Mercury's orbit quite wrong. Even where the best corroborated theory in a certain field has *no* obvious 'anomalies' we shall surely not rely on its untested predictions if these are of a *new kind* of event. For example, Fresnel's wave theory was in 1830 undoubtedly the best corroborated theory of light. In that year Hamilton showed that the theory made the startling prediction that a beam of light directed along a certain line into a crystal of a certain type would open up inside the crystal into a hollow cone of rays. In 1833 Humphrey Lloyd set up the experiment to test this prediction. Suppose (implausibly, of course, since the experiment is fairly delicate) that you were guaranteed that Lloyd's experiment would detect the effect if there was an effect to detect. Would the rational man *rely* on Fresnel's prediction being correct? Would he to make things precise, agree to be strapped into a device which fired a bullet into his brain if the light beam in Lloyd's experiment failed to open out into a cone? Well, I like to think of myself as fairly 'rational' and I can tell you that there is no way you would have got me into that contraption under those conditions. I don't myself see how we can do without some notion of rational degrees of belief in cases like this – if he *had* to bet, surely the rational man would back Fresnel, given his enormous success with other types of optical effect, or at any rate, he would be considerably readier to back Fresnel's prediction than, say, the prediction of some religious figure who predicted the end of the world in 1833. But there is surely no question here of anything even approximating absolute rational reliance.

This marks a clear difference between this type of case, exemplified by conical reflection, and the type of case exemplified by the weakened version of Galileo's law that you have been discussing. I just report it as a fact about me (and I conjecture about most people) that while there is no chance of my agreeing to be strapped into the bullet-firing contrivance in the Lloyd case, if someone asked me to be strapped into it on condition that it was absolutely guaranteed to fire *only if* our friend the floater carries out his plan and gets to the ground successfully (without tricks) then, at the expense of a donation to my favourite charity to cover my inconvenience, I'd accept.

So, while we may or may not have rational degrees of belief in them, it seems to me that we do *not rely* on a high-level theory's predictions, no matter how well corroborated that theory might be. What we *do* however rely on are the further predictions of already well tested *low level* empirical generalisations. (Of course the hitherto unsuspected observational generalisations entailed by high-level theories will standardly very quickly *be* empirically – repeatedly and independently – checked. If these checks are positive, we shall be in the position of relying at the same time on *both* the predictions of the best corroborated high level theory *and* the predictions of an exceptionless low-level observational generalisation, but it's the latter that really counts.) Many recent philosophers (some of them influenced by you, Sir Karl) have held that *all* observation statements, and therefore all observational generalisations, are 'theory laden' in a sense which makes them not only open to correction *in principle* but which also makes their corrigibility a serious *practical* concern. They seem to me to make their point, however, by treating *very* 'high level' statements as observational. (One of them, I believe, even treats 'The Brownian particle is a *perpetual* motion machine of the second kind' as directly empirical![21]) But if you go to a low enough level (to the level of what Poincaré called 'crude facts' and Duhem 'practical facts' – meter readings, angles of inclination of telescopes, digital printouts and the like) then there is, I claim, no case in the whole history of science where a once accepted, well and independently checked, low level observational generalisation turned out subsequently to be falsified.[22]

This is why our friend the floater here is such a revolutionary! He is taking his stand on the possible future falsity of an overwhelmingly confirmed empirical generalisation at the practical factual level, in precisely the area where it has been confirmed again and again.

Concerning predictions made by a theory of a type not already empirically well confirmed (even though the theory overall may be far and away the best confirmed we have), due regard to Popperian fallibilism is entirely in order – because indicated by the history of science itself. But it is precisely in the case of already well-confirmed low-level practical empirical generalisations on which we *do* rely and must rely for survival, precisely in those cases where the history of science supports an 'optimistic' rather than a 'pessimistic' meta-induction, precisely in those cases where 'rational to rely on' clearly means 'the right thing to rely on for survival purposes', that we all do make inductions and where it seems clearly irrational *not* to do so.

FLOATER: I agree that I am a true radical. I am not merely claiming that predictions about new kinds of phenomena made by the so-far best scientific theory may be false – a possibility clearly instantiated in the history of science. I am claiming that, even if there is not a single case in history of a well-tested *low-level empirical generalisation* which suddenly turned out to give an incorrect prediction, still we have no *reason* to act as though the predictions of any such generalisation will turn out correct in the next instance. Certainly I am still waiting to hear a Popperian reason for always acting in this way, and so I am still waiting to hear a Popperian reason why my proposal to float down is irrational.

SIR KARL: Your chief problem appears to be that you are still caught up in what I call the 'justificationist framework'. If you keep asking for *reasons*, keep on asking 'why', then – as the Greeks already saw – you end up in an infinite regress and hence in scepticism. You must accept certain assumptions when neither you nor any one else has produced any *good* reason to criticise them. The Greeks thought that the only way out of scepticism is through dogmatism – through the simple assertion of certain claims which are regarded as self-evident and therefore not in need of any further justification. I, however, have shown how to 'transcend' the dogmatist-sceptic controversy. We must, as I have indicated, accept certain substantive assumptions without justification; but this is not dogmatism because, should anyone subsequently raise serious criticisms of these assumptions, then we could then challenge them, and perhaps replace them – though this would of course involve taking for granted, *again temporarily*, other assumptions.[23]

FLOATER: But, as I indicated before, your notion of what counts as a 'good reason' or a 'serious criticism' is carrying all the weight here and indeed simply prejudges the debate against me. After all, there is, of course, within the framework I adopt, no serious criticism of my counter-inductive claim – only 'mere doubt'. I need hardly add that if you argued that there *is* a serious criticism of my counter-inductive policy even within my framework – namely that anyone working on this policy in the past would not have survived – then I will reply that this criticism presupposes induction. It seems to me that for all your arguments there are just the two, selfsame options that we started with: you *either* allow that my framework and yours are equally respectable and that there is no rational way of adjudicating between them (and that's just scepticism) *or* you *assert* that induction (at any rate of this very limited kind) is rational (and that's just dogmatism).

Isn't it better just to come clean and admit that you simply *assert* as part of your theory of rationality that it is rational to take the best corroborated generalisations (at any rate of this low level kind) as your guide to future action? Why not admit that this is an inductive principle and you simply assert it? You can then call this a 'solution' of the problem of induction if you like, but solutions by fiat seem perilously close to admissions that the problem is insoluble.

DAVID MILLER: But how can someone call himself a rationalist if he simply dogmatically asserts some substantive principles of rationality? The whole Popperian position was aimed at taking nothing for granted beyond *possibly* the results of low-level observations and the principles of deductive logic.

FLOATER: Yes, but the basic problem for the so-called rationalist was that if that's all you have in your rationality theory, if you settle for really hard-nosed deductive rationality then you must end up in inductive scepticism: precisely my own position. It must indeed be embarrassing for a rationalist to admit that his theory of rationality is based on substantive assumptions which, since they partially constitute rationality, cannot themselves be accredited as rational – to admit that we must, as I believe Popper himself once acknowledged,[24] make the irrational (or, better, non-rational) decision to be rational. What I am suggesting is that deductive logic itself with its great first principle that, as they say where I come from, 'you don't get owt for nowt', demands that if you *are* going to bridge the acknowledged deductive gap between past and future and thus characterise my proposed float as 'irrational', then you must make some substantive inductive principle part of your theory of rationality.[25]

DAVID MILLER (producing a gun):[26] You're right, I just have to assert an inductive principle dogmatically – now get in that lift!

FLOATER: Hang on – let me toss my coin again. Ah: this time it has landed 'tails', so I'd better go along with the prediction of the so far best-corroborated theory about what would happen to me if you actually fired that thing. I'm still doubtful about that lift, but at least if we *do* reach the ground safely, I'll be able to explain why there are no reasons, apart from inductive ones, for banning gruesome hypotheses.

2. MILLER AND POPPER ON GRUESOME HYPOTHESES

FLOATER: Well, we did get down safely. Of course, that has no consequence at all about what it will be rational to do when I ascend the tower again tomorrow. But before you go, let me explain to you my worries about your treatment of what you called 'gruesome hypotheses'.

You will remember that the alternative hypothesis to Galileo's 'law' that I originally asked you to consider was that all freely falling bodies are, let's say, uniterm – this is a new technical term: in your old-fashioned language it means "either falling before time t_0 and roughly uniformly accelerated or falling after t_0 and quickly achieving the terminal velocity of 1″ per second'. (Remember, though, that I don't actually assert that this alternative is the one to act on, but only that it has equal claims with your own favourite. It is you who wants to stick out his neck, I take a nice (pre-Thatcherite) English middle of the road position, deciding on the toss of a coin.) Now clearly you need to ban this alternative hypothesis if your general advice to prefer that theory we have least reason to think false is ever to deliver any practical advice at all. For by 'reason to think false' here, you intend, I think, what you call a '*good* reason': that, after all, was the whole appeal of the falsificationist position – that it needed no doubtful inductive principle, but only deductive logic and basic statements. But clearly we have no more 'good reason' to think this alternative false than we have for thinking that Galileo's 'law' is false. Both are equally unrefuted so far.

DAVID MILLER: Exactly – that is why it is a general principle that 'conflicting hypotheses must not be admitted to empirical science unless there are crucial tests that can eliminate at least one of them...If we are not prepared to delay until [after time t_0] a decision between [the two]...we must not admit them both into science...'.[27]

FLOATER: Well, since my original hypothesis took t_0 to be about half an hour ago, and since I just noticed that someone droppped a tennis ball from the top of the Tower and it fell with roughly uniform acceleration, if you *had* been prepared to delay a decision and had not forced me into the lift then I should have been dead by now. By the way, I of course rejoice in the refutation of this erstwhile theory of mine; and am very happy to have let the theory die in my stead. I need hardly add however that I now have a new suggested alternative theory to Galileo's 'law', which differs from my old one only in reidentifying t_0 as noon tomorrow. But let's go back to the old situation: clearly one of the two generalisations needed to be excluded if you were to give me advice on whether or not to carry out my float. But *which one?*

DAVID MILLER: 'The answer is all too obvious.'[28]

FLOATER: Well it *is* obvious which one you would *like* to be barred so that your principle to prefer the theory we have least reason to think false delivers the advice which you and everyone else apart from me seems to think is the only rational advice. But what *reason* can a Popperian give for barring the gruesome hypothesis in advance of any further evidence?

DAVID MILLER: Gruesome hypotheses like 'all freely falling bodies fall unitermly' are less falsifiable than 'natural' hypotheses like 'all freely falling bodies fall with constant (or roughly constant) acceleration'. The reason is that in order to falsify the second we need only an accepted basic statement which says that a body fell freely but not with even roughly uniform acceleration; while in order to falsify the first we would need *both* a description of a body's fall *and* a statement of the time at which that fall occurred. It is of course a basic (and *non*-inductive) principle of Popper's philosophy that more falsifiable theories are to be preferred to less falsifiable ones – unless, that is, the more falsifiable one is actually falsified. Until the more falsifiable Galileo generalisation gets refuted, its less falsifiable competitor should not even be considered.[29]

FLOATER: Goodman already showed, I believe, that any such suggestion is inevitably language-dependent. If we take his own example, if 'grue' and 'bleen' rather than 'green' and 'blue' are regarded as primitive predicates, then all we need in order to refute 'all emeralds are grue' is a non-grue-emerald, whereas in order to refute the weird hypothesis 'all emeralds are green' (equivalently: all emeralds are grue and observed before t_0) or bleen and observed after t_0, then we need both a statement of the emerald's 'colour' and a time at which it is observed.

But aside from this, it would anyway surely be right to spell out the quantificational structure of Galileo's law as "for any body *b* and for any time *t* if *b* is released above but close to the earth's surface at *t* then *b* falls to the earth with (roughly) uniform acceleration'. In order to falsify the law as thus formalised, we should need, of course, a description of a body *and* a time interval during which that body was allowed to fall freely, but in which it did not fall with constant (or roughly constant) acceleration.

Finally, and even ignoring these points, it does seem strange to appeal to the fact that we could forget information that we in fact can freely have (the motion if observed must have been observed at some time) as a justification for preferring one theory over another.

DAVID MILLER: Here is a different Popperian reason for banning the gruesome hypothesis rather than Galileo's. You see the problem only arises because so many philosophers are justificationists and *will* insist on talking in terms of 'empirical support' for theories. Since the gruesome hypothesis has been constructed precisely so that, on justificationist criteria, it will be supported by just as many facts as is Galileo's the justificationist is bound to think it mysterious that the gruesome hypothesis is not considered as standing on a par with Galileo's. But for the Popperian this problem simply does not arise. The aberrant hypothesis needs to be considered by a Popperian only if it promises to *solve some problem* not already solved by Galileo's law; and this it clearly fails to do.[30]

FLOATER: There are of course accounts of 'empirical support' which do not in fact yield Galileo's law and its gruesome rival as equally well supported.[31] However, my main reply to your argument is simply that it does no more than push the problem one stage back. As I already acknowledged several times, there is, of course, no doubt that science is against me: science does, *as a matter of fact*, sanction what everyone else seems to believe are the intuitively right predictions. As everyone knows, Hume himself accepted, indeed emphasised, this; but his question and therefore the question all along was not *whether* science sanctions the "intuitively right" actions as rational, but *why, on what grounds*, it does so.

You see I seem to lack the intuitions which everyone else shares; so you are not going to convince me by pointing out that scientists all have these intuitions – I knew that already. Galileo's law is no doubt an accepted part of science, perhaps even of "background knowledge". But we must now ask why Galileo's law *came* to occupy this privileged position on the basis of all this evidence about falling bodies, when exactly the same evidence also follows from the alternative hypothesis that I suggested. I don't suppose you would be happy just to regard this as an historical accident. After all, had my alternative been the one to be accepted initially, then your rule would have banned consideration of Galileo's law on the grounds that it solved no problems not already solved by an accepted theory. (Each theory basically solves the same set of 'problems' by describing with equivalent precision the motion of *every* freely falling body close to the earth's surface.) The problem is *not* do I have reason to take the lift rather than the air *if I accept science*, since obviously science predicts safety in the first case and disaster in the second. The problem is precisely *why* science accepts certain universal generalisations which therefore make predictions about the future when it can of necessity have only finite evidence from the past for these universal claims. Scientists do clearly intuitively prefer Galileo's law to any gruesome alternative, but what makes their intuitions better than my own? The only answer seems to be that it is the Galileo-style generalisations rather than the gruesome ones that have proved successful in the past. I don't accept this answer of course – precisely because it is inductive – but at least it's an answer: from a Popperian point of view the ban on gruesome hypotheses remains entirely arbitrary.

3. WATKINS' FAILURE TO SOLVE THE PROBLEM OF INDUCTION

JOHN WATKINS: Excuse me for butting in but I heard about your plight and immediately flew over to see if I could help. You see although I agree with you that Popper has completely failed to solve the problem, I have a 'neo-Popperian' solution which really does the trick.[32] Since you are clearly responsive to argument I think I may well be able to convince you that your new proposal to float tomorrow is irrational.

First of all let me say where I agree with Popper. My neo-Popperian, just like Popper himself, 'seeks to exclude all non-deductive inferences and insists that corroboration appraisals of hypotheses have no implications for their future performance'.[33] However, I agree with you (and of course with Hume) that the problem precisely is that deductive logic and observation results so far have no implications for future action. I also agree with you that if a philosophical system *is* going to advise a technologist to act on the prediction of the best corroborated theory, then it had better yield some *reason* for doing so.[34] Let's not pussyfoot around – to give such a reason we are going to have to incorporate certain extra *assumptions* into our theory of rational action, assumptions which, by the very nature of the problem, cannot be sanctioned either by observation or by deductive logic. In fact I propose two such assumptions which together solve the pragmatic problem of induction.

FLOATER: I am of course eager to hear your assumptions. But let me first make a general point about your whole enterprise here, since it may well turn out to clarify our later discussions. Of course it would be easy to develop a theory of 'rationality' which underwrote your and everyone else's insistence that my proposed float is irrational. For one thing you could simply *assert* that it is rational to act on the prediction of the best-corroborated theory. This 'solves' the problem 'at a stroke'. I am sure, however, that you would not expect me to be convinced by this – since it 'solves' the problem of induction simply by 'presupposing' that induction is rational, and hence simply presupposing the point at issue between us. Another, often tried way to 'solve' the problem is less obviously *ad hoc*, but equally unsatisfactory in the end. This involves invoking some assumption of the *uniformity of nature*. Now I readily concede that, suitably formulated, such a principle can sanction our crossing the deductive gap between past performance of a theory and its future predictions. But, as I understand has often been pointed out, if we ask what grounds we have for making the assumption that nature is 'uniform', then the answer can only be *inductive* – nature has seemed to be 'uniform' (in certain special respects) in the past, it seems reasonable to suppose that it *is (always)* uniform. The principle of uniformity of nature 'stands or falls' with induction itself, and cannot therefore provide a satisfactory rationale for induction. This criticism is, I believe, entirely standard, surely correct and I assume you accept it.

So, in claiming to provide a positive solution of the problem of induction, you must presumably believe that you can produce some principle or principles which (a) seem somehow plausible as general principles of rationality, (b) together sanction the step from 'best according to tests so far' to 'prefer its predictions', and (c) seem somehow *independent* of the straight inductive assumption, seem somehow *not* to 'stand or fall' with induction itself.

But even before you give me your specific proposals, it seems clear to me that you are skating on pretty thin ice. Conditions (b) and (c) give all the appearances of being contradictory. Whatever general assumptions about rationality you advocate, they cannot fulfill conditions (b) and still be *logically* independent of the inductive assumption. The only clear sense of 'presupposes' that I know of is 'entails' (that is, *A* presupposes *B* just in case *A* deductively entails *B*): so it is difficult for me to see how a set of general rationality assumptions can even be a candidate for solving the pragmatic problem *without* 'presupposing' (that is, entailing) that the best-corroborated hypothesis will be best corroborated in the future (or whatever your preferred inductive link is).

Another tack someone who proposed a positive solution of the problem of induction might take is to claim that he can produce principles of rationality which don't just entail the 'inductive link', but actually *explain* it. But I have problems with the whole idea of *philosophical* explanation. It's true, of course, that in the case of (deductive) explanations in science, it wouldn't be usual to complain that the explanans simply presupposes the explanandum, although it certainly does: if Newton's theory explains the planetary motions, then it must *entail* those motions, and hence must 'presuppose' them in the perfectly clear sense that Newton's theory itself cannot be true unless those motions occur. This, I suggest, does not worry us in science, because we only regard a general theory as an explanation of some fact or regularity if there is lots of *independent* evidence for it: we wouldn't regard Newton's theory as an explanation of the planetary motions if it weren't the case that it entails lots of other testable (and correct) claims besides. The notion of a 'philosophical explanation' on the other hand is altogether more nebulous:

certainly it hardly seems realistic to expect an explanation of induction to be independently testable in any very definite sense. That's why in the end I think you're just going to have to reach for your gun, like David Miller – that is, you're going to have to *assert* that I am irrational, and that's an end to it.

JOHN WATKINS: No, you have entirely misunderstood my whole approach to the problem. If my aim were to present rationality assumptions which entailed that the universe is and will be such as to make the success rate of predictions made by well-corroborated theories higher than that of predictions made by not so well-corrorobated theories then I would, of course, agree with you that I must implicitly presuppose an inductive principle. But all that I seek is some slight reason for preferring what I call corroborationism over any version of countercorroborationism – assuming that in order to survive we need to act on predictions using one or the other policy. I am merely trying to show that there is a reason which just tips the balance in favour of corroborationism; I am definitely *not* trying to develop a position which entails the truth (or likely truth) of the assumption about the world which underlies corroborationism. If I can give you a *reason*, no matter how marginal, for preferring corroborationism, and if no one has given any reason, no matter how marginal, for the opposite preference, then that's surely good enough. And if this reason is *not* that the world is such (or likely to be such) as to make corroborationism true, then I can surely advocate corroborationism *without* invoking any inductive assumption.

FLOATER: Oh dear I think I'm about to start to sound like a philosopher – but it does seem to me that a lot here depends on what counts as a 'reason'. If this just means something which seems intuitively relevant, is presented as an argument and makes no obvious appeal to emotion or whatever, and which perhaps as a matter of fact would persuade some people, then nothing is easier of course than to produce a reason which 'tips the balance' in favour of inductivism (or corroborationism). One such 'reason' is that following the inductivist (corroborationist) policy has always (or, at any rate, much more often than not) paid off in the past. This is indeed surely the 'reason' that most people would, if pressed, cite for their inductivist tendencies. But, it's clear that if it's considered good enough to cite *such* 'reasons' then Hume's problem need never have arisen in the first place. Suppose to make the problem sharp, I cited the previous *lack* of success of counterinductivism as a reason which tips the balance in *its* favour (after all, 'every dog has its day' and all that, and clearly inductivism really has pushed its luck so far). If you simply refused to countenance this as a reason then you are really no better off than David Miller, who in the end simply dismissed my counter-corroborationist conjecture as not serious. If you don't just dismiss it out of hand, then I can see no way out of simply asserting that while your balance tipping reason (whatever it is) really does make it more likely that inductivism (corroborationism) is true in the future, my balance tipping reason does not. But then we are back to making what you yourself admitted is an inductive assumption.

JOHN WATKINS: I propose that we postpone further consideration of this general point until after you have heard my specific proposal. You see your general point just can't be right: because my solution of the pragmatic problem of induction (as you'll see if you ever let me get it out), although it certainly uses the past success of the policy of following the best-corroborated theory, uses that past success in an entirely *non*-inductive way.

I see that this claim has stunned you into silence so let me give you my solution. It relies, as I admitted, on some assumptions – in fact just two of them. The *first* is that 'well corroborated hypotheses *have hitherto* proved better guides for practical decision makers than have hypotheses that were not well corroborated'.[35]

FLOATER: Well I am of course more than happy to grant you *that* assumption – indeed it just seems to me an obvious *fact* about the history of science and technology (very broadly construed). I would even say, as I indicated before, that if we restrict the issue, as I believe at any rate in the first place we should, to low-level empirical generalisations, then following the best-corroborated hypothesis has as a matter of fact been an infallible sure-fire method. Certainly if your second 'assumption' is as innocuous as this one, then my floating days are over.

JOHN WATKINS: Well my second assumption is simply this:

> it is rational, other things being equal, to choose that course [of action] whose success presupposes *the weaker forecast about the future*. For a weaker forecast is less likely, other thing being equal, to turn out wrong than a stronger one.[36]

So, for example, if you were on a family holiday and choosing between two ways to spend the day, and if plan *A* and plan *B* would each deliver the same number of utiles if successful and the same number of disutiles if unsuccessful, and if finally plan *A*'s success depended only on the weather not being windy, while plan *B*'s success required that it be *neither* windy *nor* rainy, then it is surely rational to choose plan *A*.

FLOATER: This certainly sounds like a very plausible (though also, I can't resist pointing out, very non-Popperian) general principle of rationality: in practical concerns be as timid, as non-bold, as you can possibly be. I would be very happy to accept it if restricted to cases of the sort you mention. After all, because of the deductive entailment relation between the two forecasts (neither windy nor rainy *implies* not windy), whenever plan *B* would have been successful so also would plan *A*, though not necessarily *vice versa* (and indeed in this case given what we know about the weather not in fact *vice versa*). However, I'm not at all clear how your second assumption is going to apply to the sort of cases over which we are in dispute. After all, the two forecasts which we make about future falling bodies, and in particular me, are logically inconsistent: I say that I shall quickly achieve a terminal velocity of 1″ per second if I jump, you say that I shall accelerate roughly uniformly. So, far from one entailing the other, *each* entails the other's negation. In what sense is your forecast about the future *weaker* than mine?

JOHN WATKINS: Well, since I want to help not just you but any others who suffer from these unfortunate irrational tendencies, let's consider the issue of which *general policy* to follow – corroborationism, which involves always preferring the prediction of the best-corroborated hypothesis, and some version of counter-corroborationism. The claim about the world which underlies corroborationism is that the overall superior success rate for well-corroborated theories is a 'constant feature' of 'the whole history (past and future) of mankind'.[37] As for the claims underlying countercorroborationism, it seems to me best to differentiate two versions. According to one version, the straight 'contrary' of corroborationism, the overall (past and future) success rate of actions guided by less well corroborated hypotheses is higher than the success rate of actions guided by the best corroborated hypotheses. The second, 'grueish' version of counter-corroborationism just says that the *future* success rate will be higher for actions guided by less well-corroborated hypotheses.

FLOATER: I'm not happy with your switch to talk of *general* policies here. One thing is that I don't hold one myself: I never claimed that there was some positive reason to prefer my proposed float to my friend's proposal of descent by lift, but only that there was nothing to choose between the two proposals from the point of view of rationality. As I said, I actually decided to float on the toss of a coin. Another difficulty is that I'm not clear exactly what your countercorroborationists say – do they say, in each instance, *which* particular non-best corroborated theory yields the rational prediction to follow? If not, they don't seem to say anything about getting down from the Eiffel Tower tomorrow. Countercorroborationism then would entail that one can't rely on all future falling bodies falling in the 'normal way', but it would *not* entail that I won't fall in the 'normal way' tomorrow. Finally it is presumably incoherent to assume that less well corroborated theories have a high success rate in the future: once they start to have any success they start to be corroborated. Presumably we are to take the available theories at some particular time, now say, divide them into best and less well corroborated *now*, and then just consider how they perform in the future, forgetting what this future performance might do to the original corroboration appraisals.

JOHN WATKINS: Well why not go along with the switch to general policies for the moment, and see where it leads? I accept your last clarification; and, as for your other point, I certainly want countercorroborationism to stand initially (that is, ahead of any evidence) entirely on a par with corroborationism, so let me clarify by assuming that countercorroborationism does pick out a *definite* prediction in each case (this may, for example, be the prediction of the fourteenth best corroborated theory in the field – the details won't in fact matter, I think). Indeed the whole idea of my solution is that the two principles should, in advance of the evidence about what has happened so far, stand entirely on a par, but that bringing in the evidence (the evidence, namely, that corroborationism has been more successful in the past) transforms the situation in favour of corroborationism: having made claims of exactly the same strength as its rival, it *now*, in the light of the evidence, makes the *weaker* prediction about the future.

FLOATER: This sounds like a classic inductivist assumption at the meta-level: the past success of inductivism makes its prediction (that it will continue to be successful in the future) weaker, 'that is, less likely...to turn out wrong'.

JOHN WATKINS: On the contrary, my claim is that the prediction that countercorroborationism is forced to make about the future, given the evidence E of the success of its rival in the past, is stronger than the corresponding prediction of corroborationism in a quite clear and *completely non-inductive sense*.

If we concentrate initially on my first version of countercorroborationism (which just says, remember, that its overall – past and future – success rate is higher than the overall success rate of corroborationism) then it entails, given *E*, 'not merely that the success-rate of actions guided by hypotheses that are not well corroborated will in the future be higher than that of actions guided by well corroborated hypotheses, but that it will be *sufficiently higher* to offset its lower success-rate in the past. No counterpart of this is yielded by the corroborationist principle when conjoined with *E*'.[38]

FLOATER: I'm not at all clear what 'counterpart' means here, but why isn't the statement that the success rate of countercorroborationism will *not* be sufficiently high to offset its past record the required counterpart? (Or equivalently the statement that the future success rate of corroborationism won't be sufficiently bad to wipe out its present advantage.)

JOHN WATKINS: Well, I agree that the formal theory of counterparthood that I developed in *Science and Scepticism* for a restricted range of cases does not straightforwardly apply here.[39] But certainly you can't just assume that the negation of a sentence can count as its counterpart. A leading intuitive idea behind my account was that counterparts should be of roughly equal logical weight. Well, if you take a statement like Newton's principle of universal gravitation, for instance, then it is clearly altogether stronger than its bare negation.

FLOATER: We may need to come back to consideration of 'equal logical strength' later, but let's now be clear: your initial attempt to specify what it means for one prediction to be stronger than another has failed. You pointed out that countercorroborationism, together with *E*, has this consequence, let's call it *C*, about its future success rate being *sufficiently* high to offset its present disadvantage, and you said that this made the countercorroborationist's claim about the future stronger, because no counterpart of *C* follows from corroborationism together with *E*. You refuse to allow the negation of *C* as the required counterpart. But since not-*C* clearly does follow from corroborationism plus *E*, and since the negation of the negation – *C* itself – cannot count as the counterpart within courtercorroborationism for not-*C*, we are back to a situation of complete symmetry between the two principles so far as consequences and counterparts go. Each has a consequence for which there appears to be no counterpart in the other.

JOHN WATKINS: Oh, but surely it is clear intuitively that the future success rate of countercorroborationism will be, not just higher than that of its rival, but *sufficiently* higher to offset its present bad record is a much stronger assertion than its negation.

FLOATER: O.K., but then you accept that your attempt to explicate 'strength' in terms of counterparts has (so far at least) failed, and you are thrown back on what are, on your account so far, *primitive* judgments of greater and less 'logical strength'. I would like to investigate the *grounds* for these judgments (I assume that you don't think they are primitive in some absolute sense.) Since they are explicitly *not* deductive judgments, what else can they rely on but *inductive* judgments about what is more or less likely, given what has happened in the past?

JOHN WATKINS: I don't really see why you have such trouble with this idea, but perhaps an analogy will help you.[40] Suppose two runners, *A* and *B*, are to race over a distance of 10 miles. You have no information in advance of the race – certainly none about the outcome of previous races. Initially then the two claims – that *A* will win, and that *B* will – surely stand on a par: they have equal logical strengths. They then start to race and at a certain stage you look to see how things are going: in fact *A* has covered 5 miles, *B* only 4. Given this evidence, the claim that *B* will win now surely has what is, other things being equal, a *stronger* consequence about the future than the claim that *A* will win. *Not* in some inductive sense, that in lots and lots of 10 mile races in the past anyone who has been 1 mile ahead at the halfway stage has won much more often than not, but in the simple non-inductive sense that, if you like, the prediction that *B* will win is, given the evidence, incompatible with more possibilities than the prediction that *A* will. The *B*-prediction now rules out, for example, more winning margins (counting this as the distance the loser is behind when the winner finishes): given the evidence the maximum winning margin for *A* is, of course, 6 miles (*B* would need to find himself unable to take another step), while the maximum winning margin for *B* is 5 miles.

FLOATER: Well, another of the books I was reading before Sir Karl arrived concerned difficulties with the 'classical theory of probability', difficulties with how to count up 'equal possibilities'. You may begin by describing the outcomes of some repeatable experiment using one set of basic events, which it seems reasonable to say are 'equally likely'. But then some one produces another description of physically the same set-up using quite different basic events, which however again seem intuitively equally likely. Following the classical account in the two cases and assigning in each case a uniform probability distribution over the 'basic events' leads to incoherence: the very same event (under of course two different descriptions) receives different probabilities in the two cases. Essentially the same problem has repeatedly come up in the recent history of inductive logic (and the recent history of the idea of verisimilitude). Now it's true of course that you yourself are not in the business of assigning probabilities (at any rate not officially). But you clearly *do* rely on partitioning the whole outcome space into some finite number of 'possibilities' each of which we are supposed to take as counting equally: so that if one prediction rules out more of these possibilities, as thus individuated, than does a rival prediction, then the first is 'stronger' than the second. But this suggestion runs into the same old problem: namely the existence of other ways of partitioning the outcome space into 'possibilities', other ways which yield different strength judgments. Why, for example, in your race case, should I not simply count the possibilities as the following: *A* wins having been ahead after 5 miles, *A* wins having been behind after 5 miles, *B* wins having been ahead after 5 miles, *B* wins having been behind after 5 miles? Given *that* partitioning into 'possibilities' (and who's to say that they aren't equally likely?) then the prediction that *A* will win and the prediction that *B* will are both consistent with two possibilities initially and the evidence rules out *one* possibility in each case. Thus on this partitioning, the two predictions will, on your account, remain equally 'strong' after the evidence is brought in.

Indeed, it seems to me that if, as you do, you rule out inductive considerations then there is no way of distinguishing between various different ways in which the race outcome space might be partitioned (even ones in which the prediction that the front runner will win turns out to rule out *more* possibilities[41]). And this means that the strengths of the two predictions should just on your account be regarded as incomparable. I think most people's intuitions (which, as I keep saying, do seem unfortunately to be inductivist) would tell them that (a) if they have seen lots of races of this or similar type in which someone so far ahead at half way eventually wins, and few, if any, in which he loses, then the prediction that *A* wins is indeed, given *E*, 'weaker', that is, more likely (inductively) to be correct; (b) if, on the contrary, they've seen lots of races in which someone so far ahead eventually 'blows up' and loses and few, if any, in which he wins then the prediction that *A* wins would, given *E*, actually be 'stronger' – less likely to be correct; and (c) if, as in your case, all inductive considerations are excluded so that no previous race outcomes can be taken into account, then they would (or at any rate ought to) say that the evidence fails to tell one way or the other (who knows, if he never saw a race before, whether a halfway lead is an advantage or a disadvantage?). It might seem natural to translate this failure to tell either way into the assertion that the two post-evidence predictions are equally strong – but this would be the analogue of the classical probability mistake: the only safe assertion seems to be that they are of non-comparable strength.

JOHN WATKINS: I'm still not sure that you have grasped the main point of my position; but, while I'm thinking about that, let me hear what you think about my treatment of the 'grueish' variant of countercorroborationism. This says, remember, that although following the best-corroborated theory in the past was more successful, following less well corroborated theories will be more successful in the future. Again it seemed to me that, in view of corroborationism's past success, this grueish view makes a stronger forecast about the future than any made by the corroborationist. It 'implies that a time will come when success rates change over, and, moreover, that *this time has come now*'. Corroborationism, on the other hand, 'says nothing corresponding to this'.[42]

FLOATER: Well, Goodman, of course, showed that this idea of a 'change' in the course of nature is language-dependent: native grue/bleen speakers would spot a 'change' tomorrow (or whenever) if emeralds, in our terms, continued to be green (for them, emeralds would have changed from grue to bleen). But laying this point aside for the moment, my reply is essentially the same as before. The corroborationist *does* say something corresponding to this change-over prediction, namely that the time has *not* now come for any change-over.

JOHN WATKINS: But, as I told you before, my notion of counterparthood explicitly excludes automatically counting the negation of a sentence as its counterpart. Surely you must admit that the statement that the time has *now* come for the change-over is much stronger intuitively than its negation. The negation is of course a very weak statement compatible not only with a change-over occurring at *any other* time than now, but also compatible with no change-over ever occurring.

FLOATER: Well again your original explication of 'strength of prediction' won't in fact deliver the judgment you want. Given that you disallow the negation from counting as counterpart to the 'time is now' prediction, then the corroborationist also makes a prediction for which there is no counterpart in countercorroborationism: the prediction that the time has *not* now come for a 'change-over' in success-rates. So far as your counterpart argument goes the two principles remain on a par after E is brought in: each entailing a consequence for which there is no counterpart entailed by the other.

So again you are forced back on intuitive, for you primitive, judgments about the relative logical strength of two contradictory propositions. You just take it as obvious that 'the time has *not* now come over for a change-over' is 'weaker' than 'the time *has* now come for a change-over'. You will not be surprised to learn that I don't share this judgment and that I shall argue that, although most people might well accept it, that acceptance is informed precisely their inductive preconceptions.

However, in order to develop my reply in detail I need to raise a few formal niggles so as to get clearer on what exactly your 'grueish' counter-corroborationist is saying. The talk of changing success rates is unclear. The success rate of countercorroborationism relative to corroborationism can't just change in an instant – in the clear cut case where we assume corroborationism has been uniformly successful so far, then even if in the very next case of a prediction we had a 'change' in that finally the best-corroborated theory got it wrong, and some hitherto badly corroborated theory got it right, then countercorroborationism would still have a long way to go to catch up. Moreover, assuming that we are not in this clear cut case – which your talk of success *rates* seems to imply – so that we have already some variability of outcomes, then we can't even talk of the success rates *starting* to change, since the success rates then change every single time a trial is performed. If we are just talking straight *percentages* of success (and therefore assuming, as seems plausible, that only finitely many predictions will ever be made in the whole past and future history of mankind) then although it may be true that the success rate for corroborationism, say, in the subpopulation of *past* predictions is different from the success rate in the subpopulation of *future* predictions, you can see that it is still not possible to talk of success rates 'changing' just now. Almost certainly in such a case the success rate for corroborationism in the subpopulation of predictions made earlier than 1985, say, will be different from the success rate in the subpopulation of predictions from 1985 onwards up till now *and* into the future. Why say that the success rates changed *now* rather than in 1985?

If you really want to have the countercorroborationist say that there will be a 'change' *now* then, so far as I can see, you have two options. First you could talk in *probability* terms rather than success rate terms; second, you could take it that we are talking only about well-corroborated low level empirical generalisations and take advantage of my concession that the corroborationist policy would in these cases have paid off in the past, not just more often than not, but 100% of the time – in that case you could have the grueish countercorroborationist say that the very next case was going to be the first exception.

So, consider the first possibility where the two rival principles are making different claims about the probability that a randomly selected prediction of a so far best-corroborated theory will turn out true. Let's simplify by assuming that there are only two possible outcomes – the best-corroborated theory's prediction turning out false, just in case the prediction of whatever theory the countercorroborationist favours is true. The 'grueish' countercorroborationist then allows that the probability *in the past* that the well-corroborated theory would be successful was at any rate greater than a half, but claims that a change is occurring exactly now in the probabilistic set-up so as to favour the less well-corroborated theory. (Of course how these, essentially *theoretical*, claims about probabilities are to be established is a tricky matter – all that we can 'observe' are your success rates. But let's not worry about that now.) The corroborationist, on the other hand, predicts no change in the probabilities.

JOHN WATKINS: Suppose I accept this account, then surely my case is made. You must agree that the claim that there will be a change in probabilities exactly now is stronger – other things being equal – than the claim that they will continue unchanged.

FLOATER: Well the same old point comes up again: the corroborationist principle *does* imply something about the present time, namely that the same probabilities will obtain and hence, if you like, that no change will occur in these probabilities. I just don't see how, in the explicit absence of any inductive considerations, you can say that one of these claims is 'stronger' than the other. If we look at the situation overall, each of the two positions entails equally precise claims about the probability of a certain outcome of a certain 'experiment' when that experiment is performed *at any given time*. Whether or not one of the claims seems 'stronger', that is, less likely to be true, can only be judged from the evidence: do we seem to live in a chancy, inconstant world or in a more regular world of seemingly fixed probabilities?

I need hardly say that this is an inductive consideration.

JOHN WATKINS: Suppose then I took your second sense in which the grueish countercorroborationist might be predicting a change just now. He allows that the success-rate for corroborationism so far is 100% but that the first exception is going to be the next instance (perhaps from there on in there are only 'exceptions'). Surely the uniformity claim is weaker than this claim that something different will happen in the very next instance.

FLOATER: The first part of my reply is that, as Goodman showed in his original paper and as I mentioned before, the talk of a 'change' occurring here is language-dependent (speakers of a language with 'grue' and 'bleen' as primitive predicates see a change if the next emerald is green). But secondly and more centrally, any talk of a 'change', even with respect to a given language, is derivative. Basically the two positions describe in equivalent detail what will happen in each observational trial. Say that the next action to depend on the prediction of some theory is my proposed descent from the top of the Eiffel Tower. The corroborationist and countercorroborationist make entirely analogous, but different predictions about what will happen: the corroborationist predicts roughly uniform acceleration, the counter-corroborationist predicts, let's say, a gentle float. Since it is this latter prediction which, given *E*, entails the change claim, it must be at least as 'strong' as that change claim (given *E*). But then in the end you're going to have to say that the statement that 'the next freely falling body will achieve terminal velocity of '1″ per second' is, given the evidence, 'stronger' than the statement 'the next freely falling body will fall with roughly constant acceleration'. And the fact that *that* judgment of comparative strength has exclusively inductive credentials is surely patent.

JOHN WATKINS: It is important for me to speak about *overall* success and failure of the two policies and not to concentrate simply on the next instance. Let me have the assumption that the future will be finite so that I can avoid the difficulties with probabilities and stick to success-rates. I accept that I can't legitimately talk about success-rates 'changing now'. So let me amend the position of my grueish countercorroborationist: he now says simply that the success-rate of countercorroborationism within the (finite) set of future predictions will be higher than that of corroborationism within the same set.

FLOATER: Fine: but then – since you have dropped your idea that the grueish countercorroborationist claims that a change occurs now – I don't see even a *prima facie* case for his prediction about the future being any 'stronger', given the past evidence, than that of his corroborationist rival. Unlike the 'straight' countercorroborationist, he makes no claim about his future success rate being sufficiently high to offset his present bad record, he simply discounts the past. He and the corroborationist simply make equally precise but conflicting predictions about success rates in the future. Given that we are banned from making any inductive use of the past evidence to ground claims that one prediction is more likely than another, the two surely stand on a par.

JOHN WATKINS: Well, not quite. My 'grueish' chap says, remember, that his policy will definitely pay off in the future – he says that the success rate of countercorroborationism will be, not just greater than *or equal to* the success rate of corroborationism, but strictly greater than it. The past evidence allows the corroborationist, on the other hand, to rest on his laurels: he claims that the *overall* success rate of his policy is strictly greater than that of his rival, but his past success means that he need only claim that his success rate in the future will be *no worse* than that of his rival, he need not claim that his future success-rate will actually be higher. I agree that this means that the corroborationist's prediction is weaker only by a whisker: but then I always told you that my promise was only just to tip the balance in his favour.

FLOATER: Well in fact, if we go along with the finite universe assumption, your corroborationist – since he claims only that corroborationism will be ahead 'at the end of time' – could in fact allow that his success rate in the future will actually be *less* than that of the countercorroborationst, so long as it is only marginally less so that the big crunch will come before corroborationism's present advantage is finally eroded. But I always felt that your 'first past the post' formulation of the two positions didn't come to grips with the real problem.[43] I want to know why my proposed float is 'irrational'. Now even if you convinced me that there is some reason to think that the general corroborationist policy will have outscored its rival at the end of time, would this be enough to categorise my proposal to float as irrational? Surely not.

If corroborationism's prediction about the future really is compatible with less than 50% of *future* predictions from best-corroborated theories being correct, then how could it deliver the required irrationality verdict? If the corroborationist allows that it *may* be true that I have a slightly better than 50% chance of getting down safely by floating and that it *may* be true that there is a slightly worse than 50% chance that the predictions underlying the construction of the lift will hold in future instances, then how can it possibly deliver the verdict that everyone else seems so sure is the right one? Even if we turn back to probabilities and have your corroborationist say that there is at least a 50% chance of any future prediction of any best-corroborated theory being correct, then my proposal to float down can hardly be regarded as 'irrational' – remember that I am quite happy on every occasion to float or not, depending on the toss of a coin. In general, the intuitions of my opponents would surely be that, where an action's success depends on the outcome of a random process with equally likely outcomes, then it would not be irrational to base one's action in a given case on the prediction of *either* outcome. If in the future, playing 'corroborationist' or 'counter-corroborationist' is equally likely to pay off on average, then what seems to be indicated as 'most rational' is the 'mixed strategy' that I proposed, Certainly always playing 'corroborationist' is not indicated as the uniquely rational strategy (and indeed would surely stand on a par rationalitywise with the strategy of always playing 'countercorroborationist'). What you make the corroborationist say now seems to me so 'weak' that it fails to solve the problem even if established. To deliver the verdict you want, you need to make some positive (and not merely non-negative) claim about the future. Your corroborationist needs to assert *at least* that there is a better than 50% chance of his favoured predictions turning out correct in the future. But then we are back to complete parity between him and the grueish counter-corroborationist.

JOHN WATKINS: What you say would be correct, of course, if we *knew* that corroborationism would be correct only 50% (or marginally less than 50%) of the time in the future. But the whole problem was that we *don't* know what the future success rates will be. Any claim to know them, or even any claim that one set of future rates is more rationally credible (or whatever) than another, would be an acknowledgedly inductive assumption. And my position, remember, is 'neo-Popperian': I hold that *at most* we know accepted observation statements and deductive logic. Corroborationism's future success rate may, for all we know, be much higher than 50%, perhaps even 100% or close to it; but it is the (now known) fact that this future success rate *could* be only 50% (or slightly less) and the policy still be overall champion which, I claim, just tips the balance in favour of acting on the corroborationist principle.

FLOATER: Well, it *might* tip the balance in favour of the claim that corroborationism will be overall champion. (Although even this relies, as I said before, on how you count up possibilities.) But I just don't see at all how it tips the balance in favour of *applying corroborationism in future instances*, and this after all was the problem all along. If we go back to your race analogy, the analogue of the question about whether to take the lift or to the air is *not* the question of who will eventually win the race, but something like who will cover the next yard, say, the faster. And even if I accepted (which I don't) that his 1 mile lead at halfway tips the balance in favour of the prediction that A will win, this lead is, as I am sure you will accept, entirely irrelevant to the question of whether he covers the next yard faster.

This really brings me back to the general question of 'balance tipping'. As I said before, it is easy to cite 'balance tipping' factors in favour of inductivism or corroborationism. Almost everyone believes it, for example. Or why not come right out and say that what tips the balance in its favour is its past success (what you do say seems to amount in the end to not much more than this anyway)? But then the whole problem always was *why* this past success should tip the balance. In general I don't believe that you can really claim to have solved the problem of induction by citing a 'balance tipping' factor, unless that reason tells in favour of inductivism or corroborationism being correct at least more often than not *in the future*. Otherwise Popper would already have solved the problem which you yourself accept that he did not. Let's take it that Popper's system of corroboration contains no inductive element. What tips the balance in favour of the lift rather than my float? Well: the fact that Galileo's law is best-corroborated. This 'solution' does not presuppose induction, since the corroboration appraisal contains no inductive element. But Salmon's point (and others') was exactly that if this is meant to deliver the verdict that it is rational to apply one theory rather than others in the future, then Popper must presuppose that the so far best corroborated will continue to be better corroborated than its known rivals in the future. That is, he must presuppose the very point at issue. You yourself quote Salmon's argument at one point, I believe, and comment 'Game, set and match to Salmon'.[44] It seems that in the end he scores the same victory over you.

JOHN WATKINS:[45] Perhaps you're right: I've been banging my head against the same intellectual brick wall as Popper. Once it has been accepted that there is a deductive gap between past performance of a theory and its future performance, it has to be further accepted that, if we are going to deliver the verdict that your proposed float is irrational, we need to beef up our theory of rationality. We need, that is, to accept that neo-Popperianism or deductive rationality (observation statements plus deductive logic) is too weak to do the job. But if we are forced to incorporate substantive assumptions into our theory of rationality, we are not likely to find any more intuitively convincing or more plausible than the straight inductive assumption that, in certain restricted circumstances, it is reasonable to suppose that the future will be like the past. I just solve the problem by fiat and declare you irrational. By the way it strikes me that, although I've all along just taken *deductive* logic as given, if someone (I hear you have a *really* strange brother) took a similar line on deductive rationality, he would in the end force me into a similar apparently dictatorial 'irrational' position.[46] So maybe I shouldn't feel too bad about it.

But what are we going to do about you? I'm absolutely amazed that you have survived so far and quite convinced that you won't for much longer if you don't mend your ways. Pyrrho, the founder of one version of scepticism, is alleged to have actually lived the life of a sceptic as well as defended the philosophical doctrine. But I notice that he did take the precaution of surrounding himself with lots of disciples. My conjecture has always been that he would have made sure that he did not fully convince these disciples, so that whenever he headed for a cliff edge or whatever, claiming that there was no reason to think that he would come to any harm, there were always a few acolytes around whose 'emotions' got in the way of their 'reason' and who therefore pulled him back. Perhaps we can get a few apprentice philosophers to look after you – they can have fun arguing with you in the meanwhile.

London School of Economics and Political Science

NOTES

[1] See Miller (1982) both for references to Popper's early and late critics, and for Miller's attempt to argue that these criticisms are invalid.

[2] Since this volume was to be kept as a surprise, I faced the prospect of publishing the present paper without the benefit of John Watkins's critical comments. Since this would have been both an unwelcome and entirely novel experience for me – he has been kind enough to read and helpfully criticise drafts of all my other work prior to publication – I resorted to subterfuge: I am glad to say that this paper too has been (substantially) modified as a result of his comments on an earlier version of Sections 1

and 2 delivered at a conference on Popper's philosophy and on an earlier version of Section 3 delivered at his seminar. Did anyone ever really need to be assured that a critic of an earlier draft is 'not responsible for the mistakes that remain'? If so, it is hardly likely to be in this case – I have no doubt that John will have at least as many disagreements with the new version as he did with the old. As always, though, I thank him for his comments (and for the help and encouragement he has given me over the years). Thanks for comments on an earlier version are also due to Colin Howson and Elie Zahar.

[3] Popper (1972), p. 1.

[4] Miller (1982), p. 18.

[5] I don't, of course, deny the possibility of the lift cable breaking or some other disaster. But should such a disaster occur it would not be put down to our having an incorrect appreciation of the laws of nature but instead to some 'natural' (and independently checkable) cause such as metal fatigue. This means that, had the equipment been checked just in advance of the disaster, then disaster could (at any rate in principle) have been predicted and hence averted. This is, then, quite different from our friend the floater's proposal which, if it led, as we all expect, to disaster would indeed be attributed to his incorrect appreciation of the laws of nature. Let's suppose, then, that some important dignitary is due to ascend the Eiffel Tower tomorrow, and that therefore the lift equipment has been exhaustively checked. Those who just can't get the fallibility of lifts out of their head should substitute slowly walking down the emergency stairs in thick rubber-soled shoes with various safety-harnesses attached.

[6] Russell (1948), p. 9: 'Scepticism, while logically impeccable, is psychologically impossible, and there is an element of frivolous insincerity in any philosophy which pretends to accept it'.

[7] Popper, *op. cit.*, p. 21.

[8] *Op. cit.*, p. 22.

[9] Salmon (1981), p. 122.

[10] Although I have given some of my characters real names, they are of course my own 'rational reconstructions'. Except where they have been deliberately caricatured for 'dramatic effect' (in places that should be fairly obvious), my reconstructions are intended to be accurate, as indicated by the quotations and references. But readers must judge for themselves how far my intentions have been realised; and should certainly remember that, except when explicitly quoting, I am indulging in the dangerous business of putting words in other people's mouths.

[11] Miller, *op. cit.*, p. 40.

[12] *Ibid.*

[13] *Op. cit.*, p. 38–40.

[14] The question of whether there is some cogent Popperian reason to discount gruesome hypotheses is the subject of Section 2.

[15] *Op. cit.*, p. 39.

[16] *Op. cit.*, p. 41: 'if we are sufficiently finicky, and *insist boringly* on the eternal possibility that nature will entirely change its course just before our action is performed, then no proposal [for action] whatever will get rejected [in the course of critical debate].' In case this seems to give the game away, I should add that Miller goes on to make it clear that one should be less 'finicky' and less 'boring' and

therefore reject all proposals *except for* the intuitively rational ones.

[17] Miller, quoting Larry Briskman, *op. cit.*, p. 11.

[18] This is, of course, one of the points made by Goodman (see below); for Miller on verisimilitude, see his (1974) and (1975).

[19] '[Popper's] answer manifestly involves no recourse to any principle of induction. There is no suggestion that the proposal that best survives criticism is one that we have any reason to expect to be successful; but this hardly means that it will not be successful' (Miller (1982), p. 40). Cf: 'The crux of the matter is that in order to provide genuinely interesting knowledge of the world inductivism needs to *assume* that there is some order and regularity in the world, whilst falsificationism *requires only* that there is some order and regularity in the world – but it does not need to make any sort of assumption to this effect'. (*Op. cit.*, p. 33; emphasis added.) I cannot for the life of me see what credit accrues from abstaining from an assumption, when one's theory requires the world to be such as to make the assumption true.

[20] In fact the only defensible thesis here, I believe, is that the (mathematical) structure of a rejected theory (though not its ontology) standardly lives on as a limiting case. For at any rate a partial articulation and defence of 'structural realism', see my (1989).

[21] See Feyerabend (1975), and for criticism (and elaboration of the distinction between 'crude' and 'scientific facts') my (1978) and (1981).

[22] For an instructive case, concerning the pressure of light, see my (1982).

[23] See, for example, Popper (1963), Chapter 10.

[24] 'The rationalist attitude is characterized by the importance it attaches to argument and experience. But neither logical argument nor experience can establish the rationalist attitude; for only those who are ready to consider argument or experience, and who have therefore adopted this attitude already, will be impressed by them...We have to conclude from this that no rational argument will have a rational effect on a man who does not want to adopt a rational attitude...But this means that whoever adopts the rationalist attitude does so because he has adopted, consciously or unconsciously, some proposal, or decision, or belief, or behaviour: an adoption which may be called irrational ["non-rational" would surely have been better].' (Popper (1945) II, pp. 230–1). Bartley's Comprehensively Critical Rationalism was for a while regarded by Popperians as providing an escape from having to admit that their rationality, like other forms, is based on a non-rational decision. But what exactly would it mean to have a criticism of the critical method? No one ever said – at any rate not coherently. It is surely clear that Popper was right that: 'a comprehensive rationalism is untenable' (*ibid.*)

[25] It will be clear that this is what I think should be said about the problem of induction – though it is not at all clear that it constitutes a 'solution' of the problem as opposed to an honest recognition of its insolubility. Again the position is hardly new; it was held for example by Russell:

What [Hume's] arguments prove – and I do think the proof can be controverted – is, that induction is an independent logical principle, incapable of being inferred either from experience or from other logical principles, and that without this principle science is impossible. (1946), p. 647.

[26] I need hardly, perhaps, say that at this point the amount of reconstruction (rational or irrational) becomes extreme!

[27] Miller, *op. cit.*, p. 39.
[28] *Ibid.*
[29] See Miller, *op. cit.*, p. 38 and note 85: 'Actually I think that this claim [that the 'natural' and 'gruesome' alternatives are equally corroborated by the evidence on Popperian criteria] is false, since a test for grueness involves not only a color test but also a date test'.
[30] See Miller, *op. cit.*, pp. 39–40. Miller refers to a similar position adopted earlier by Bartley (1981).
[31] For example, the 'Zahar–Worrall account' (for details see my (1985)).
[32] See Watkins (1984), pp. 342–8; and the 'simpler...and better version' in Watkins (1988).
[33] Watkins (1988), p. 18.
[34] 'Can a neo-Popperian like myself...give [a technologist] any reason why he should prefer the best-corroborated hypothesis?' *Ibid.*
[35] Watkins (1988), p. 18.
[36] *Op. cit.*, p. 19.
[37] *Ibid.* There are difficulties with Watkins' use of 'success rates' which, except in extreme cases, cannot of course be constant (see *below*). But laying these aside, John Watkins' claim here is ambiguous between (a) both the success rate in the past and the success rate in the future (assuming this notion is well-defined (*below*)) are higher for corroborationism and (b) taking both past and future successes into account corroborationism has the higher rate. John Watkins is, I think, inclined to plump for the latter construal. As Mr Floater points out (*below*), this means that, even if it were accepted that there is a reason to prefer corroborationism as thus construed, then the problem of induction would still not in fact be solved. The whole question, as normally understood, is about what it is rational to do *in the future* (or less grandly, in the next instance).
[38] *Ibid.*
[39] (1984), pp. 171–7. Watkins distinguishes 'congruent' and 'incongruent counterparts'. The former are simply pairs of *identical* consequences of two different theories and raise no problem. Our concern is of course with *incongruent* counterparts.
[40] Watkins, personal communication.
[41] Thus, for example, suppose the 'possibilities' are taken to be: (i) A wins having been ahead at halfway; (ii) A wins having been, at halfway, both behind and whistling 'Dixie'; (iii) A wins having been, at halfway, both behind and not whistling 'Dixie'; (iv) B wins having been ahead at halfway; (v) B wins having been, at halfway, both behind and whistling 'Dixie'; (vi) B wins having been, at halfway, both behind and not whistling 'Dixie'; (vii) deadheat. The prediction that A will win starts off entirely on a par with the prediction that B will – both excluding four possibilities. But bringing in E, which is just, remember, the information that A is 1 mile ahead at halfway (if one is allowed to bring in extra evidence then it is even easier to make the A prediction 'stronger' given the evidence), now makes the prediction that A will win 'stronger' on the Watkins account. This is because the 'A wins' prediction together with E is, of course, compatible with only one of the seven original 'possibilities' (possibility (i)), while the 'B wins' prediction, together with E, is compatible both with possibility (v) and with possibility (vi). (The dependence of the measures of logical strength developed in Watkins' [1985] on how the 'possibilities' are parti-

tioned was pointed out at several LSE seminars by Colin Howson.)

42 Watkins (1988), p. 19.

43 The ambiguity in the formulation of the corroborationist position, pointed to above n. 37, becomes significant here.

44 Watkins (1984), p. 341.

45 Once again the amount of reconstruction becomes extreme at this point.

46 See Lewis Carroll's famous (1895) dialogue between Achilles and the Tortoise.

REFERENCES

Bartley, W.W.: 1981, 'Eine Losung des Goodman-Paradoxons' in Radnitzky and Andersson (eds.): *Voraussetzungen und Grenzen der Wissenschaft*. Tubingen: Mohr.

Carroll, L.: 1895, 'What Achilles said to the Tortoise', *Mind*, **14**, 278–80.

Feyerabend, P.: 1975, *Against Method*. London: New Left Books.

Miller, D.: 1974: 'Popper's Qualitative Theory of Verisimilitude', *British Journal for the Philosophy of Science*, **25**, 166–77.

Miller, D.: 1975, 'The Accuracy of Predictions', *Synthese*, **30**, 159–91.

Miller, D.: 1982, 'Conjectural Knowledge: Popper's Solution of the Problem of Induction' in Levinson (ed.): *In Pursuit of Truth*. Brighton: Harvester.

Popper, K.R.: 1945, *The Open Society and its Enemies*. London: Routledge.

Popper, K.R.: 1963, *Conjectures and Refutations*. London: Routledge.

Popper, K.R.: 1972, *Objective Knowledge*. Oxford University Press.

Russell, B.A.W.: 1946, *A History of Western Philosophy*. London: Allen and Unwin.

Russell, B.A.W.: 1948, *Human Knowledge: Its Scope and Limits*. London: Allen and Unwin.

Salmon, W.C.: 1981, 'Rational Prediction', *British Journal for the Philosophy of Science*, **32**, 115–25.

Watkins, J.W.N.: 1984, *Science and Scepticism*. Princeton University Press.

Watkins, J.W.N.: 1988, 'The Pragmatic Problem of Induction', *Analysis*, **48**, 18–20.

Worrall, J.: 1978, 'Is the Empirical Content of a Theory Dependent on its Rivals' in Niiniluoto and Tuomela (Eds.): *The Logic and Epistemology of Scientific Change*. North Holland.

Worrall, J.: 1981, 'Feyerabend und die Fakten' in H.-P. Duerr (Ed.): *Versuchungen – Aufsatze zur Philosophie Paul Feyerabends*. Frankfurt: Suhrkamp.

Worrall, J.: 1982, 'The Pressure of Light: the Strange Case of the Vacillating "Crucial Experiment"', *Studies in the History and Philosophy of Science*, **13**, 133–71.

Worrall, J.: 1985, 'Scientific Discovery and Theory Confirmation' in Pitt (Ed.): *Change and Progress in Modern Science*. Dordrecht: Reidel.

Worrall, J.: 1989, 'Structural Realism: the Best of Both Worlds?', *Dialectica*, forthcoming.

ALAN MUSGRAVE

SAVING SCIENCE FROM SCEPTICISM

This paper was to have had a motto, but John Watkins got in first. One of the two mottoes of his *Science and Scepticism* is "We have to try to adopt a highly critical attitude towards those theories which we admire most" (Karl Popper). I admire many of the theories of *Science and Scepticism:* difficult problems are confronted head-on and considerable ingenuity goes in to solving them. But I must say no more in that vein. Instead I must follow the exhortation in Watkins' motto, which was to have been mine too.

Science and Scepticism aims "to succeed where Descartes failed" and to save science from scepticism. The rescue-act proceeds in two stages: first, the rationality of theoretical science is defended by offering a new theory of rational acceptance in science; second, the rationality of practical science is defended by offering a new solution to the so-called 'pragmatic problem of induction'. In both cases Watkins proceeds under tight constraints. He accepts the central negative thesis of Humean scepticism. There is to be no appeal to 'inductive logic' or to 'probabilism' to escape from it. Nor is there to be any recourse to Kantian synthetic a priori principles or to any other metaphysical assumption. Finally, science is not to be saved by emasculating it in any positivist or instrumentalist or phenomenalist way: the science to be saved is thoroughly realist science.

Watkins' project is a bold one. Does he succeed in carrying it out? I think not. He saves scientific rationality by emasculating, not science, but the aim of science; as a result he provides no real answer to Humean irrationalism. His solution to the pragmatic problem of induction turns out to involve inductivism, in much the same way as Watkins and others say that Popper's solution does. Or so I shall be arguing.

Does this mean that there is no saving science within the constraints that Watkins imposes? Perhaps not. I will explain what I think Watkins ought to have said in order to solve his problems within his self-imposed constraints. It is what Popper, as I interpret him, did say in order to solve them. This interpretation is hotly disputed by several loyal Popperians and perhaps even by Popper himself. It is also hotly disputed whether the resulting position is a defensible one.

Fred D'Agostino & I. C. Jarvie, Freedom and Rationality: Essays in Honor of John Watkins, 297–323.

1. HUMEAN SCEPTICISM AND IRRATIONALISM

The scepticism from which science is to be saved is Humean scepticism. This Watkins defines as the valid argument from:

(I) there are no synthetic a priori truths about the world;
(II) any genuine knowledge we have of the external world must ultimately be derived from perceptual experience;
(III) only deductive inferences are valid

to the sceptical thesis that there can be no genuine knowledge of the external world (1984, p. 3). Let us set aside the exegetical question of whether Humean scepticism is Hume's scepticism. Let us also set aside the fact that, if genuine knowledge is of synthetic truths, (II) entails (I). Humean scepticism, thus defined, is clear enough. If it is correct, science would seem to lie in ruins.

Watkins considers and rejects attempts to avoid Hume's sceptical conclusion by rejecting (I) or (III). His own solution rejects (II): following Popper, he says that we can have genuine *conjectural knowledge* of the external world which is not derived from perceptual experience. (More on 'conjectural knowledge' at the end.)

Watkins also considers and rejects various 'antisceptical strategies' which *accept* (I)–(III). What is *anti*sceptical about a strategy which accepts (I)–(III) and the sceptical conclusion that they entail? To answer this question we must consider Watkins' discussion of the 'probabilist strategy' and his distinction between 'probability scepticism' and 'rationality scepticism'.

The probabilist strategy is to accept Hume's sceptical conclusion and say that it does not matter. When Hume concludes that there can be no *genuine* knowledge of the external world, he merely means that there can be no *certain* knowledge of the external world. Hume is right. No statement about the external world (or at least no such statement which is not deductively entailed by the available evidence) can be known with certainty to be true. Call this *certainty-scepticism.* Certainty-scepticism does not matter because probability is the guide of everyday and scientific life. Statements which transcend available evidence can be shown to be more or less probable (well-confirmed) in the light of that evidence. The reasonable person accepts that hypothesis (if there is one) which is most probable or best-confirmed. We can have rationality without certainty – provided we have high probabilities.

As well as certainty-scepticism there are:

> *Probability-scepticism*: There is no empirical confirmation or justification in any probabilist sense. *Rationality scepticism*: We never have any good cognitive reason to adopt a hypothesis about the external world. (Watkins 1984, p. 58)

Earlier the formulations had been "all empirical statements about the external world are equally uncertain" and "we never have good reason for preferring one empirical statement about the external world to another" (Watkins 1968, pp. 278–9; italicised in the original). These various formulations are not quite the same and one could quibble with all of them. I will quibble with none, for their intent is clear enough. The probabilist strategy, then, is to accept certainty-scepticism but to reject both probability-scepticism and rationality-scepticism.

Watkins is no infallibilist – he too accepts certainty-scepticism. Nor is he a probabilist – he accepts probability-scepticism too. But he rejects rationality-scepticism. Is this coherent? Do not certainty-scepticism and probability-scepticism together *entail* rationality-scepticism? Watkins (rightly) thinks not. He formulates a 'principle (IV)' (1984, p. 57) which states or implies that:

> it is not rational to accept, or to have some positive degree of confidence in, a proposition unless it is either certainly true or, if uncertain, empirically justified or confirmed to some degree. (1984, p. 59)

Certainty-scepticism and probability-scepticism *together with* (IV) entail rationality-scepticism. Watkins can accept certainty-scepticism and probability-scepticism, yet reject rationality-scepticism, because he also rejects (IV).

Let us take stock. Watkins's rejection of (IV) commits him to the view that it is sometimes rational to accept a proposition which is neither certainly true (entailed by the evidence) nor probably true (well-confirmed by the evidence). We still need to know exactly when it is rational to do this. But Watkins' key idea is clear: we can have rationality (rational acceptance) without either certainty or probability.

Most philosophers find this key idea absurd. For 'accept' means 'accept as true'. How could it ever be rational to accept a proposition as true which the evidence shows neither to be certainly true nor probably true?

A suspicion arises: perhaps Watkins uses 'accept' to mean something other than accept as true. If so, he is not so much *disagreeing* with the Humean irrationalist as *changing the subject*. The Humean irrationalist claims, in line with (IV), that it is only rational to accept as true a proposition which is either entailed by the evidence or 'probabilified' by it. Watkins says this is wrong because it is sometimes rational to accept uncertain and improbable propositions *as* X for some X other than truth. This is changing the subject by equivocation on 'accept' and its cognates. It is as if the assertion that no sceptic is to be taken seriously were rebutted with "You are wrong – some antisceptics are to be taken seriously". Sometimes it is best to change the subject. Always it is best to do so openly, declaring one problem insoluble and switching to another.

Does Watkins change the subject in this way? Does he think it sometimes rational to accept a proposition which is neither certain nor probable? If so, when is it rational to do this? And does 'accept' here mean accept as *X* for some *X* other than truth?

2. CERTAIN TRUTH, PROBABLE TRUTH, AND POSSIBLE TRUTH

Watkins's conjecturalist strategy for defeating rationality-scepticism is a complex one. What is is reasonable or unreasonable to do depends upon our aim. So Watkins seeks to formulate a nonarbitrary *optimum aim for science*. Then he outflanks rationality-scepticism by saying that we have "a very good reason to accept a theory, inductive support for which is zero, if that theory is the one in its field that best fulfils the optimum aim for science" (1984, p. 337). And which theory does this? The theory "that is the best corroborated, in essentially Popper's sense, is the one that best satisfies" the optimum aim for science (1984, p. 279). The suspicion of a change of subject mounts.

If different scientists had different aims, then it might be rational for one of them to accept something (given her aim) and not rational or positively irrational for the other to accept it (given his aim). Hence:

> If we are to defeat rationality scepticism we need a nonarbitrary aim for science, an aim to which all members of the republic of science could subscribe. (1984, p. 123)

To ensure nonarbitrariness, Watkins states adequacy requirements on any proposed aim for science, requirements which he hopes will be uncontroversial. There are five of them (I paraphrase his 1984, pp. 124–6):

Any proposed aim for science must:

1. be *coherent* – must not contain components that sometimes or always pull in opposite directions;

2. be *feasible* – must not be such that we know that it cannot be fulfilled;
3. *serve as a guide* in choices between rival theories or hypotheses;
4. be *impartial* – must not favour one side in the scientific battle because of a bias towards theories that endorse a particular metaphysical view of the world;
5. involve the idea of *truth*.

It is sanguine to think all of these uncontroversial. Idealistic metaphysical views seem to be ruled out by the way Watkins poses his problem, though to be fair, he had criticised some of them earlier (1984, pp. 14–17 and 84–91). Since I share his impatience with all forms of idealism, I will not harp on this. More controversial is the requirement that the aim "involve the idea of *truth*". Instrumentalists deny in various ways that scientific theories are true or false in the first place. Other antirealists deny that true theories are any part of the aim of science. Watkins mentions Laudan as one of the latter and cannot disguise his impatience:

> ...to say that truth is no part of the aim of science is on a par with saying that curing is no part of the aim of medicine or that profit is no part of the aim of commerce. (1984, p. 126)

I share Watkins' impatience with those who "give up on truth". The trouble is that he himself does just that – but I anticipate.

Having formulated his adequacy requirements, Watkins sets them aside. He also sets aside Humean scepticism. He takes us back to the pre-Humean glad confident morning of modern science and formulates a utopian aim for science called the *Bacon-Descartes ideal*: science aims to render all phenomena "explainable or predictable by deduction from true descriptions of initial conditions together with universal principles that are certainly true, ultimate, unified and exact" (1984, p. 129). Next we have a more feasible 'progressivised' version of this, in which the aim is to progress towards the final goal with theories that are ever more certain or probable, ever deeper, ever more unified, ever more exact, ever more predictively powerful (1984, p. 130).

But the Bacon-Descartes ideal is incoherent. Aiming for ever more certain or probable theories pulls against aiming for ever deeper, more unified, exact, and predictively powerful theories. Something has to give. Watkins follows Popper: what has to give is certainty rather than depth etc. However, we cannot simply *eliminate* the demand for certainty. A truncated depth-only aim would not involve the idea of truth: theories can get ever deeper etc. without any of them being true. We must replace the demand for certainty with something that "involves the idea of truth". What shall this be?

Well, *truth* involves the idea of truth. Given Watkins's impatience with those who give up on truth, one would expect that truth would replace certainty in his optimum aim for science. But one would be wrong. He distinguishes truth from *possible truth*, which is not just internal consistency but also consistency with the available evidence. But not even possible truth replaces certain truth and probable truth in Watkins's optimum aim. The "strongest feasible core that can be retrieved from" the Bacon-Descartes aim for certainty or high probability is rather:

> (A*) Science *aspires after* truth. The system of scientific hypotheses adopted by a person X at any time should be possibly true *for him*, in the sense that, despite his best endeavours, *he has not found* any inconsistencies in it or between it and the evidence available *to him*. (1984, pp. 155–6; italics mine)

Truth has become something hoped-for (aspired-after) rather than aimed-for. Actually, (A*) says nothing about the aim (as opposed to the hope) of science. The rest of (A*) states an epistemic principle of acceptance (adoption) which is couched not in terms of truth, nor even in terms of possible truth, but in terms of possible truth for a person at a time. What is the epistemic principle doing in a statement of (part of) the aim of science? Why not aim for truth? Is possible truth for a person at a time really part of the optimum aim for science? Let us take these questions in reverse order.

Imagine a historical sequence of ever-deeper etc. theories, all possibly true (in Watkins's sense), and furthermore *all true*. Imagine a second sequence of ever-deeper etc. theories, again all possibly true (in Watkins's sense), yet sadly *all false*. Surely, the former sequence fulfils the optimum aim and the latter does not. Imagine now a third sequence of ever-deeper etc. theories, all possibly true for the folk involved, in that none of those folk ever succeeded despite their best efforts in finding any inconsistencies in them or between them and the available data, yet all *inconsistent* either internally or externally, and hence *utterly false*. Surely, this third sequence is even worse than the second as far as attaining the *optimum* aim for science is concerned.

Why not aim for truth (as well as depth etc.)? Perhaps to do so is to violate Watkins's adequacy requirements on any aim for science. Let us see. Such an aim certainly *involves the idea of truth*. Such an aim is *impartial*, in that no particular metaphysic is favoured by it. Such an aim *serves as a guide* in theory-choices: it is because we are after truth that we will reject theories we have reason to think false. Such an aim is *coherent*: certainty pulls against depth etc., but truth does not. Finally, such an aim is *feasible*: we do not know that it cannot be fulfilled, that science is somehow constitutionally incapable of obtaining a theory that is both true and deep etc.

Some dispute this last point. They appeal to the so-called *pessimistic induction*: all theories *so far* proposed in the history of science have been false; therefore (or therefore probably) all theories *ever* proposed in the history of science will be false. This argument is as bad as an argument can be: its premise is preposterous and its conclusion does not follow. It is preposterous to say that science has come up with no theoretical truths. So lunar eclipses are not caused by the moon passing into the earth's shadow? So water is not a compound of hydrogen and oxygen? So freely falling bodies do not accelerate? So puerperal fever is not brought about by micro-organisms? So limestone is not formed by sedimentation? So bloodletting does cure some diseases? So some multi-cellular organisms do arise by spontaneous generation from non-living matter? The list might be extended indefinitely.

Faced with such a list, pessimistic inductivists try to save their premise by confining it to *big* theories. (What is a big theory? To find out, replace the little word 'big' by your favourite big word or phrase from the following list: paradigm, disciplinary matrix, scientific research programme, conceptual scheme, global theory, research tradition, world-view, etc.) Pessimistic inductivists now owe us a distinction between big and little theories, and they had better not say that the big theories are all those that have turned out to be false! Suppose (which I doubt) that 'big theory' can be defined in a non-question-begging way. Still, confining the pessimistic induction to big theories leaves little ones unscathed. And whether confined to big theories or not, the pessimistic induction is invalid: it does not show that science can never attain a true theory.

Watkins could never appeal to the pessimistic induction to show that truth is not a feasible aim for science. He is a deductivist, who eschews all inductive inferences, including this one. Yet he does think truth an infeasible aim. Why? Because his feasibility requirement is actually much stronger than the officially stated one. For an aim to be feasible it is not enough that we *do not* know that it *cannot* be achieved. It must also be the case that we *do* know, and *know for certain*, when it *has been* achieved. He says:

We could defeat rationality-scepticism if we could know, at least in a good many

> cases, which one of a set of competing hypotheses, given the present evidence, best satisfies [the optimum aim for science]; for we surely have the best possible reason to adopt the one that best satisfies the optimum aim for science. (1984, p. 279).

Truth is not a feasible aim in this stronger sense. Certainty-scepticism in science translates to meta-science: if no theory *T* can be known for certain to be true, then no meta-statement of the form "Theory *T* is true (and deep etc.)" can be known for certain to be true either. This is why Watkins jettisons truth from his optimum aim for science.

He jettisons possible truth too. He even jettisons possible truth for a person at a time, defined thus: "a system of scientific hypotheses *S* is possibly true for *X* at time *t* if *X* knows that *S* is: (i) internally consistent and (ii) consistent with the total evidence available to *X* at *t*". Even this is an infeasible aim, because it is "well-nigh certain" that a person could not *know* that *S* is both internally consistent and consistent with all the available evidence. The most we can aim for is that a person has looked for inconsistencies and not found any. This is a feasible aim because a person can be certain that "despite his best endeavours, he has not found any inconsistencies in [*S*] or between [*S*] and the evidence available to him" (1984, pp. 155–6; see also his 1987, pp. 77–8).

The other features of Watkins' optimum aim for science, depth, unity, exactness, and predictive power, are all analysed as *logical* features. Presumably, Watkins thinks them epistemically unproblematic, in that a person can know for certain whether and to what extent a theory possesses them. This is optimistic. Watkins' pursuit of meta-scientific certainty should lead him to do for depth etc. what he does for truth, and substitute 'possible depth etc. for a person at a time' as the aim of science. But let us not dwell on this point. Let us return to truth, which is my main concern.

Watkins' optimum aim for science is an *ignoble* aim. Out goes truth. Out goes consistency. Out goes consistency with the data. The aim of the activity which Watkins admires so much is to get scientists into the position where they have looked for inconsistencies and not (yet) found any. (Notice that the question "Does this theory fulfil the optimum aim for science?" has become ill-formed – we must add "for Scientist *X* at time *t*".) For this is the only *feasible* aim, the only aim that scientists can *know for certain* that they have achieved. Antirealists like Kuhn, Laudan, and van Fraassen all jettison truth from the aim of science in favour of something like empirical adequacy. This is bad enough for realists. How much worse is Watkins' proposal that we jettison truth in favour of empirical adequacy *as far as we now know it*!

At the end of my first section I raised the suspicion that Watkins does not so much reject rationality-scepticism as change the subject. You *reject* rationality-scepticism if you think that sometimes we do have good cognitive reason to adopt *as true* a hypothesis neither entailed nor rendered probable by the evidence. You change the subject if you think that sometimes we do have good cognitive reason to adopt such a hypothesis *as X* for some *X* rather than truth. Watkins changes the subject:

> When I speak of a scientist *X adopting* (or accepting or preferring) one of a set of competing theories, I...mean that he regards it (i) as *possibly true* in the sense indicated, and (ii) as the *best* of the theories in its field, as they now stand. (1984, p. 156)

Here 'adopt' does not mean 'adopt as true'. Watkins does not think we ever have good cognitive reason to adopt a hypothesis which is neither certain nor probable as true. He tacitly endorses rationality-scepticism and talks about something else.

The reason why Watkins changes the subject is that he hankers after meta-scientific certainty. He is happy to eschew certainties (and near-certainties or high probabilities) in science. (This is not quite right. Watkins also hankers after certainty at the experimental or rather experiential level: he has a 'level$_0$' of first-person experiential reports that are also supposed to be certain. But this is another story.) In science 'conjectural knowledge' is the order of the day, and Watkins is presumably quite happy with the locution "Scientist *S* conjecturally knows that *P*". (We have still to find out what this *means*, however.) Oddly, Watkins is unhappy with the meta-locution "Scientist *S* conjecturally knows that *P* is true" or with the equivalent meta-locution "Scientist *S* conjecturally knows that *P* fulfils the truth part of the aim of science, which is truth". In meta-science certainty is the order of the day: meta-scientific appraisals of whether some theory fulfils the aim of science must be certain. (Indeed, when Watkins talks of the scientist having good cognitive reason to adopt a theory as fulfilling the truth part of the optimum aim for science, this is as certain as anything can be, for it is *analytic*: "That scientist *S* has tried and failed to detect internal or external inconsistencies in a theory is a good cognitive reason for *S* to adopt that theory as fulfilling the (truth part of) the optimum aim for science, that is, as being possibly true for him, that is, as being such that he has tried and failed to detect internal and external inconsistencies in it".) Watkins's hankering after meta-scientific certainty leads him to change the subject in the face of rationality-scepticism and to downgrade truth from aim to aspiration. As

Lakatos once said, "did Pyrrho or Hume have anything against...entertaining 'hopes'?" (Lakatos 1974, p. 161; cited by Watkins 1984, p. 287).

It is odd for a self-professed 'conjecturalist' to hanker after certainty anywhere. It is odd for a self-professed realist to downgrade truth from aim to aspiration. It is odd for a self-professed rationalist to be tacitly endorsing rationality-scepticism. However, our discussions make it perfectly clear what Watkins *ought* to have said in order to save science from scepticism within his self-imposed constraints. It is what Popper's critical rationalism, as I understand it, *does* say.

3. CRITICAL RATIONALISM

Let truth resume its place as part of the optimum aim for science. Science aims for theories that are ever-deeper, more unified, more exact, more predictively powerful, *and true*. On the epistemic side we are to endorse certainty-scepticism and probability-scepticism, yet deny rationality-scepticism. So we must say that sometimes we have good cognitive reason to adopt a hypothesis as true which is neither entailed, nor rendered probable, by the available evidence. And we must, of course, give an account of when it is reasonable (when we have good cognitive reason) to do this. The only price paid for all this is that meta-scientific appraisals of whether or not some theory fulfils the optimum aim for science will be as conjectural and fallible as the theory itself.

Watkins's account of the method of science is a thoroughly Popperian one. Theories are subjected to criticism in order to eliminate the mistaken ones. In science the main (but not the only) way to try to criticise a theory is to subject it to severe experimental tests. A theory which passes such a test is *corroborated*. There can be *degrees* of corroboration. With luck one of the competing theories will emerge as the *best* corroborated. (More generally, with luck one of the competing theories will emerge as the one which best withstands all the criticisms, empirical and non-empirical, directed at it.) To defeat rationality-scepticism (as opposed to changing the subject) the critical rationalist proposes a principle such as the following:

(CR) It is reasonable to adopt a theory as true at time t if that theory is the best-corroborated theory at time t.

(The proposal is, then, that a theory's being the best corroborated at a certain time is a good cognitive reason to adopt that theory as true at that time.)

Given some such principle, rationality-scepticism collapses: sometimes we do have good cognitive reason to adopt a hypothesis as true which is neither certain nor probable given the available evidence.

This proposal, unlike Watkins's proposal, is not trivial. On the contrary, it will be met with disbelief if not derision. Yet it is the proposal or something like the proposal which Popper makes – *as I understand him.* It is also the proposal which Watkins should have made in order to solve his problem (defeat rationality-scepticism) within his self-imposed constraints (endorse certainty-scepticism and probability-scepticism).

This proposal means that a scientist might reasonably adopt a theory as true which turns out to be false. Well, what is wrong with that? If an accepted theory does turn out to be false, we will say that what we accepted was wrong, not that it was wrong of us (unreasonable of us) to have accepted it. Any reasonable theory of reasonable belief (acceptance as true, adoption as true) must allow for reasonable beliefs in falsehoods.

But it is much worse than that (a critic might say). Not only might what we 'reasonably adopt' turn out to be false, but also we have been given no reason, conclusive or inconclusive, in favour of its truth. For a certainty-sceptic, no degree of corroboration however high establishes that the corroborated theory is true or provides a conclusive reason for it. For a probability-sceptic, no degree of corroboration however high establishes that the corroborated theory is probably true or provides an inconclusive reason for it. More generally, the failure of our best efforts to show a theory false, however these best efforts are to be explained or defined, provides neither a conclusive nor an inconclusive reason for the truth of that theory. So how can it be reasonable to accept it as true?

This question takes us to the heart of the matter. Popper drew a distinction which is crucial if we are to understand (I do not say 'accept') the critical rationalist response to Humean scepticism-cum-irrationalism. Talk of justifying or giving reasons for beliefs (acceptances as true, adoptions as true, etc.) is ambiguous between justifying or giving reasons for *what is believed* and justifying or giving reasons *for the believing of it.* Popper's contention is that we cannot justify the things we believe but *can* sometimes justify our believing them; that we can give no good reason for what we accept as true, but can sometimes give a good reason for our accepting it (tentatively of course). Critical rationalism is the view that the failure of our best efforts to show a theory false is a good reason for *us to accept* it tentatively as true, without being any reason at all *for the theory itself.*

This is, I admit, only one *interpretation* of Popper's position. Textual support for it is ambiguous and has become more so with the passage of time. This is partly because Popper developed a phobia about the term 'belief', partly because for a while increasing verisimilitude supplanted truth as his preferred aim for science, partly because of a growing number of unexplained qualifications of various kinds (including 'scare quotes'). But here are some of the passages on which this interpretation is based:

> Hume was right in stressing that our theories cannot be validly inferred from what we can know to be true – neither from observations or from anything else. He concluded from this that our belief in them was irrational. If 'belief' here means our inability to doubt our natural laws...then Hume is again right. ... If, however, the term 'belief' is taken to cover our critical acceptance of scientific theories ... then Hume was wrong. In such an acceptance of theories there is nothing irrational. There is not even anything irrational in relying for practical purposes upon well-tested theories, for no more rational course of action is open to us. (1963, p. 51)

> I saw that what has to be given up is the *quest for justification*, in the sense of the justification of the claim that a theory is true. *All theories are hypotheses*; all *may* be overthrown.
> On the other hand, I was very far from suggesting that we give up the search for truth: our critical discussions of theories are dominated by the idea of finding a true (and powerful) explanatory theory; and *we do justify our preferences by an appeal to the idea of truth*: truth plays the role of a regulative idea. *We test for truth*, by eliminating falsehood. (1972, pp. 29–30)

> Although we cannot justify a theory ... we can sometimes justify our *preference* for one theory over another; for example if its degree of corroboration is greater. (1976, p. 104).

> ... we can never justify a theory. But we can sometimes 'justify' (in a different sense) our preference for a theory, considering the state of the critical debate; for a theory may stand up to criticism better than its competitors. (1976, p. 149)

> If two competing theories have been criticised and tested as thoroughly as we could manage, with the result that the degree of corroboration of one of them is greater than that of the other, we will, in general, have *reason to believe* that the first is a better approximation to the truth than the second. (1982, p. 58)

> But the critical discussion can, if we are lucky, establish sufficient reasons for the following claim:
> ...'This theory seems at present, in the light of a thorough critical discussion and of severe and ingenious testing, by far the *best*...' (1982, p. 82).

Passages like these could be multiplied. But multiplying them would not, I admit, establish that Popper proposes the view I attribute to him. That view was, recall, that a theory's having fared best in the critical discussion does not justify the theory, but does justify tentative adoption of the theory as true.

Most philosophers find this view absurd. Most philosophers find the following *justificationist principle* self-evident: *to justify adopting a theory as true one must justify the theory adopted – a reason for believing a theory or accepting it as true must be a reason for the theory believed or accepted.* This principle lies behind Watkins's 'Principle (IV)' (see above, p. 299). This principle, combined with certainty-scepticism and probability-scepticism (which together show that there are no conclusive or inconclusive reasons for theories), leads straight to rationality-scepticism.

If rationality-scepticism (irrationalism) is to be denied, then there are only three options: (1) deny certainty-scepticism; (2) deny probability-scepticism; (3) deny the justificationist principle. (Here I assume, of course, that changing the subject is not a real fourth option.) Critical rationalism takes the last of these options. And to repeat, those who think the justificationist principle self-evident should think critical rationalism absurd.

The critical rationalist can motivate his rejection of the justificationist principle. Our epistemic predicament is that there just are no good reasons for theories which transcend the evidence. Inductive inferences being invalid as Hume said, there are no conclusive reasons for them and the certainty-sceptic is right. Probabilism having foundered, there are no inconclusive reasons for them either and the probability-sceptic is right. Accept the justificationist principle, and irrationalism is unavoidable. In this predicament we can only reject the justificationist principle. The best we can do is provide reasons for our tentative acceptance of theories which are not reasons for the theories accepted. Now there is nothing more rational than a thorough and well-conducted critical discussion. So the best reason we can have for accepting a theory is that it has survived such a discussion.

We find something like this motivating argument in the following passage:

> ... since we *have* to choose, it will be 'rational' to choose the best-tested theory. This will be 'rational' in the most obvious sense of the word known to me: the best-tested theory is the one which, in the light of our *critical discussion*, appears to be the best so far, and I do not know of anything more 'rational' than a well-conducted critical discussion. (Popper 1972, p. 22; 'best tested' obviously means 'best-corroborated'.)

However, I admit again that this is a matter of interpretation (I do not know, for example why Popper always puts scare-quotes around the word 'rational').

We also find something like this motivating argument in David Miller's paper 'A Critique of Good Reasons'. Miller argues that there can be no good reasons for the things we beileve, for the theories we accept or adopt or prefer (or as he prefers, classify as true). The argument is subtle and ingenious. Conclusive reasons are not good reasons because they beg the question; inconclusive reasons are not good reasons because they beg part of the question and are reasons *against* the other part of the question (for details see Miller's 1987, pp. 346–351). I do not know whether this argument is sound. If it is sound, it reinforces our motivating argument for critical rationalism, and it reinforces the need to distinguish reasons for what we believe from reasons for our believing those things.

Miller, however, draws no such distinction. Instead, he tacitly adopts the justificationist principle that a reason for believing a theory must be a reason for the theory believed. For having argued that there are no good reasons of the latter kind, he straightaway concludes that there are none of the former kind also:

> ... there are no such things as good reasons; that is, sufficient reasons for *accepting* a hypothesis rather than *rejecting* it, or rejecting it rather than accepting it, or *anything like that*. (1987, p. 343, italics mine)

> ... no reason at all is needed for *classifying* an hypothesis as true ... (1987, p. 352, italics mine)
> ... statements [which contradict it] will not supply us with a reason *for rejecting* "All swans are green". (1987, p. 356, italics mine)

> ... the critical rationalist's answer to the question 'Why do you *think that h* is true? or 'Why do you *think that* action *a* should be performed?' will be 'Why not?' (1987, p. 356; italics mine)

This is incredible. In Miller's critical hands rationalism has turned into wholesale irrationalism. Let H_1 and H_2 be the only hypotheses in the field. Let H_1 be as soundly refuted as a hypothesis can be, and H_2 as strikingly corroborated as a hypothesis can be. No doubt Miller's 'critical rationalist' will classify H_1 as false and H_2 as true. But according to Miller he has no good reason for these classifyings; they are unreasonable; they are just as unreasonable as those of the madman who classifies H_1 as true and H_2 as false. (For more details, see my 1989, section 3.)

Miller attributes this incredible position to Popper. In doing so, he has to set aside passages from Popper like the ones cited earlier in this paper. So far as I know Popper has nowhere repudiated Miller's interpretation of his 'critical rationalism'. The exegetical question is insoluble (at least by me). Setting it aside, I will merely say that if Miller's interpretation is correct, then so-called 'critical rationalism' is another name for irrationalism.

Miller denies that a high degree of corroboration is any reason to accept a hypothesis as true. As we saw earlier, Watkins denies it too. Both endorse rationality-scepticism, despite their protestations to the contrary. Why do Miller and Watkins, travelling very different routes, arrive at the same disastrous destination? My guess is that it all has to do with the pragmatic problem of induction.

4. THE PRAGMATIC PROBLEM OF INDUCTION

Recall Popper's statement that "There is not even anything irrational in relying for practical purpose upon well-tested theories, for no more rational course of action is open to us" (1963, p. 51). This statement, and others like it, have prompted the most frequently urged criticism of Popper's entire philosophy. In "relying for practical purposes upon well-tested theories", Popper must be smuggling in some kind of *inductive inference* or *inductive assumption*. He must be assuming that predictions from well-corroborated theories will be true, while predictions from refuted theories will be false. Or at least, he must be assuming that predictions from well-corroborated theories will be more likely to be true (more probable) than predictions from refuted theories. Some tried to answer Hume and validate induction by appealing to metaphysical principles like "Nature is uniform" or "The Future will be like the Past" or "Unobserved cases resemble observed cases". Hume showed this would not do, and Popper says that he agrees. But Popper himself solves the pragmatic problem of induction by tacitly appealing to some metaphysical principle of the same kind.

This cricitism, in one form or another, has been levelled at Popper by Salmon, Lakatos, Ayer, Putnam, Grünbaum, and O'Hear (references are given in Watkins 1974, p. 279), and also by Good, Hesse, Warnock, Levison, Feigl, Cohen, Niiniluoto and Tuomela, Howson and Worrall (references are given in Miller's 1982). A careful statement of it is to be found in John Worrall's paper in the present volume. Watkins, having once tried to rebut Salmon's formulation of the criticism, came to endorse it too and concludes his recounting of that debate with "Game, set and match to Salmon!" (1984, p. 341). Watkins then proceeds to have another stab at the problem (his 1988 is a "simpler and better" version of his 1984 solution).

The criticism can be rebutted. Popper's solution to the pragmatic problem of induction involves no inductive inferences and invokes no metaphysical inductive principles. (Perhaps I should say that a *Popperian* solution can be presented which does neither of these things, since I do not know whether

Popper would subscribe to what follows.) On the other hand, Watkins's allegedly non-inductivist alternative solution does need such an inference or such a principle if it is to be acceptable. Or so I shall argue.

Taking Watkins's solution first, he poses the problem thus:

> Let a technologist confront a practical problem upon which several rival hypotheses bear, each indicating a different solution. One of these is better corroborated than its rivals. Can a neo-Popperian like myself, who seeks to exclude all non-deductive inferences and who insists that corroboration appraisals of hypotheses have no implications for their future performance, give him any reason why he should prefer the best corroborated hypothesis? (1988)

Watkins's solution rests on two assumptions: (i) "well-corroborated hypotheses *have hitherto* proved better guides for practical decision-makers than have hypotheses that were not well-corroborated"; (ii) "it is rational, other things being equal, to choose that course whose success presupposes *the weaker forecast about the future.* For a weaker forecast is less likely, other things being equal, to turn out wrong than a stronger one" (1988). Next we have a *corroborationist principle*: the *overall* success-rate of actions guided by well-corroborated hypotheses is higher than that of actions guided by hypotheses that are not well-corroborated. Finally, we have two versions of a *counter-corroborationist principle*: first, the *overall* success-rate of actions guided by hypotheses that are not well-corroborated is higher than that of actions guided by hypotheses that are well-corroborated; second, this will be the case *from now on.*

Watkins now adds the historical assumption (i) to each of these principles to see what each of them entails (*deductively* entails) about the future. The corroborationist principle entails that corroboration-guided actions will continue to have a higher success-rate. (Actually, it does not entail this – but let us let this pass). The counter-corroborationist principle (first version) entails that counter-corroboration guided actions will have a higher success-rate in the future and that this rate will be sufficiently higher to offset their lower success rate in the past. The counter-corroborationist principle (second version) entails that a time will come when 'success-rates change over' *and* that this time has come *now.* Next the crucial step: the two counter-corroborationist forecasts about the future are *stronger* (hence less likely to be true) than the corroborationist forecast; hence it is rational, by assumption (ii), to go in for corroboration-guided actions. The trick is, Watkins claimed, that *non-inductive use* is made of the historical assumption (i).

Clearly, everything hinges of Watkins' claims about the relative strengths of the various forecasts:

(F_1) Corroboration-guided actions will have a higher success-rate than counter-corroboration-guided actions.

(F_2) Counter-corroboration-guided actions will have a sufficiently higher success-rate than corroboration-guided actions to outweigh their lower rate in the past.

(F_3) Counter-corroboration-guided actions will have a higher success-rate than corroboration guided actions.

Watkins insists that F_1 is weaker (hence more likely to be true) than either F_2 or F_3. These judgments are intuitive ones and I have to report that I do not share Watkins's intuitions. My intuition in such cases is, I confess, impoverished. What impoverishes it is my *deductivism.* When *A* deductively entails *B* and not *vice versa*, I have a clear intuition that *A* is stronger and less likely to be true than *B*. Almost everywhere else my intuition fades – and Watkins's forecasts are somewhere else. His intuitions seem to depend largely on the way things are *formulated.* The attentive reader will have noticed that my formulation of F_3 differs from that given in the preceding paragraph, which is Watkins' preferred fomulation. I admit that the first formulation pushes intuition in the direction Watkins wishes. I submit that my F_3 is just another way of saying the same thing. I have no clear intuition that F_3 is stronger and less likely to be true than F_1.

But one can do better than trade intuitions or the lack of them. Watkins's judgements of the relative strengths of these forecasts (and hence of their relative probabilities) are either *a priori* judgements or *a posteriori* judgements. Suppose they are *a priori* judgements. Then they are synthetic *a priori* judgements. Watkins' judgement regarding F_1 (that it is weaker and more likely to be true than F_3) tells us that we are more likely than not to have a higher success-rate with corroboration-guided actions. This assertion is not analytic. (Indeed, it is a synthetic metaphysical inductive principle.) But appeal to any synthetic *a priori* judgement is ruled out by Watkins.

The other alternative, that Watkins's judgements are *a posteriori*, will be equally unpalatable to him. On what experience could they be founded? The only experience in the matter is that reported in the historical assumption (i). But then we would have a straightforward *inductive argument*: given that well-corroborated hypotheses have *hitherto* proved better guides for practical decision-makers than have hypotheses that were not well-corroborated, it is more probable than not that they will *continue* to do so. (I suspect that any who share Watkins's intuitions in this matter tacitly appeal to this 'Straight Rule' argument to get the relative *probabilities* of the forecasts and then read off their *strengths* from these probabilities.) Needless to say, appeal to an inductive argument such as this is also ruled out by Watkins.

So Watkins's solution to the pragmatic problem of induction rests *either* on synthetic *a priori* judgements *or* on inductive reasoning. And when you think about it, it could not be otherwise. He needs to show that one forecast about the future is more probable than another. The entire thrust of his own previous discussions had been that showing such things requires either metaphysical inductive principles (synthetic *a priori* assumptions) or inductive inferences. He has simply failed to heed his own lesson. (There are many other things to be said about Watkins's solution. For example, is he really entitled to his assumptions (i) and (ii)? But since John Worrall's paper in the present volume says most of these things, I shall say no more.)

Instead, let us return, somewhat chastened and perhaps even a little softened-up, to the critical rationalist solution as I understand it. The widespread accusation is that it too involves tacit inductive arguments and/or tacit metaphysical inductive principles. Let us see. Earlier we formulated a Popperian principle of rational acceptance in science:

(CR) It is reasonable to adopt a theory as true at time t if that theory is the best corroborated at time t.

(This principle will not quite do as it stands, but never mind for the moment. Besides it is confined to corroboration and does not mention non-empirical critical considerations; but never mind that for the moment either.) Is any inductive argument involved in applications of this principle? NO. The argument involved is a straightforward deductive one:

(CR) Theory T is the best corroborated theory at time t.
Therefore, it is reasonable to adopt theory T as true at time t.

But we are also told that we can rely (or 'rely') for practical purposes on well-corroborated theories: we are told, in other words, that it is also reasonable to accept as true the *predictions* which well-corroborated theories make. Surely an inductive argument is involved here.

Well, again NO. Here is a (hopefully uncontroversial) principle of *applied deductive logic*:

(ADL) If it is reasonable to adopt P as true, and if P deductively implies C, then it is also reasonable to adopt C as true.

The argument involved in applying rationally-adopted scientific theories is another perfectly straightforward deductive argument:

(ADL) It is reasonable to adopt theory T as true at time t. (The conclusion of the first deductive argument.)
Theory T deductively implies prediction P.
Therefore, it is reasonable to adopt prediction P as true at time t.

So only deductive arguments are involved in rationally accepting theories as true (in accordance with CR) and in applying accepted theories in practice (in accordance with ADL).

Suppose this is accepted (it seldom is). It will immediately be objected that an *inductive principle* must figure in the premises of these arguments. The obvious candidate is CR. Is this not an inductive principle? Well, CR is a principle of a quite different kind from the usual inductive principles. They were *metaphysical* principles, CR is an epistemic principle. Let us not quarrel about the word 'inductive'. Let us agree to call CR an *epistemic inductive principle.*

Epistemic inductive principles of one kind or another are ubiquitous. *Total irrationalism* is the view that it is unreasonable to accept anything as-anything (as true or as-X for any X other than truth). Nobody is a total irrationalist. That is, everybody accepts *some* epistemic inductive principle or other, some principle stating that it is reasonable to accept something as-something. Watkins' Humean sceptic can be viewed as one who accepts the spartan principle "It is only reasonable to accept as true reports about one's immediate experiences and statements deductively implied by such reports." The probabilist can be viewed as one who accepts the principle "It is reasonable to accept as true hypotheses that can be shown to be more probable than not in the light of the available evidence". Those who accept so-called *inference to the best explanation* can be viewed as accepting the principle "It is reasonable to accept as true the best available adequate explanation of any body of facts". (Inference to the best explanation is commonly regarded as a species of inductive inference; it is best seen, however, as a deductive pattern of inference with this principle as its major premise (see my 1989). Thus, there is nothing to stop a *deductivist* from accepting inference to the best explanation, as Watkins does in his 1984.)

All epistemic inductive principles are 'ampliative' in the sense that they enable you to obtain conclusions which do not follow from the other premises of the deductive arguments in which they figure. For example, CR is not a *redundant* premise in the argument-scheme given a few paragraphs back. To be ampliative in this sense is simply to be *non-analytic.* As deductivists see it, there are no ampliative epistemic inductive *principles.* And everybody who is not a total irrationalist (that is, everybody) can be seen as accepting one or more of them.

Even if this is accepted (which it seldom is), it will immediately be objected that so far CR is just an arbitrary proposal, no reason has been given for it, it is completely unjustified. *Given* CR, the problem of rational theory-acceptance may be solved; given ADL as well, the pragmatic problem of induction may be solved. But can we be 'given' CR with no reason for it? CR is not an analytic truth, as has just been admitted, so we *need* a reason for it. And once we start to think of possible reasons for it, possible justifications of it, we will see immediately that any such reason or justification will be a *metaphysical* inductive principle of the sort we are supposed to be avoiding. *Why* do you think it reasonable to adopt well-corroborated theories? Obviously, it must be because you think that corroboration is a good guide to truth (or verisimilitude, or high probability). Induction is smuggled in after all! And it is really quite obvious that it *must* be smuggled in. Corroboration is 'backward-looking': it concerns past predictive success. By itself, a corroboration-report can say nothing about future performance. So we must supplement corroboration with some principle linking past success to future performance – that is, with some metaphysical inductive principle. (Reports on the more general state of the critical discussion are just as 'backward-looking' as corroboration-reports, and the same argument applies to them. I will talk for a while in terms of corroboration alone, as is customary.)

I have stated this objection as forcefully as I can, since it takes us to heart of the issue. I will consider it 'back-to-front'. *Of course* corroboration-reports are 'backward-looking': *one cannot report on future tests.* But we need not supplement a corroboration-report with any metaphysical inductive principle for it to have 'forward-looking' implications. Two simple deductive argument-schemes were given earlier. A corroboration report appeared as the second premise of the first argument. The assertion that it is reasonable to adopt a certain *prediction* appeared as the conclusion of the second argument. I defy the reader to locate a metaphysical inductive principle among the premises of either. We need no such principle to forge a past-future link – rationally-adopted scientific theories do that job for us.

But does not some metaphysical inductive principle *lie behind* the epistemic principle CR? This brings us back to the first part of the argument. We need a reason for CR and only a metaphysical inductive principle could provide one. Now the demand for a *reason for* CR assumes in meta-science precisely what critical rationalism denies in science. It assumes that it only reasonable to *adopt* CR if a reason can be given *for* CR. The consistent critical rationalist must refuse to give a reason for CR: any such reason would be (as the objection rightly states) a metaphysical inductive principle. But this does not mean that CR must be *arbitrarily* adopted. For the consistent critical rationalist could give a reason for (or justify) his *adoption of* CR. He could say that it is reasonable to adopt CR because it has withstood philosophical criticism better than rival epistemic principles. Such a reason or argument has, in effect, already been given, when we described the *epistemic predicament* in which we find ourselves (see *above*). That argument might now be strengthened by adding Miller's negative result (see *above*) about the non-existence of good reasons for what we believe (assuming, of course, that this negative result is correct).

Perhaps these considerations will shed some light on the following passage from Popper:

> But every action presupposes a set of expectations; that is, of theories about the world. Which theory shall the man of action choose? Is there such a thing as rational choice? This leads us to the pragmatic problems of induction:
>
> Pr_1 Upon which theory should we rely for practical action, from a rational point of view?
>
> Pr_2 Which theory should we prefer for practical actions, from a rational point of view?
>
> My answer to Pr_1 is: From a rational point of view, we should not 'rely' on any theory, for no theory has been shown to be true, or can be shown to be true.
>
> My answer to Pr_2 is: But we should prefer as basis for action the best-tested theory.
>
> In other words, there is no 'absolute reliance'; but since we have to choose, it will be 'rational' to choose the best-tested theory ...
>
> Of course, in choosing the best-tested theory as a basis for action, we 'rely' on it, in some sense of the word. It may therefore be described as the most 'reliable' theory available, in some sense of this term. Yet this does not say that it is 'reliable'. It is not 'reliable' at least in the sense that we shall always do well, even in practical action, to foresee the possibility that something may go wrong with our expectations. (1972, pp. 21–2)

Then again, perhaps what I have been saying sheds no light on this, because *nothing* can.

We had the critical rationalist giving as a reason for *adopting* CR (but not as a reason for CR itself) the fact or alleged fact that CR has withstood criticism better than alternative epistemic principles. It will immediately be objected that this is *circular*. The critical rationalist says that it is reasonable to adopt CR *by CR's own standard of when it is reasonable to adopt something*. Quite so. At this level of abstraction such circularity is unavoidable (any general epistemic principle will be subject to it) and *alternatives to it are even more unacceptable*. One alternative is simply to postulate CR and give no reason for adopting it. "The method of "postulating" what we want has many advantages; they are the same as the advantages of theft over honest toil" (Russell 1919, p. 71). The other alternative is to try to give a reason for adopting CR (or for CR itself) which is not of a critical rationalist kind. A reason for CR would be a metaphysical inductive principle, anathema to critical rationalists, and no use anyway, since if such a principle is proffered as a reason for CR, we will immediately need a reason for it, and so on *ad infinitum*. A reason for adopting CR of a non-critical-rationalist kind would mean that critical rationalism was not self-contained or comprehensive, in that the critical rationalist would no longer be a critical rationalist *regarding his critical rationalism*.

Still, a self-contained or comprehensive critical rationalism is not an unmixed blessing. Is anyone sceptical of CR going to be *convinced* that he ought to adopt it by being told that it withstands criticism better than alternative epistemic principles? It is not true that there are no arguments for adopting critical rationalism, since there are circular ones. It is true that none of these arguments will convince anyone who rejects critical rationalism in the first place. (Similarly, it is not true that there are no valid arguments for the practice of arguing validly. It is true that none of these arguments will convince anyone who does not believe in the practice of arguing validly.)

Let us descend from these lofty regions. We had the critical rationalist arguing that it is reasonable to adopt CR because it has withstood criticism better than alternative principles. Is even the *premise* of this (curiously circular) argument correct? CR has been compared to justificationist alternatives like naive inductivism and probabilism. Suppose that it does *better* than these (Watkins would presumably accept this much). This does not mean that it does *well* in the critical debate. Here is an *objection* to critical rationalism: it will encourage a *reckless profligacy* in belief or theory-acceptance. The critical rationalist (so-called) thinks that one or two striking corroborations of the latest and most speculative theory in some field will make it reasonable to accept that theory as true. This will discourage further critical investigation and further testing of a theory. Worse than that, the critical rationalist (so-called) also thinks it reasonable to employ this accepted speculative theory *in practice*. This is a recipe for (technological) disaster. We need a much more sober and cautious theory of rational acceptance than this, both in theoretical science and especially in practical science.

This is a serious objection which must be directly rebutted. The critical rationalist might first emphasise that his beliefs or theory-acceptances are always *tentative* ones. Such beliefs do not discourage further critical investigations, nor do they encourage practical recklessness (especially where the costs of making a mistake are high). But second, and more important, the critical rationalist should *deny* that one or two striking corroborations of some speculative theory at the frontier of science are enough to justify even its tentative acceptance. At the frontier of science, where the critical battle is still raging, no theory will have yet established its superiority. The epistemic attitude critical rationalism will recommend here is 'suspension of judgement'. The theories the critical rationalist tentatively accepts as true have *shown* their superiority at withstanding criticism. They really are *well-corroborated* theories. (Not merely the *best*-corroborated theories, for the *best*-corroborated theory in a field may be *poorly* corroborated. That is why I said earlier that the formulation of CR, which was inherited from Watkins' discussions, would not quite do.) They are theories like those listed above against the pessimistic induction, theories which justificationists describe as 'established theories'. The critical rationalist would claim that the listed theories and others like them are all *true*, all *accepted as true* by current scientists, and all *rationally* accepted as true. (They are not, however, *known* to be certainly or probably true.) Moreover, these are the sorts of theories that we use without qualm in practice without incurring technological disaster.

But (Philip Pettit objected) what if we *have to choose* for practical purposes between theories at the frontier, concerning which the critical debate still rages? The scenario is unlikely, I replied. Unlikely maybe but possible, Pettit rightly responded. Then we are like the person lost in the bush, with two tracks to choose from, and no reason to prefer one track to the other. Inaction being a disastrous form of action, we *flip a coin*. We *act on* the hypothesis "This track will get me out of the bush" without *believing* it (accepting it as true). No more rational course of action is available to us, in this unlikely scenario.

Here is another objection to critical rationalism: it cannot cope with the *grue-problem*. Let theory T be as highly-corroborated by a body of evidence E as a theory can be. The critical rationalist enjoins us to accept T tentatively as true and to employ it in practice. But we can always concoct a grueish variant T^* of T which is also consistent with E and equally well corroborated by it, but which makes utterly different predictions about the future. Suppose for simplicity that T's only untested prediction is "If conditions C obtain, E will result". Then T^* might be "Until now the world has been as if T were true, but henceforth if conditions C obtain, E will not result". For the critical rationalist it is just as 'reasonable' to accept T^* as T, and it is no more reasonable to expect E to occur than to expect E not to occur in conditions C. This difficulty is quite general: for any theory T, any body of evidence E, and any prediction P made by T, you can concoct a grueish variant T^* which will be just as well corroborated by E as T is and which predicts not-P. Thus critical rationalism is empty: it cannot ever single out one of a set of competing theories as the one which it is reasonable to adopt, or any prediction as the one which it is reasonable to make. It cannot vanquish Humean irrationalism after all.

This objection too must be met head-on. The critical rationalist must simply *deny* that grueish variants like T^* emerge as well from the critical discussion as well-corroborated scientific theories. Perhaps T^* is not so well-corroborated by E as T is; true, T^* was concocted so as to be consistent with E, but corroboration requires more than mere consistency with available evidence. (What more it requires depends upon the detailed theory of corroboration a critical rationalist defends.) Or there may be *non-empirical* criticisms of grueish variants like T^*. Perhaps T^* is not even a candidate *universal* theory in the first place, since it postulates a time (now) when there will be a 'change in the course of nature'. Perhaps T^* is not even a candidate *explanatory* theory in the first place, because "Up until now the world has been *as if* T were true" does not really explain E as T *was* assumed to do. Perhaps it can be shown to be *less simple* than T (and perhaps critical rationalists can show that they are not precluded from appeals to simplicity). Perhaps it can be criticised as a theory which is addressed to no specific explanatory problem in science. Without developing any of these responses in detail (for this paper is getting overly long), I will merely say that I think that critical rationalism has ample resources for eliminating grueish variants of well-corroborated theories. (There is more on this in my 1988, p. 223–5).

Here is another objection to critical rationalism: it is *incoherent* because it talks of 'conjectural knowledge' and this concept is incoherent. This is a weak objection. The critical rationalist *need* not talk of 'conjectural knowledge' at all, if the concept really is incoherent. But is the concept incoherent? 'Conjectural knowledge' is, of course, Popper's term: remember 'Conjectural Knowledge: My Solution of the Problem of Induction' (the title professing an aversion to defining things (an aversion he has managed to overcome in other contexts). But our preceding discussions suggest a perfectly natural definition which will serve to establish the consistency of the notion.

The traditional *justified true belief* account of knowledge runs as follows:

A knows that *P* if and only if
(1) *A* believes *P*,
(2) *P* is true,
(3) *A* can justify (give a reason for) *P*.

My suggested definition of conjectural knowledge differs in only one way, though it is a crucial way, from this:

A conjecturally knows that *P* if and only if
(1) *A* believes *P*,
(2) *P* is true,
(3*) *A* can justify (given a reason for) his believing that *P*.

The crucial difference is, of course, that the traditional account demands a reason for what is believed (in accordance with the justificationist principle), while the conjecturalist account demands a reason for the believing of it (in accordance with the critical rationalist's rejection of that principle). 'Conjectural knowledge', thus defined, is a perfectly consistent notion: 'false knowledge' is a contradiction in terms, but 'conjectural knowledge' is not. But do we need such a notion? Why not confine ourselves to saying "*A* conjectures that *P*" or even "*A* reasonably conjectures that *P*" instead of "*A* conjecturally knows that *P*"? Well, these are not the same thing at all: one can conjecture that *P*, even reasonably conjecture that *P*, even when *P* is false and one does not conjecturally *know* that *P*. It is the truth-condition (2) which entitles conjectural knowledge to be regarded as a species of *knowledge*.

I doubt that Popper will like my definition of his term 'conjectural knowledge', couched as it is in terms of belief. He says that he does not believe in belief (1972, p. 25). (This turns out to mean that he does not believe in dogmatic, irrationally held beliefs, *faiths, commitments*. Which sensible person does?) He prefers different locutions: 'prefer' (with respect to truth or verisimilitude), 'choose' (when one's aim is truth or verisimilitude), 'accept' or 'adopt' (tentatively); David Miller adds 'classify as true'. I find these terminological fads peculiar coming from philosophers who say that "words do not matter". As has been suggested throughout this paper, I can see no significant *difference in meaning* between the following locutions (for simplicity I leave verisimilitude out of them and omit the word 'tentatively'):

A believes that *P* is true.
A prefers *P* with respect to truth.
A chooses *P* while aiming at truth.
A accepts *P* as true.
A adopts *P* as true.
A classifies *P* as true.

More important than terminological fads is the idea of 'Epistemology without a Knowing Subject' (to cite the title of the third chapter of Popper's 1972). *There can be no such thing* – or at least, there can be no epistemology without an *epistemic* subject. As Popper has taught us, there can be epistemology without a knowing-for-certain subject (forgive the barbarism). Popper's own epistemology is an epistemology of the conjecturally-knowing-subject (forgive the further barbarism). But believing, preferring, choosing, classifying as true, or whatever, are all things that epistemic subjects do – there are none of these things in Popper's (or Frege's) 'third realm' of the objective contents of thoughts. The epistemological *problem* is whether we can know anything for certain, or conjecturally, or have reasonable beliefs, and so forth. You do not even address that problem if you fly to 'world 3' and confine yourself to 'third-world considerations' – instead, you change the subject. Hume thought that any belief which went beyond the immediate evidence of the senses was unreasonable. You merely change the subject if you say (*meaning* it) that you "do not believe in belief".

The positive part of this paper has been more concerned to explain what critical rationalism as I understand it *asserts* than to defend its assertions as *correct*. But I have not tried to disguise the fact that I think critical rationalism is correct. You see, I am one of the few who agree with Popper when he says (rather immodestly) "I think that I have solved a major philosophical problem" (1972, p. 1). That problem is, of course, the problem of induction (in all its facets). Popper's solution is so frequently misunderstood, by its defenders as well as its detractors, that I have been chiefly at pains in the positive part of my paper to say what the solution *is*. Its *acceptability* is another matter.

University of Otago

REFERENCES

Lakatos, I.: 1974, 'Popper on Demarcation and Induction', in P.A. Schilpp (Ed.), *The Philosophy of Karl Popper*, Open Court, La Salle, pp. 241–273; reprinted in Lakatos (1978), vol. I, pp. 139–167.

Lakatos, I.: 1978, *Philosophical Papers*, ed. J. Worrall and G. Currie, Cambridge University Press, Cambridge.

Miller, D.: 1982, 'Conjectural Knowledge: Popper's Solution to the Problem of Induction', in P. Levinson (Ed.), *In Pursuit of Truth: Essays in Honor of Karl Popper's 80th Birthday*, Humanities Press and Harvester Press, pp. 17–49.

Miller, D.: 1987, 'A Critique of Good Reasons', in J. Agassi and I.C. Jarvie (Eds.), *Rationality: The Critical View*, Nijhoff, Dordrecht, pp. 343–358.

Musgrave, A.: 1988, 'Is There a Logic of Scientific Discovery?', *LSE Quarterly*, **2**(3) (Autumn 1988), pp. 205–227.

Musgrave, A.: 1989, 'Deductivism versus Psychologism', in M.A. Nottorno and J. Wettersten (Eds.), *Psychologism*, Martinus Nijhoff, The Hague.

Popper, K.R.: 1963, *Conjectures and Refutations*, Routledge and Kegan Paul, London.

Popper, K.R.: 1972, *Objective Knowledge*, Clarendon Press, Oxford.

Popper, K.R.: 1976, *Unended Quest: An Intellectual Autobiography*, Fontana/Collins, Glasgow.

Popper, K.R., 1982, *Realism and the Aim of Science*, W.W. Bartley III (Ed.), Hutchinson, London.

Russell, B.: 1919, *An Introduction to Mathematical Philosophy*, George Allen and Unwin, London.

Watkins, J.W.N.: 1968, 'Hume, Carnap, and Popper', in I. Lakatos (Ed.), *The Problem of Inductive Logic*, North-Holland, Amsterdam, pp. 271–282.

Watkins, J.W.N.: 1984, *Science and Scepticism*, Hutchinson, London.

Watkins, J.W.N.: 1987, 'A New View of Scientific Rationality', in J.C. Pitt and M. Pera (Eds.), *Rational Changes in Science*, Reidel, Dordrecht, pp. 63–82.

Watkins, J.W.N.: 1988, 'The Pragmatic Problem of Induction', *Analysis*.

ELIE ZAHAR

JOHN WATKINS ON THE EMPIRICAL BASIS AND THE CORROBORATION OF SCIENTIFIC THEORIES

In Chapter 7 of his *Science and Scepticism* on 'The Empirical Basis', John Watkins addresses one of the most serious difficulties confronting Popper's philosophy of science. As is well-known, Popper's methodology rests on a demarcation criterion according to which only empirically refutable propositions count as scientific.[1] This criterion arises from the asymmetry, in regard to scientific hypotheses, between verification and falsification. While such hypotheses are taken to be universal statements subsuming infinitely many states of affairs, their potential falsifiers (basic statements) are singular sentences. One is thereby led to believe that the impossibility of verification has to do with the (physical) impossibility of performing infinitely many operations; while the possibility of falsification rests on the possibility – at least in principle – of making finitely many observations in order to decide the truth-value of a basic statement. However, both in his (1930/31) and in his (1934), Popper goes out of his way to assert that observation, as commonly understood, namely as a process grounded in perception, bears no epistemological relation to basic statements.[2] Sense-experience may motivate or cause us to accept a falsifier but it provides no *reason* for doing so. Potential falsifiers can be and in fact usually are objective statements about the external world. Not only are they theory-laden, they often consist of low-level hypotheses containing dispositional terms; their actual verification is therefore impossible and it certainly has nothing to do with the experimentalist's perceptions. According to Popper, a basic statement is acceptable if and only if there is intersubjective agreement about its truth. Such an acceptance can be revised and is therefore non-dogmatic; but, when carefully examined, such 'revision' merely means that the previously accepted basic statement, taken in conjunction with another low-level hypothesis, can be rejected as a result of a fresh *agreement* about some other falsifier. At *no* point does this potentially infinite process bring in epistemological considerations, i.e., considerations which might link the truth-value of the basic statement either to the act of observation or to that of reaching a consensus.

This view of the empirical basis completely destroys the presumed asymmetry between verification and falsification; for we now have simply one proposition, which we call 'theory *H*', confronting another proposition

Fred D'Agostino & I. C. Jarvie, Freedom and Rationality: Essays in Honor of John Watkins, 325–341.

called 'basic statement B'. Despite these labels, all we are entitled to assert is that H and B are logically incompatible. I.e. $H \Rightarrow \neg B$, or, equivalently, $B \Rightarrow \neg H$; so the relationship between H and B is perfectly symmetric. True, Popper often describes B as a *low-level* hypothesis, but this is of no help; for since $(H \not\Rightarrow B)$ and $(B \not\Rightarrow H)$, the levels of H and B are not comparable, except in some vague intuitive sense. More seriously: nothing in Popper's analysis tells us that the theories impregnating B are less risky than H; yet we arbitrarily take H rather than B to be under test. All we can say is that, over the last 400 years or so, science has, as a matter of fact, pursued a largely empiricist policy; i.e. that physicists have often reached agreement over statements of the same form as B rather than over statements resembling H; but, as Popper himself admitted in his (1930/1), we can offer no scientific explanation for the social success of this empiricist strategy.[3] Worse still: we cannot explain why the pursuit of such a policy issued in great *technological* breakthroughs. These breakthroughs are taken to depend, at least partially, on the truth, on the probable truth or on the approximate truth of some consequences of our theories; yet truth-considerations have so far played no role in our methodology. Sustained technological progress becomes a continuing miracle.

Popper's initial position thus leads to scepticism about everything save logic and mathematics. All we can claim to know is that H and B cannot hold good together.

As mentioned by John Watkins, Popper subsequently relented and (implicitly) went along with Ayer in holding that observation not only motivates our acceptance of a basic statement like B but also provides reasons for it.[4] Watkins correctly pointed out that Popper was thereby making an inductivist concession to Ayer: sense-experience provides inductive support for the proposition B which, according to Popper, does not by itself refer to such experience. Once available, this support tips the epistemological balance in favour of holding B to be true, hence of regarding H as having been refuted.

CRITIQUE OF WATKINS'S SOLUTION

John Watkins rightly emphasizes that perceptual experience and the propositions describing it – which he calls level-0 reports – have an important role to play with regard to physical science. I deliberately chose the expression 'with regard to' in preference to 'within'; for, as will presently be seen, level-0 statements are held by Watkins not to belong to science proper. He divides propositions into various levels: 0, 1, 2, ... As already mentioned, level-0 reports describe autopsychological processes while level-1 propositions typically assert the existence of physical objects. Higher-level sentences may refer to more abstract, i.e. to theoretical, entities and may contain additional quantifiers. I think that only one basic dichotomy is called for; namely the one separating the zero level from the rest; where the rest consists of propositions referring to transcendent entities or containing at least one universal quantifier. (An entity will be called transcendent if it goes beyond the present contents of the mind; i.e. if it is non-immanent). Naturally, all sentences are partially ordered by the logical consequence relation, which gives rise to infinitely many layers; but, from the epistemological viewpoint, only the one dichotomy explained above is fundamental.

For the sake of argument, I shall go along with Watkins's assumption that level-1 statements describe external objects which, in some intuitive sense, may, but need not, correspond to perceptual experiences. For example: 'I think (or it seems to me) that there is a red house in front of me' is a level-0 report of which the level-1 correlate is the proposition that there exists, in some spatio-temporal region, a physical house reflecting light of a certain frequency.

Watkins puts forward the following three theses:

(1) Perceptions constitute explananada; these can be hypothetico-deductively explained by means of level-1 statements conjoined with certain physical, or psycho-physical, theories. Watkins calls such conjunctions mini-theories.

(2) No matter how scientific theories might be augmented or beefed up, they can never be tested against pure level-0 reports. So we might just as well test our hypotheses against (fat rather than thin) level-1 statements.

(3) As already indicated in (1), these level-1 propositions, taken together with certain theoretical assumptions, can be checked against sense-experience, i.e. they can be tested by means of level-0 reports. Should they withstand such tests, then they can be 'quasi-rationally' accepted. Thus: by enabling us 'quasi-rationally' to accept the level-1 propositions which test scientific theories, perceptual experience plays a significant methodological role with regard to science, without actually belonging to science proper.

Through showing that the three theses above are mutually inconsistent, I hope to vindicate my own view, which I called the phenomenological thesis and for which I already argued against Popper's conventionalist view of basic statements.[5]

Let T be a physical theory which is to be tested by some propositional function p of level-1 statements. We shall distinguish three cases according to whether p follows from, confirms, or else refutes T. In what follows, each e_i $(i = 1, 2, 3, \ldots)$ will represent a single level-1 sentence. Thus, according to Watkins's 3rd thesis, there exist a theory H_i and two level-0 reports q_i and $\bar{q}_i$, such that q_i confirms, while $\bar{q}_i$ refutes, the conjunction $H_i \wedge e_i$.

(i) If p logically follows from T, then p will usually be of the form

$$p \equiv (e_1 \rightarrow e_2),$$

where

$$\vdash T \rightarrow (e_1 \rightarrow e_2),$$

e_1 describes boundary conditions and e a prediction. In fact, e_1 typically asserts the existence of a physical object (or process) in some spatio-temporal region; for example, e_1 could be Leverrier's assumption of the presence of an extra planet within the solar system. Whatever the case may be, the conjunction $T \wedge H_2 \wedge e_1$ of theories and boundary condition is refuted by the level-0 report $\bar{q}_2$; this contradicts Watkins' second thesis, according to which T cannot be strengthened so as to become refutable by level-0 reports alonge.

(ii) If p confirms T, then p will be of the form $e_1 \wedge e_2$ where

$$\vdash T \rightarrow (e_1 \wedge e_2).$$

In this case $(T \wedge H_1 \wedge H_2 \wedge e_1)$ is confirmed by

$$q_1 \wedge q_2$$

while, as above,

$$(T \wedge H_2 \wedge e_1)$$

would still be refuted by $\bar{q}_2$. Both these results contradict Watkins's 2nd thesis.

(iii) Now suppose that p falsifies T. Thus p is of the form $p \equiv (e_1 \wedge e_3)$, where both

$$\vdash (T \rightarrow (e_1 \rightarrow e_2))$$

and

$$\vdash (e_3 \rightarrow \neg e_2).$$

In this case, Watkins is prepared to say that, were q_1 and q_3 to be verified, this would make it 'quasi-rational' for us to accept both $H_1 \wedge e_1$ and $H_3 \wedge e_3$ and thus regard T as having been refuted (strictly speaking, only the conjunction $T \wedge H_1 \wedge H_3$ has been really undermined). This quasi-rationality is however nothing but Ayer's inductive rationality: the 'consequent' $q_1 \wedge q_3$ provides inductive support for the 'antecedent' $H_1 \wedge e_1 \wedge H_3 \wedge e_3$ or at least for the conjunction of level-1 statements: $e_1 \wedge e_3$. It is precisely this support which then enables us to say that, of the two incompatible propositions T and $e_1 \wedge e_3$, the former rather than the latter has been falsified.

Watkins's account is, to my mind, a correct description of what we often do in actual test situations; but then, we ought openly to admit that our inference needs to be underwritten by some inductive principle. I, for one, am not averse to using inductivist arguments but, like John Watkins, think that induction should be kept at bay for as long as possible. Induction is like a powerful but poisonous (because strictly invalid) drug which should be administered only in extreme situations. The phenomenological thesis, according to which conjunctions of theories, auxiliary assumptions and boundary conditions can be tested against perceptual experience, enables us to dispense with induction longest. It defers the use of inductive reasoning up to the point where we need to say that those hypotheses which have so far withstood the severest perceptual tests are also the most reliable in technological applications.

Anyway, even in this third case, Watkins faces the same objections as in the first one; for we still have a situation in which a conjunction of theory and boundary conditions, e.g. $H_1 \wedge e_1$ and $H_3 \wedge e_3$, can be refuted by a level-0 report, e.g. by $\bar{q}_1$ and $\bar{q}_3$ respectively; and this thesis (2) above declares impossible. True, Watkins calls $H_1 \wedge e_1$ and $H_3 \wedge e_3$ mini-theories; but they are still *theories* and Watkins's objections to theories being tested by level-0 reports are so general that they also apply (or ought to apply) to the cases of $H_1 \wedge e_1$ and $H_3 \wedge e_3$. This is why I shall now turn to these objections and try to show that they are untenable.

In other words: I claim to have demonstrated that theses (1)–(3) are mutually inconsistent and I now propose to show that thesis (2) is the real culprit. (Let me, already at this stage, say that I fully accept thesis (1); as for thesis (3), most people act on it in experimental situations, although they may remain unaware of its inductive presuppositions).

WHY LEVEL-0 REPORTS CAN TEST CONJUNCTIONS OF THEORIES AND BOUNDARY CONDITIONS

A major criticism of the view that level-0 reports should form the empirical basis of science is that neither Newton's nor Maxwell's, Einstein's or Schrödinger's theories refer to anybody's psychological states of mind. So every such theory, T say, is testable by means only of propositions of the form $e_1 \rightarrow e_2$ where e_1 and e_2 are level-1 statements. The problem is that e_1 and e_2 refer to transcendent states of affairs and are therefore conjectural. Thus John Watkins rightly claims that T does not, by itself, entail any level-0 report. I propose however to show that he is wrong in maintaining that T cannot be beefed up so as to become testable by combinations of pure level-0 statements.

Watkins's critique runs as follows. Let O_1 and O_2 be the level-0 correlates of e_1 and e_2 respectively. Our aim, which he deems unattainable, is to strengthen T into T^* in such a way that

$$\vdash T^* \rightarrow (O_1 \rightarrow e_1)$$

and

$$\vdash T^* \rightarrow (e_2 \rightarrow O_2).$$

Since T^* is moreover chosen to be stronger than T, we have

$$T^* \Rightarrow T \Rightarrow (e_1 \rightarrow e_2);$$

hence

$$\vdash T^* \rightarrow (O_1 \rightarrow O_2).$$

T^* would, in such a case, be refutable by the perceptually decidable proposition $(O_1 \wedge \neg O_2)$. The reasons why Watkins thinks the condition $T^* \Rightarrow (O_1 \rightarrow e_1)$ unfulfillable are best explained by an example. Let O_1 be the level-0 report describing my visual impression of a palm-tree; and let e_1 be the proposition that there exists an actual palm-tree in some spatio-temporal region. The problem is so to beef up T that I am thereby enabled to infer e_1 validly from O_1. Now, one might simply try adding to T the crude hypothesis H_1 which says that, whenever an ego thinks it sees a palm-tree, then there exists a corresponding physical object out there. However, at least since Descartes, we know H_1 to be false; so we replace H_1 by another assumption H_2 consisting of two conjuncts: the first says that whenever a person of sound body and mind experiences an O_1-like sensation (i.e. thinks he sees a palm-tree), then an e_1-like situation obtains (i.e. there exists a palm-tree out there).

The second conjunct says that whenever a person feels fine, cannot remember either taking drugs or drinking to excess for the last six months and moreover thinks that his ideas are clear and distinct, then he/she is actually sound in mind and body. I now add H_2 to T and thus obtain $T_2 \equiv T \wedge H_2$. Such a process could of course be pushed further, issuing a sequence of hypotheses: $H_1, H_2, \ldots, H_n$.

Watkins's point seems to be that hypotheses like $H_1, H_2, \ldots$ are all suspect and, in this, he is of course right. It can even be granted that none of these hypotheses is as respectable as Newtonian Mechanics or Einsteinian Relativity. However, every scientific theory is dubious in the strict sense of being probably false. (Even relativity yields, for the subatomic domain, results which are known to be false.) H_2 may be false because healthy people may experience mirages. We can however weaken H_2 by strengthening the antecedents of one of its conjuncts through the proviso: '... and there is no desert around ...'; or more precisely: '... and the person in question perceives no large yellowish expanse around him ...'. This illustrates the trivial point that, possibly through monster- or exception-barring, an auxiliary hypothesis H_n can be reached such that: H_n has not so far been falsified and:

$$\vdash (T \wedge H_n) \rightarrow (O_1 \rightarrow e_1).$$

Similarly, a mirror-image operation should yield a theory H'_n such that

$$\vdash (T \wedge H'_n) \rightarrow (e_2 \rightarrow O_2).$$

Hence, in view of

$$\vdash T \rightarrow (e_1 \rightarrow e_2),$$

we obtain:

$$\vdash (T \wedge H_n \wedge H'_n) \rightarrow (O_1 \rightarrow O_2).$$

As already mentioned, it remains true that the hypotheses H_n and H'_n are dubious; so that, in case of refutation, i.e. if $O_1 \wedge \neg O_2$ should be ascertained, the falsity of

$$T \wedge H_n \wedge H'_n$$

could be imputed to that of H_n or of H'_n; that is: the auxiliary assumptions, rather than the core theory, are blamed for the failure of the whole system. In fact, this is what usually – and quite properly – happens; which also explains why dramatic confirmations play more important a role in the progress of science than do falsifications: should $O_1 \wedge O_2$ be verified, then

$$T \wedge H_n \wedge H'_n$$

is taken to have been confirmed; this, partly because the chances that the 'falsities' of the disparate hypotheses T, H_n and H'_n should have exactly compensated for one another are felt to be slim (though non-zero).

Thus I cannot accept the claim that level-0 statements do not test scientific theories. As explained above, such a claim in any case runs counter to Watkins's – correct – thesis that perceptual experiences constitute explananda. How could level-0 sentences be explained *except* through being derived from physical and psycho-physical theories conjoined with level-1 statements acting as boundary conditions?

JOHN WIDEAWAKE AND THE INCORRIGIBILITY OF LEVEL-0 REPORTS

John Watkins rightly accepts that we can infallibly know the truth of our autopsychological statements – at least while asserting them. The following question arises: is there any way we can argue for the truth of our level-0 reports apart from pointing to their so-called self-evidence?

According to Brentano[6], this self-evidence constitutes a fundamental intuition which we have to accept, but which we cannot justify without falling either into a circle or into an infinite regress. Brentano must ultimately be right: level-0 propositions are not true by virtue of ordinary logic alone; their negations are not self-contradictory; so they cannot be justified without recourse to some synthetic principle. Yet the latter – and this is a possible answer to Brentano – may be more intuitively convincing or immediate than the alleged self-evidence of the level-0 reports themselves. Anyway, such talk of self-evidence smacks of the sort of psychologism which Popper strongly condemned[7]: is such 'self-evidence' anything more than the strong *feeling* of conviction we experience while affirming level-0 propositions; i.e. the feeling that the opposite of what we assert is (psychologically) unthinkable? But it is well-known that *feelings* of conviction can be misleading: while dreaming, for example, I may have a very strong feeling of conviction about the existence of some threatening object which later turns out to have been illusory. Could we not be just as mistaken about the alleged truth of level-0 reports? The answer to this objection is that our insight into the truth of such statements directly flows, not from any *feeling* of conviction on our part, but from the reflexive character of the level-0 sentences themselves. This is clearly indicated by Descartes in his Meditations: I can doubt, but I cannot doubt that I doubt; I may not be hearing a real (physical) noise, but it certainly seems to me that I hear a noise, etc. Brentano has in fact given what seems to be the best account of this reflexive character.[8] First, he underlined the intentional structure of consciousness which is, in his view, always a consciousness of something; i.e. it always intends, or is directed towards, some object. Secondly however, since this object need not exist independently of consciousness in order for the latter to be directed at it, Brentano called such objects intentional; these are, by definition, strictly immanent entities, i.e. they do not exist 'outside' the mind. It does not of course follow that *no* transcendent entity exists over and above the corresponding intentional one; only that the latter is always present, whether or not the former is. But can the intentional object be said to *exist* in any legitimate sense?

On this point, Brentano changed his mind over the years. Initially, he ascribed to intentional entities a kind of existence which he misleadingly called their inexistence in consciousness.[9] But Brentano grew dissatisfied with the shadowy existence of his intentional realm. This is why he put forward his second and definitive solution,[10] which is best illustrated by a concrete example. Consider the proposition 'I see a red house' and assume that there exists no physical house *out there* at all; i.e. suppose that I am hallucinating. If 'red house' is given its ordinary realist meaning, then the proposition above will be false. It could conceivably be turned into a true sentence if 'red house' were to denote an *existing intentional object*. However, should one deny the existence of intentional entities, then the truth-value of our proposition would become problematic. Fortunately, the sentence 'I think I see a red house', or 'It seems to me that there is a red house in front of me', has a truth-value independent of the existence or non-existence of any physical building reflecting light of a certain frequency.

According to Brentano's final solution, all forms of consciousness are of the kind described by statements like 'I think I see a red house'. I.e.: consciousness always possesses two indissolubly connected aspects. There is first a primary consciousness directed at a primary object, e.g. the red house. However, in no sense can a primary intentional object, i.e. the red house as *pure phenomenon*, be said to exist; there is simply *no immanent red house*; only the ego with its directional conscious activity exists; hence all statements about the intentional red house are shorthand for pronouncements about the conscious acts themselves. Secondly, this primary consciousness is always accompanied by a so called secondary consciousness which takes the primary conscious activity itself as its intentional object; such an intentional entity, which consists of certain primary conscious *acts*, necessarily exists and, through it, secondary consciousness indirectly intends the (non-existent) primary intentional object (the red house qua pure phenomenon). That is: whenever I primarily (seem to) see a red house, I am at the same instant secondarily aware of seeing it. Finally, this secondary consciousness is simultaneously self-awareness; i.e. secondary awareness is aware both of itself and of primary consciousness. Thus, according to Brentano's definitive view, every conscious act is such that:

(a) There exists a primary conscious activity directed at a non-existent intentional object. (This does not, of course, automatically imply that no corresponding *real* primary object exists.)
(b) There also exists a secondary conscious activity which directly targets the (certainly existing) primary activity and, through this primary consciousness, indirectly targets the (non-existent) primary intentional object.
(c) Finally, secondary consciousness is also aware of itself. Hence, tertiary and all higher levels of consciousness collapse back onto the secondary level.

There is a lot wrong with Brentano's final solution; but it can, in my opinion, be altered into an acceptable form. First note that John Watkins's level-0 reports are equivalent to statements of the same form as 'I think (or it seems to me, or I believe) that I see a red house'. We are therefore dealing with expressions of secondary acts of consciousness; thus, as already mentioned, we can abstain from discussing the existence or non-existence of primary intentional objects; this can remain an open question which in no way affects the truth-value of level-0 reports. Secondly, we find Brentano defending the view that every conscious act must be accompanied by secondary consciousness and is therefore self-conscious or, as he calls it, self-aware. (See (c) above). I find this thesis unconvincing. Not only does it exclude subconscious and even semi-conscious activities; it also entails that all animals either are capable of self-consciousness or else have no consciousness at all. There is in fact direct evidence that, in everyday life, we are not always reflexively conscious of perceiving; we simply perceive, i.e. we intend certain primary objects. It remains true that the primary acts themselves *can be reflexively made* into the – necessarily existing – intentional objects of higher-order mental activities, e.g. of secondary consciousness. This gives rise to the level-0 reports: 'I think I see a red house' or 'I believe I see a red house', whose intentional object, namely the perception of a primary entity, is both immanent and under the scrutiny of secondary consciousness. Having admitted that one level of conscious activity can occur without necessarily involving a higher level, we no longer need Brentano's assumption (c), which was introduced mainly in order to stop an infinite regress: if every conscious level presupposes a higher one *distinct from the first*, then we would have to accept an actually infinite number of strata in every conscious act. Brentano arrested this regress at the secondary stage through identifying the tertiary with the secondary level. We can however accept that every level could (but need not) be made into the actually existing intentional object of a higher and distinct level. Everyday experience occurs at the primary level and issues in statements like 'I see a red house', which remains fallible if 'red house' is given a transcendent meaning. All levels above the primary one have immanent, hence necessarily existing, intentional objects; and these give rise to infallible propositions. Of this kind are Watkins's level-0 reports which describe the operations of secondary consciousness. (Why such propositions are infallible will shortly be explained). Thus, if I am right, then the testing of all scientific theories takes place, at the level of secondary consciousness, through statements of the form 'I think I see a red house'. As for philosophical reflection, it is carried out at least at the tertiary level, since it discusses both primary and secondary conscious activities.

In his *Science and Scepticism* John Watkins introduces an important character, namely John Wideawake (J.W.), who achieves a large measure of self-awareness. J.W. in fact – reflexively and through an act of will – raises himself to the level of secondary consciousness, he can thus issue level-0 statements which he knows to be incorrigibly true.

Before accounting for the infallibility of level-0 reports, let me try to dispel the myth according to which propositions can be justified only by means of other propositions. Of course, once a statement q has been deduced from another, p say, then we know that the truth of p entails that of q. After all, by definition, deduction transmits truth. The above-mentioned myth affirms the converse of this trivial deductive principle: it maintains that all we can really know is the transmissibility of truth, not the truth of any single proposition. Of course, if one subscribes, as John Watkins and I certainly do, to the correspondence theory of truth, then such a correspondence can never be directly ascertained between a realist hypothesis and the transcendent state of affairs it refers to. After all, the external world is, by definition, something we do not directly apprehend; we have, not knowledge by acquaintance, but conjectural knowledge of it through theories which cannot be factually verified. However, this conclusion does not apply to the relationship between level-0 statements and the immanent states of affairs they refer to. Here subject and object coincide, or rather the object forms part of the subject. More precisely: through reflexion, consciousness acquires direct, *privileged* access to itself; it can simultaneously grasp its own states and examine the level-0 statements purporting to describe these states; in this way it can ascertain whether any correspondence obtains between the two components in its grasp. Doubts have been raised about the meaning of the universals occurring in level-0 reports: how do we know that they subsume what we actually perceive? But these meanings are *as intended* by, and during, the conscious activity and do not extend beyond it. In other words: there is *no* guarantee that other people, or I at a later date, should attach the same meaning to 'redness' as I do now. With all these provisos however, the statement 'I think I see a red house' can, at the moment it is uttered, be infallibly known to be true – or so I hold.

WATKINS'S THEORY OF CORROBORATION

I now move to a different criticism of Watkins's account of science – one which is indeed quite independent of any claim about the empirical basis. In the following, then, it will not matter whether we consider the empirical basis to consist of level-0 or of level-1 statements.

John Watkins defines the empirical content Ct (T) of a proposition T as the set of all the singular predictive implications (SPI's) of T; i.e. Ct (T) consists of all sentences of the form $e_1 \rightarrow e_2$ where:

$$T \vdash e_1 \rightarrow e_2,$$

e_1 expresses boundary conditions and e_2 a certain prediction. (As already mentioned, I shall remain neutral as between the view that e_1 and e_2 are level-1 statements and the view that e_1 and e_2 express level-0 reports.) Suppose that $(e_1 \rightarrow e_2) \in$ Ct (T) and that an experiment has been actually set up in which e_1 was taken to have been realized. (Note that the boundary conditions referred to by e_1 may have to be expressed by universal, hence strictly unverifiable, propositions.) Should $\neg e_2$ be ascertained, then $T \wedge e_1$ will have to be refuted. Suppose however that e_2 is observed. Can we, without further ado, conclude that this experimental outcome provides genuine support for $T \wedge e_1$? The answer must surely be in the negative; for, as already pointed out by Popper, conventionalist stratagems can always bring experimental results into line with any core hypothesis. Thus, a central problem facing any methodology is how to demarcate between genuine and spurious support.

According to the so called Zahar–Worrall view[11] (henceforth referred to as ZW), an SPI strongly confirms a hypothesis only if the latter has *not* been adjusted to yield the SPI in question. This may sound like a psychologistic, or at least a historical, criterion requiring the methodologist to analyse the scientist's states of mind as they actually unfolded during the construction of a theory. This is however not so: 'to adjust' typically means to introduce some parameters: $\lambda_1, \lambda_2, \ldots, \lambda_n$ into T, then to solve a system of equations ensuring that $(T \wedge e_1) \Rightarrow e_2$. If at most as many independent equations are available as there are parameters, i.e. if the parameters are underdetermined or, at worst, uniquely determined, then the ensuing theory has been doctored, in an objectively *ad hoc* manner, in order to yield the fact e_2.

It has to be admitted that, despite its objective character, ZW has an unwelcome feature: it concerns itself not simply with the relationship between theory and fact but with the relationship between theory, fact and the method of construction of the theory. John Watkins understandably tried to find an alternative demarcation between genuine and spurious confirmation which avoids the disadvantages of, while yielding most of the singular value judgements flowing from, ZW.

Watkins seems to insist on two main conditions which an experimental result must fulfil in order to count as a hard (positive) test of T:

[a] the result must be subsumed by an empirically novel experimental generalisation g implied by T. (g is said to be empirically novel if *no* test carried out before T was proposed is relevant to g, i.e. is either incompatible with, or else subsumed under, g). All other results which follow from T are said to provide soft or, at best, medium support for T. However, the difference between these two adjectives is explicated in terms of incongruent counterparts; since the latter have been defined only within simple (monadic) languages, the distinction between 'soft' and 'medium' confirmations will not bear on the examples adduced below. Only the divide between hardness and non-hardness is of any real significance.

[b] $(e_1 \rightarrow e_2)$ must belong to the corroborable content Co (T) of T, where Co (T) is defined as follows. First assume that each of the primitive predicates occurring in T has been labelled either 'theoretical' or else 'observational'. Secondly, let T_H be the conjunction of all the axioms of T which contain only theoretical predicates and let A be the conjunction of the rest; i.e. T_H represents the theoretical core of T, while A is the conjunction of those axioms which essentially involve at least one observable relation. Thus $T \equiv (T_H \wedge A)$. Now define Co (T) $=_{\mathrm{Def}}$ Ct(T)\Ct(A) = Ct ($T_H \wedge A$)\Ct(A). (I.e. Co (T) is the complement of Ct (A) relatively to Ct (T)).

As it stands, this definition depends on the way T has been axiomatized. After all, T can be regarded as the single axiom of T; in which case T_H would be non-existent and Co (T) empty. We are thus driven to differentiate between allowable and forbidden (i.e. unnatural) axiomatisations and to stipulate that the above definition should apply only to the former.

An allowable axiomatization is defined as consisting of a sequence of distinct, non-analytic axioms: B_1, ... B_N, such that: no proper segment of B_i, for any i=1, 2, ..., N, is a logical consequence of the theory, i.e. of $B_1 \wedge B_2 \wedge \ldots \wedge B_N$. (It can be effectively shown that every axiomatisation can be turned into an allowable one.)

Let me immediately say that I consider condition [b] to be of the utmost importance. Not only does it naturally lead us to distinguish between permissible and impermissible axiomatizations; it also constitutes a necessary condition for the hardness and even for the bare relevance of all tests. However, I shall now explain why condition [b] remains insufficient while [a] is, in my opinion, unacceptable. I.e. I propose to show that [a] is too stringent and [b] too lax to solve the problem at hand.

Let me start with [a], which is predicated on a strict distinction between theories and experimental generalisations. Such a distinction is essential if [a] is to apply at all; for any theory T, no matter how revolutionary, has to explain some empirical successes of some of its predecessors; hence, were T itself to be regarded as a generalisation, then T could not be novel.

I can see no way of distinguishing between a theory T and a generalisation g on the basis either of logical form or of relationship with the facts; for both T and g could be of the form $(\forall x)[F(x) \rightarrow G(x)]$ and both are experimentally refutable. To my mind, one can demarcate between T and g only genetically, that is through the different modes of their construction, i.e. of their genesis. As its name indicates, a *generalisation* is obtained by generalising from certain facts, e.g. through curve-fitting, while a *theory* emerges in hypothetico-deductive fashion.

On this criterion, Kepler's first law about the planets' elliptical paths turns out to be a generalisation obtained by curve-fitting. As for Galileo's equation $v = gt + v_0$, since it was proposed as the next simplest hypothesis after the relations v = constant and $v = ks$ had been refuted, it must be considered a theory rather than a generalisation. (Anyway, velocity could not be directly measured; so $v = gt + v_0$ had to be tested via $s = 1/2\ gt^2$ or via $s = 1/2\ g \cos \theta\ t^2$.) The case of Kepler's equal areas law is ambiguous: on the one hand, it might well have been obtained through observation; on the other, it could have been deduced from Kepler's dynamics, according to which the transverse components of both force and velocity are inversely proportional to the distance from the centre of force.

As an adherent of ZW, I have nothing against criteria based on (objective) heuristic considerations; but, in the present case, the criterion entails that all generalisations, being founded on past facts, must be non-novel. This implies that no confirmation of any theory could possibly be hard.

There exists another and more serious objection to [a]. Suppose that some past hypothesis, or empirical law L, got the facts right in some small interval (a,b), where L traces a curve Γ connecting certain observables; and suppose a new revolutionary theory T implies a law giving rise to a curve Γ^*, where Γ^* extends from $-\infty$ to $+\infty$, but happens to be observationally equivalent to Γ in (a,b). By virtue of condition [a], no test carried out on Γ^*, *no matter how far away from* (a,b), e.g. no matter how near the tail ends of Γ^*, can ever provide hard support for T; and this despite the fact that L, unlike T, may have been obtained in a completely ad hoc way. Such a conclusion is to my mind completely unacceptable: the relationship of T to the evidence, *especially to facts lying right outside the domain of* L, should have nothing to do with the accident that L was proposed before T. It is no fault of T that someone noticed certain facts in (a,b) and took the bother of framing L before T was produced.

Let me again resort to a favourite example. For a long time, the precession of Mercury's perihelion was taken to provide the strongest form of support for Einstein's theory, even though Mercury's anomalous behaviour had long been noticed and subsumed under a low-level generalisation.[12] (Even Einstein's law concerning the bending of light-rays, given the smallness of the predicted magnitudes, had already been implicitly tested by every experiment which failed to detect deviations from Euclidean geometry: Einstein's theory predicts, or rather retrodicts, a bending angle below the then available observational level. Hence, had such bending been *observed*, Einstein's hypothesis would have been born refuted. Let us however go back to Mercury's case, which has the added advantage of showing that [b] constitutes no sufficient condition for hardness of support.)

Obviously, the precession of Mercury's perihelion cannot be explained on the basis of the initial conditions alone; so Einstein's theory satisfies [b]. But, as is well-known, we could modify the boundary conditions, more particularly the mass distribution within the sun, so as to obtain a classical account of Mercury's anomalous motion. Naturally, such a Newtonian account, exactly like Einstein's, fails condition [a]; but, also exactly like Einstein's explanation, the classical one passes [b]; for the modified boundary conditions alone do not yield Mercury's motion but, in order to do so, have to be conjoined with the theoretical core of Newtonian gravitational theory. *Thus, Watkins's two conditions [a] and [b] treat Einstein's and Newton's accounts on a par. This conclusion, which ZW naturally negates, is to my mind intuitively unacceptable.* Be that as it may, one thing remains certain: [a] and [b] do not always yield the same singular value-judgments as those of ZW. Let me end by venturing a bold conjecture: no methodology *M* which takes account only of facts, theories and their comparative dates of discovery, can come up with the same singular pronouncements as ZW; and this for a very simple reason: one may arrive at the same hypothesis *T* along different routes which make use either of different facts or of no facts at all. ZW judges, not an isolated theory, but also the path issuing in that theory; i.e. ZW assesses a heuristic as well as its end-products; the *same theory* may therefore receive *different appraisals* because it is paired with different heuristics. As long as the assessments of *M* remain functions of facts, theories and their respective dates alone, they cannot always coincide with those of ZW.

London School of Economics and Political Science

NOTES

[1] Popper, K.R.: 1934, *Logik der Forschung* (Springer, Wien), Chapter 1.
[2] Popper, K.R.: 1934, p. 29; also: Popper, K.R.: 1930/31, *Die beiden Grundprobleme der Erkenntnistheorie* (Mohr-Siebeck, Tubingen), p. 11.
[3] Popper, K.R., 1930/31, p. 132 (p. 11).
[4] Watkins, J.W.N.: 1984, *Science and Scepticism* (Princeton University Press, Princeton), p. 72.
[5] Zahar, E.G.: 1982, 'The Popper-Lakatos Controversy' (*Fundamenta Scientiae*, **3**, No. 1), pp. 21–54. Especially p. 11.
[6] Brentano, F.: 1925, *Versuch über die Erkenntnis* (Felix Meiner, Hamburg), Section 2.
[7] Popper, K.R.: 1930/31, p. 121.
[8] Brentano, F.: 1924, *Psychologie vom empirischen Standpunkt* (Felix Meiner, Hamburg), Vol. I, Book 2, Chapter 2.
[9] Husserl was later to adopt Brentano's early view. As is well-known, Husserl wanted all logical operators to be both objective and man-made; so he felt he had no option save to look upon all logical and mathematical entities as existent, though ideal, intentional objects. Cf. Husserl, E.: 1948, *Erfahrung und Urteil* (Felix Meiner, Hamburg), Section II.
[10] Brentano, F.: 1924, Vol. 3, Chapters 4 and 5.
[11] Zahar, E.G.: 1973, 'Why did Einstein's Programme Supersede Lorentz's?' (*British*

Journal for the Philosophy of Science, **24**), p. 11; also Worrall, J.: 1978, 'Methodology of Scientific Research Programmes', in Radnitzky, G. and Andersson, G. (Eds.): *Progress and Rationality in Science* (Reidel, Dordrecht).

[12] Zahar, E.G.: 1980, 'Einstein, Meyerson and The Role of Mathematics in Physical Discovery' (*British Journal for the Philosophy of Science*, **31**), pp. 1–43.

GRAHAM ODDIE

THE UNITY OF THEORIES

John Watkins is a critical rationalist. I can think of no more appropriate way of celebrating his contribution to philosophy than by taking one of his conjectures and subjecting it to critical scrutiny: his account of the unity of theories.[1] But criticism is only part of critical rationalism, and it is only part (Section 1) of this paper. In Section 2 I rework this conjecture utilising a possible-worlds framework and a thoroughgoing realism concerning causation, law and theory. Most of the desiderata which Watkins lays down for a concept of unity appear to be satisfied within the new account. Moreover, it yields the serendipitous consequence that unified theories must also possess another virtue for which Watkins argues at length: depth. In Section 3 I offer an account of the aim of scientific inquiry which unifies the distinct elements of Watkins's account.

1. UNITY AND DEPTH IN *SCIENCE AND SCEPTICISM*

1.1. Security versus Depth

The central thesis of *Science and Scepticism* is that the 'Bacon–Descartes ideal' of inquiry contains two radically different poles: a depth pole and a certainty pole.

> The utopian aim for science may be summarized thus: all empirical phenomena rendered explainable or predictable by deduction from true descriptions of initial conditions in conjunction with universal principles that are certainly true, ultimate, unified, and exact. (p. 120)[2]

Stated in this way the aim does appear hopelessly utopian. Nevertheless, while it may be impossible to achieve the aim, it may be possible to make progress towards achieving the aim. A 'progressivised' version of the Bacon–Descartes ideal requires that an inquiry:

Fred D'Agostino & I. C. Jarvie, Freedom and Rationality: Essays in Honor of John Watkins, 343–368.

progress with explanatory theories that are ever

(A) more probable;
(B1) ever deeper,
(B2) more unified;
(B3) more predictively powerful;
(B4) more exact. (p. 130)

The *security* pole is encapsulated in component A; the *depth* pole in components B1–B4. And as the metaphor of poles suggests, security and depth lie in opposite directions – we cannot make progress towards both at once. Watkins argues that we should go for depth, accept the Bs and reject A. However, the Bs concern purely logical features of a theory and only component A contains a reference to truth. Thus if truth is to feature as part of the aim of inquiry we cannot *simply* jettison A in favour of the Bs. Watkins's proposal is to weaken the first component to something that is compatible with the remainder but which nevertheless incorporates something to do with truth: theories are not required to be probably true (or ever more probable) but merely *possibly* true. In effect, this is tantamount to requiring that they be unrefuted by the observational data.

A different way of incorporating truth in the aim, while eschewing certain or ever more probable truth, would be to make use of the concept of truthlikeness. Popper resorted to truthlikeness and it seems that he did so partly because without it his account of inquiry seemed too pessimistic. If the aim of inquiry is truth, but we can never know that we have a true or probable theory, and the best we can do is to show that our favourite theories are false, things look pretty bleak. Science could never contribute anything positive. The concept of truthlikeness provided some hope. If one false theory can be closer to the truth than another then it is at least possible that the succession of falsehoods which science issues may constitute progress towards the truth. Truthlikeness allows the possibility of progress. But, for one thing, the concept of truthlikeness requires logical and metaphysical commitments which Watkins hoped to avoid. For another, if truthlikeness is to be epistemically accessible we almost certainly require an account of inductive probability.[3]

1.2. Unity and Organic Fertility

Watkins introduces the problem of unity in the following passage which it is worth quoting at length:

> Now we may think of the content of a theory as encapsulated in its axioms; and its axioms must be logically independent of each other ... Then why should not the conjunction of *any* logically independent statements constitute a theory? For instance, why should not 'All cows are herbivorous' plus 'All crocodiles are thick-skinned' plus 'All crows are black' constitute a zoological theory no less than the conjunction of Newton's logically independent axioms constitutes a cosmological theory? And why should not the conjunction of Newton's axioms with, say, those of Bohm-Bawerk's *Capital and Interest* constitute one theory superior to either of its two halves taken separately?
>
> Is is rather remarkable that, although scientific theories are taken as the basic units by many philosophers and nearly all historians of science there is no extant criterion, so far as I am aware, for distinguishing between a theory and an assemblage of propositions which, while it may have much testable content, remains a rag-bag collection. (p. 204).

Watkins suggests that theories are *unified* collections of statements: despite being logically independent the statements 'mesh together', in some sense, to generate more 'content' than these statements generate separately. If we take the content of a statement, or class of statements, to be the class of its consequences then it is hard to see how this idea can be captured. However, if we take the 'content' at issue to be *testable content* it appears that some progress can be made. Suppose we could characterise T's testable content (T^{test}) in some satisfactory way. (Watkins's characterises it as the class of T's *singular predictive implications*.) Intuitively, the testable content of the union of Newton's laws and Bohm-Bawerk's laws is just the union of the testable content of Newton's laws and the testable content of Bohm-Bawerk's laws. So the two sets of laws don't mesh together to generate new testable content. By contrast the testable content of Newton's theory of gravitation, N, exceeds the union of the testable content of the law of gravitation, g, and the testable content of the remainder: $N - \{g\}$. By itself the law of gravitation has no testable content, $\{g\}^{\text{test}} = \varnothing$, and clearly N^{test} is greater than $(N-\{g\})^{\text{test}}$. Moreover, the same holds for any other partition of N into two sub-classes of postulates. In any such partition the two parts mesh together in an 'organically fertile' way to produce more testable content than is contained in the union of their separate testable contents.

This suggests Watkins' *organic fertility requirement* as a necessary condition for the unity of T: for T to be unified there must be no partition of T into two sub-theories, T_1 and T_2, such that T^{test} is just the union of $T_1{}^{\text{test}}$ and $T_2{}^{\text{test}}$.

1.3. The Problem of 'Natural' Axiomatisations

The organic fertility requirement applies to a theory considered syntactically, as a set of statements, not considered semantically, as the proposition expressed by the conjunction of those statements. The proposition so expressed will, of course, be axiomatisable in infinitely many different ways if it is axiomatisable at all. If intuitively disunified theories could always be reaxiomatised to satisfy the organic fertility requirement, or intuitively unified theories could always be reaxiomatised so as to fail the organic fertility requirement, then the criterion would be empty. Certainly the former trivialization result can be proved immediately for finitely axiomatisable theories: for any such theory can be axiomatised by a single sentence. What Watkins proposes in order to deal with this difficulty is a criterion of 'naturalness' of axiomatisations. And it is implicit that a theory, considered semantically, is unified just in case any natural axiomatisation of the theory is unified.[4]

Here I will not rehearse the motivation for, but simply state each of the five conditions Watkins lays down for a natural axiomatisation.

1. *Independence requirement:* each axiom in the axiom set must be logically independent of the conjunction of the others.
2. *Nonredundancy requirement:* no predicate or individual constant may occur inessentially in the axiom set.
3. *Segregation requirement:* if axioms containing only theoretical predicates can be separately stated, without violating other rules, they should be.
4. *Wajsberg's requirement:* an axiom is impermissible if it contains a (proper) component that is a theorem of the axiom set, or becomes one when its variables are bound by the quantifiers that bind them in the axiom.
5. *Decomposition requirement:* if the axiom set can be replaced by an equivalent one that is more numerous (though still finite) without violating the preceding rules, it should be. (pp. 208–9).

Clearly the task of characterising a 'natural' axiomatisation is not a straightforward one, and this particular set of rules is the product of a number of revisions by counterexample.

I will present two simple examples of propositional 'theories' which are, intuitively, naturally axiomatised. The definition judges neither to be so. Theory *A* is probably a unified theory (at least it ought to be if such judgements make sense in the propositional domain and the motivation behind Watkins' theory is on the right lines); and theory *B* (a variant of Watkins' rag bag of zoological axioms) is clearly not unified. The definition yields results contrary to the intuitions concerning unity: that A is not and that B is unified. A and B correspond to two first-other theories, UA and UB, which share the same intuitive features. Unfortunately, it becomes well nigh impossible to apply the definition of unity even to these simple first-order examples.

EXAMPLE 1. Let P and Q be two singular observation statements and let T be a singular theoretical statement. A has just two axioms: $\{P \supset T, T \supset Q\}$.

Neither $P \supset T$ nor $T \supset Q$ has any testable consequences in isolation:

$$\{P \supset T\}^{\text{test}} = \{T \supset Q\}^{\text{test}} = \varnothing.$$

A itself entails $P \supset Q$, which has the form of a 'singular predictive implication'. Thus A^{test} is not the union of

$$\{P \supset T\}^{\text{test}} \text{ and } \{T \supset Q\}^{\text{test}}.$$

A seems to a natural enough axiomatisation of its total content, and it also seems that if any theory in this framework is unified A is. Now consider a reaxiomatisation of A, A^*:

$$\{((P \wedge \sim Q) \supset T), ((P \wedge \sim Q) \supset \sim T), ((\sim P \wedge Q) \supset T), ((P \wedge Q) \supset T)\}.$$

As is easily checked, A^* satisfies requirements 1–4. But it also satisfies the decomposition requirement. Using only the resources P, Q, and T the weakest statements (those that cannot be further weakened without becoming tautologous) are *minimal disjunctions*. That is, statements of the form:

$$(\pm)P \vee (\pm)Q \vee (\pm)T$$

where in the place of $(\pm)$ there is either a negation sign or else nothing at all. A^* is tantamount to the following set A^{**} of minimal disjunctions:

$$\{(\sim P \vee T), (\sim P \vee Q \vee \sim T), (P \vee \sim Q \vee T), (\sim P \vee Q \vee T)\}.$$

A^{**} does not satisfy Wajsberg's requirement, because $(\sim P \vee Q)$ is both a theorem of A^{**} and a proper part of one of its axioms. Hence A^{**} is not natural.[5] And because it is not natural its existence is irrelevant to the unity of A. Since every statement is uniquely decomposable into a conjunction of minimal disjunctions, and A^* is a four-membered natural class of minimal disjunctions, by the decomposition requirement, any other natural reaxiomatisation of A will contain at least four axioms, say $C_1, \ldots C_4$. It follows from this (and the independence requirement) that $C_1, \ldots C_4$ must be logically equivalent to the members of A^{**} and A^*.[6]

Do all the reaxiomatisations of A which satisfy conditions 1–5 also satisfy the organic fertility requirement? Since the organic fertily requirement does not depend on the syntactic form of the axioms, but only on their logical strength, either all these natural axiomatisations satisfy the organic fertility requirement or none do. There is only one singular predictive implication of A and that is $(P \supset Q)$.[7] Consider the partition:

$$A^*_1 = \{((P \wedge {\sim}Q) \supset T), ((P \wedge Q) \supset {\sim}T)\}$$
$$A^*_2 = \{((P \wedge Q) \supset T), ((P \wedge Q) \supset T)\}$$

$(P \supset Q)$ follows from A^*_1 alone, whilst the testable content of A^*_2 is empty. It follows that all natural axiomatisations of A fail the organic fertility requirement, and so A itself, no matter how it is axiomatised, lacks unity.

Even though A is a simple propositional example it corresponds to a more interesting lawlike theory: that being P is sufficient for being T, and that being T is sufficient for being Q. Within the first-order framework of Watkins' account such laws have to be identified with regularities. The lawlike theory corresponding to A is thus the universally quantified theory, UA, involving two observational predicates, P and Q, a theoretical predicate T, and two axioms:

$$\{(\forall x)(Px \supset Tx), (\forall x)(Tx \supset Qx)\}$$

The decomposition requirement becomes extremely unwieldy even in this simple first-order framework. First-order correlates of minimal conjunctions can be defined and every statement is equivalent to a conjunction of these.[8] Not surprisingly, UA stated in such normal form involves massive violation of Wajsberg's requirement. What we can say, however, is that the axiomatisation UA is not natural, since the universally quantified version of A^*, (UA^*) is more numerous. That's counterintuitive. And if UA^*, like its propositional correlate, is natural, then all natural reaxiomatisations of theory A fail the organic fertility requirement, and A is not unified. If, as Watkins assumes, the axioms of UA do have the logical form of a law then this judgement is counterintuitive. For surely two such laws of the kind exemplified by UA mesh together in an organically fertile way. If UA does not count as a unified first-order theory then it would be hard to think of one that would. This aside, the difficulty of actually applying the definition to such simple cases suggests it cannot be on the right lines.

EXAMPLE 2. Let P, Q, R and S be singular, logically independent observation statements. Theory B also has two axioms:

$$\{P \supset Q, R \supset S\}.$$

(The universally quantified correlate of B, UB, is rather like Watkins' zoological example: $\{(\forall x)(Px \supset Qx), (\forall x)(Rx \supset Sx)\}$.)

Firstly, B is, intuitively, a natural axiomatisation of its content. Moreover, B is not unified since its two axioms do not mesh together to generate testable content that is not already in the two axioms taken separately.

In order to satisfy the decomposition requirement once again we have to go to the minimal disjunctions or their logical equivalents. Reaxiomatisation B^* satisfies the decomposition requirement, and again any other axiomatisation satisfying the requirement will have axioms individually equivalent to the axioms of B^*.

B_1 $((P \wedge \sim Q \wedge R) \supset S)$
B_2 $((P \wedge \sim Q \wedge R) \supset \sim S)$
B_3 $((P \wedge \sim Q \wedge \sim R) \supset S)$
B_4 $((P \wedge \sim Q \wedge \sim R) \supset \sim S)$
B_5 $((P \wedge Q \wedge R) \supset S)$
B_6 $((\sim P \wedge Q \wedge R) \supset S)$
B_7 $((\sim P \wedge \sim Q \wedge R) \supset S)$.

B has two singular predictive implications: the two axioms $(P \supset Q)$ and $(R \supset S)$. It is not hard to see that no subset of B^* yields $(P \supset Q)$ unless it contains the first four axioms: B_1, B_2, B_3, B_4. Similarly no subset of B^* yields $(R \supset S)$ unless it contains the axioms B_1, B_5, B_6, B_7. Since B_1 occurs in both sets no partition of B^*, $\{B^*_1, B^*_2\}$ can be such that the testable content of B^* is the union of the testable contents of B^*_1 and B^*_2.[9] Hence every natural reaxiomatisation of B satisfies the organic fertility requirement, and B emerges as unified.

Similar remarks apply to the universally quantified version of B, UB, as apply to *UA*. It may be that to satisfy the decomposition requirement the universally quantified correlates of the axioms of B^* are not weak enough. That is counterintuitive in itself, since *UB* looks natural enough. But if they are weak enough, then UB^*, and by implication the zoological rag-bag, turns out to be unified.

2. UNITY AND DEPTH REDEFINED

2.1. A Sketch of the Proposal

Let us take seriously three theses:

(i) Theories are collections, or conjunctions, of causal laws.
(ii) Causal laws involve a relation of natural necessitation.
(iii) Natural necessitation is an irreducible relation.

When these three theses are spelt out with a little logical precision it will be apparent that each theory T is a conjunction of a *unique* set of laws: T^L. This fact enables us to avoid all the problems of 'natural' axiomatisation which bedevil Watkins' account. Once we have sharpened up the concept of observational content we can then take over Watkins' criterion of organic fertility more or less intact. Where P is a proposition, or class of such, let P^{Ob} be the observational content of P. A theory T is organically fertile just in case every partition T^L into two subtheories, T_1 and T_2, has the following feature: the observational contents of T_1 and T_2, taken separately and conjoined, do not exhaust the total observational content of the whole theory T. That is to say, $(T_1{}^{Ob} \wedge T_2{}^{Ob})$ is (strictly) logically weaker than T^{Ob}. With this account it is possible to prove that there are non-trivial and intuitively acceptable examples of unified theories.

2.2. Propositions, Attributes, and Inquiries

Treating theories as syntactic items, even interpreted syntactic items, has well known limitations. It often turns out to be heuristically advantageous to treat theories rather as items of information, and, as is well known, information can be explicated in terms of classes of possible worlds or of world-times.[10] These information-theoretic entitities are often identified with propositions.

By a possible world I don't mean anything ontologically exotic. In particular I don't think of possible worlds as agglomerations of *things*. Rather, a possible world is a way things might be. With Wittgenstein, I think of the world as the collection of facts, not things. Worlds can be thought of as distributions of some attributes through a domain of entities. The basic attributes which generate the logical space need not all be of the first-order, and in what follows higher-order attributes will play an essential role.[11]

In what follows I lay down a few assumptions about attributes (that is, properties and relations in the broadest sense) and then offer a familiar framework which explicates these assumptions. An attribute is a condition which an entity of a certain kind (or an *n*-tuple of entities of various kinds) may either have or lack, and the class of those entities which have it is the attribute's actual extension. Properties are monadic attributes, relations are dyadic, triadic, etc., attributes.[12] Eschewing super-essentialism, some attributes are had by some entities contingently: a particular entity might have lacked some of the properties it has in fact, or borne different relations to other individuals. So at least some attributes might have had different extensions from the extensions they have in fact.

These relatively uncontroversial theses can all be translated into the idiom of possible worlds. A possible world is a completely specific state (relative to some class of generating attributes and a domain of entities). In each possible world an attribute has an extenstion: the subclass of those entities, or *n*-tuples of such, which have the attribute in that world. It may be that in a world *w* the extension of the attribute is empty (nothing has the attribute in *w*). Attributes are thus associated with functions from worlds to classes of entities (or *n*-tuples of entities) of the right sort. A property of individuals induces a function from worlds to classes of individuals. A relation between individuals induces a function from worlds to classes of couples of individuals. A property of properties of individuals induces a function from worlds to classes of properties of individuals, and so on.[13]

Suppose we have a logical space in which the attribute of being a planet features, and consider the following questions:

Q_1 What is the remainder upon division by three of the number of planets?

There are three possible complete answers to Q_1: 0, 1, 2. And corresponding to these are the following propositions:

A_1 The remainder upon division by three of the number of planets is 0.

A_2 The remainder upon division by three of the number of planets is 1.
A_3 The remainder upon division by three of the number of planets is 2.

$\{A_1, A_2, A_3\}$ form a partition of the logical space and it is easy to see that question Q_1 is tantamount to asking which cell of the partition contains the actual world. While a precise answer to Q_1 specifies just one of these cells, imprecise answers are also possible. For example, one imprecise answer is: A_1 or A_2. In general an imprecise answer is a disjunction of cells of the partition. Some propositions entail a precise answer to Q_1 while others entail, at best, an imprecise answer. The propositions:

P_1 The number of planets is 9.
P_2 The number of planets is either 8 or 9.
P_3 The number of planets is either 7 or 8 or 9.

entail respectively the following answers to Q_1:

A_1; (A_1 or A_2); (A_1 or A_2 or A_3).

A_1 is what P_1 says *about* Q_1: (A_1 or A_2) is what P_2 says *about* Q_1. And (A_1 or A_2 or A_3), what P_3 says *about* Q_1.

These observations can be generalised. A question, or more generally an inquiry, induces a partition of the class of worlds, and each such partition is tantamount to a question: the question, *which cell of the partition contains the actual world?* The cells of the partition are just the different possible answers to the question. A *precise* answer to a question (or partition) Q specifies a particular cell C of the partition. Imprecise answers to the questions give a range of alternative specific answers. The limiting case of imprecision is the whole collection of cells. If this tautologous answer is counted as a degenerate case (the vacuous answer) then every proposition gives some kind of answer to any question. For any question/partition Q, proposition P, and precise answer C, if P shares a world in common with C then P does not rule out the possibility that C is the correct answer to P: hence C is in the range of alternatives allowed by P. Thus P's *answer* to Q, what P says *about* Q is the union or disjunction of all cells not excluded by P: that is, the set of all those cells which share at least one world in common with P. If P gives the vacuous answer to Q then it says *nothing* about Q.[14]

Questions can stand in interesting logical relations. One such relation is *logical reducibility*. For consider the following questions:

Q_2 What is the remainder upon division by six of the number of planets?
Q_3 What is the remainder upon division by four of the number of planets?
Q_4 What is the number of the planets?

Intuitively Q_1 is logically reducible to Q_2 but not to Q_3 and the first three are logically reducible to Q_4. What lies behind this idea of reducibility is that any precise answer to Q_2 logically entails a precise answer to Q_1 whereas some precise answers to Q_3 do not. And any precise answer to Q_4 logically entails a precise answer to all four questions. Quite generally, a question Q is logically reducible to another $Q\#$ just in case any precise answer to $Q\#$ logically entails a precise answer to Q. It is not difficult to see that this is tantamount to the following:

> Question Q logically reduces to question Q# just in case the partition of the logical space induced by Q can be embedded in the partition induced by $Q\#$.

An attribute M can be associated with a question, or inquiry, namely: *What is the truth about the extension of M?* The partition associated with M is thus the set of all precise answers to the question: *what is the extension of M?* The different possible answers to this question specify one of the possible extensions of M. Call this the *M-question* or *M-partition*. Just as a single attribute is associated with a question, so too is a class M of attributes. If N is $\{N_1 \ldots, N_m, \ldots\}$ the relevant question is: What are the respective extensions of $N_1, \ldots, N_m, \ldots$? The set of complete answers to this question in turn induces a partition of the logical space. These different possible answers specify, completely precisely, the respective extensions of $N_1, \ldots, N_m, \ldots$ Call this question the N-*question* or N-*partition*.

Note that this connection between classes of attributes and their associated inquiries yields an account of the logical reducibility of one class of attributes to another. For this is clearly tantamount to the logical reducibility of the associated questions. Attribute class M is logically reducible to attribute class N just in case the M-question logically reduces to the N-question. That is

> M is *logically reducible* to N just in case the M-partition is embedded in the N-partition.

Two features of this account of logical reducibility are worth noting. Firstly, it has little to do with *definability in a language* or the expressive powers of any particular language. It is framed entirely in terms of non-syntactic items. Secondly, the account is wider than the traditional account, in that it allows application to cases in which the attributes at issue are of different orders.

Logical reducibility is one extreme relation between attribute classes. At the other extreme a class of attributes M may have nothing to do with the attribute class N. Suppose that none of the answers to the N-question leaves you any the wiser concerning the right answer to the M-question (and hence none of the answers to the M-question leaves you any the wiser concerning the right answer to the N-question). This occurs just in case every cell of the N-partition overlaps with every cell of the M-question. In that case the attribute classes M and N are *logically independent*.

2.3. *Observational, Theoretical, and Mixed Attributes*

With Watkins I will assume that we can divide basic attributes (those which generate the set of possibilities) into two classes: observational and non-observational. Let **Ob** be the class of basic observational attributes and let **Th** be the class of basic non-observational attributes. Thus **Ob** $\cup$ **Th** is the class of all the generating attributes. Since a world is a complete state relative to the class of generating attributes the partition of the space effected by **Ob** $\cup$ **Th** is maximally fine. Every partition of the space can be embedded in it. Thus every attribute reduces logically to **Ob** $\cup$ **Th.** The class of *observational attributes* is just the class of those attributes logically reducible to **Ob.** The class of *theoretical* attributes is the class of those attributes logically reducible to **Th.** We can call those attributes which are neither observational nor theoretical *mixed.*

2.4. *Laws and Theories*

One positivist dream was that all attributes dealt with in science should reduce, logically, to the class of basic observational attributes; at bottom all attributes are observational. There are two connected, though distinct, motivations for this: one epistemological (it was difficult to see how we could verify propositions involving non-observational attributes); the other semantic (it was difficult to see how we could connect predicates with non-observational attributes). The epistemological problem already infects the observational attributes, for there are many attributes reducible to **Ob** for which a verification procedure cannot be carried out by us, or any creature with finite resources in an infinite universe. As Watkins has demonstrated so elegantly, the desire for epistemic security cannot be satisfied even by truncating the logical space to one generated solely by observational attributes. And the semantic problem just seems unavoidable. We have witnessed the repeated failure of philosophers to reduce the concept of causal necessitation to the class of observational attributes, and yet we all perfectly well grasp the concept of causal necessitation.[15]

It is plausible that the proposition that x's having a natural property P *causes* x to have property Q, entails the proposition that *anything's* having P causes that thing to have Q. The particular identity of x is irrelevant: it is the kinds, P and Q, which are causally relevant. The universalizability of causal connections can be explained by the thesis that there is a second-order, contingent relation of natural necessitation between first-order properties. The causal relation between the events, x's having P and x's having Q, holds in *virtue* of the fact that natural necessitation links the causally relevant properties themselves. The causal connection between x's having P and x's having Q is thus derivative. I take Armstrong's proposal to be that natural necessitation should be regarded as one of the *basic* attributes, one of those which generate the logical space of scientific inquiry: in particular natural necessitation is not logically reducible to the class of observational attributes. Natural necessitation entails, but is not entailed by, the regularity relation *Reg*: the second-order relation which holds between first-order properties P and Q just in case there is no P which is not a Q.

It is a virtue of the account of reducibility given above that it applies to cases in which M is of an order different from those of the attributes in the class N, and it turns out that the intuitive judgements are precisely what the account delivers. Consider the class of all first-order properties, $\mathbf{P}^1$. Associated with this there is the $\mathbf{P}^1$ question, *what are the extensions of the first-order properties?* and the corresponding $\mathbf{P}^1$-partition. Clearly the *Reg*-partition can be embedded in the $\mathbf{P}^1$-partition, and so *Reg* is always logically reducible to the class of first-order properties, $\mathbf{P}^1$. Further, ontological misers sympathetic to Humeanism will be sympathetic to the thesis that all attributes are observational. In that case $\mathbf{P}^1$, the class of first-order attributes, would itself be logically reducible to **Ob**, and thereby *Reg* itself would logically reduce to **Ob.** Indeed, it is easy to see that if **Ob** exhausts the class of basic attributes then *all* higher-order properties and relations are reducible to the observational.

I follow Watkins in assuming that there are attributes that are not so reducible, and I follow Armstrong in assuming that natural necessitation is one of them. Natural necessitation, or *Nat* for short, is thus a member of **Th.** A *law* is a proposition to the effect that property *P* naturally necessitates property *Q*, for some properties *P* and *Q*.[16] Where *P* and *Q* are the properties connected by law *L* we will call *L* the *P*-*Q* law. The *P*-*Q* law entails what we might call the *P*-*Q* regularity, *RegPQ*, but is not entailed by it. And, in line with the desiderata (i)-(iii), I take a *theory* to be a class of laws.

Note that for each pair of properties *P* and *Q* there is a unique *P*-*Q* law, and each theory is a unique class of laws. Since a law is a proposition (a set of possible worlds) we are not troubled by the problems which trouble a syntactic approach to theories. There will normally be different ways of expressing a law within a language, or in different languages, (because there will normally be different ways of expressing the component properties). But any two expressions of the one law will pick out one and the same proposition.[17]

The *P*-*Q* law *L* is *observational* if both *P* and *Q* are observational properties, *theoretical* if both *P* and *Q* are theoretical, and *mixed* otherwise.

2.5. *Observational, Theoretical, and Mixed Content*

We can now combine these elements into an account of observational and theoretical components of both propositions and theories. The maximal observational inquiry can be identified with the question: *what is the observable structure of the world?* That is, *what are the extensions of the attributes in* **Ob**? Different propositions give different answers to this question. What a proposition says *about* the observational facts is just its answer to this question. What a proposition says *about* the observational facts is just its answer to this maximal observational question. Thus the observational content of proposition *P*, P^{Ob}, can be identified with the disjunction of all cells of the observational partition which overlap with *P*.[18] Similarly the theoretical content of *P*, P^{Th}, can be identified with *P*'s answer to the question: *What are the extensions of the theoretical attributes?* It is the disjunction of all cells of the theoretical partition which overlap with *P*. *P*'s *mixed* content is the weakest proposition which in conjunction with P^{Th} and P^{Ob} yields *P*.[19] It is the remainder after subtraction from *P* of its observational and theoretical components. The same considerations can be applied to a theory *T* (that is, a class of laws) by taking the conjunction of the laws in *T*.[20]

Propositions themselves can now be classified into those that are observational, theoretical, and mixed. A proposition *P* is *purely observational* (or just *observational* for brevity) if it is identical in content to its observational content: $P = P^{Ob}$. Thus an observational proposition is one which does not breach the cell divisions of the observational partition. Similarly, a *purely theoretical* proposition (or just *theoretical* for brevity) is one which is equivalent to its theoretical content: it does not breach the cell divisions of the theoretical partition. Propositions which are neither purely observational nor purely theoretical are *mixed*.[21] The observational (respectively, theoretical) content of a proposition is, of course, an observational (respectively, theoretical) proposition itself. Moreover, the class of observational (respectively, theoretical) propositions is closed under infinite Boolean operations.

As a result of the lack of logical reducibility of *Nat*, even an observational law is logically stronger than its own observational content. *NatPQ* entails *RegPQ*. (That's part and parcel of the concept of natural necessitation). Since *Nat* is a non-observational attribute we can stipulate that the observational content of any *P*-*Q* law, $(NatPQ)^{Ob}$, is just the observational content of the corresponding *P*-*Q* regularity, $(RegPQ)^{Ob}$.[22] As is easily checked, if *P* and *Q* are purely observational then *RegPQ* itself is a purely observational proposition ($RegPQ = (RegPQ)^{Ob}$).[23] Thus the observational content of an observational law is just the corresponding observational regularity. Of course, even if *P* and *Q* are themselves purely observational, the *P*-*Q* regularity does not entail the *P*-*Q* law. Hence even purely observational laws are not purely observational propositions, but this is just what one should expect on a realist, non-reductionist account of laws.

An observational proposition may have mixed consequences, or even theoretical consequences. Similarly, theoretical propositions may have mixed, or even observational consequences. This is because the two classes of attributes, theoretical and observational, are not logically independent.[24] Natural necessitation is a theoretical attribute. Hence the partition of the logical space effected by {*Nat*} is embedded in the theoretical partition. A particular causal law, say the *P*-*Q* law, does not breach the boundaries of that partition, and so is itself a theoretical proposition. Hence all causal laws are theoretical propositions. However, if *P* and *Q* are observational and logically independent the *P*-*Q* regularity is a non-trivial observational proposition. And so every observational law is a theoretical proposition which entails a non-trivial observational proposition. A theoretical law, however, may be *observationally vacuous:* if its observational content is tautologous.

2.6. *Unity*

We are now in a position to give an account of the unity of theories. Note that for any partition $\{T_1,T_2\}$ of theory T, the observational content of T, that is, $T^{\mathbf{Ob}}$, is always at least as strong as the conjunction of the observational contents of T_1 and T_2 taken separately: $(T_1{}^{\mathbf{Ob}} \wedge T_2{}^{\mathbf{Ob}})$. T is organically fertile, or unified, just in case for any such partition $T^{\mathbf{Ob}}$ exceeds $(T_1{}^{\mathbf{Ob}} \wedge T_2{}^{\mathbf{Ob}})$.

> A theory T is *unified* just in case for any partition $\{T_1, T_2\}$ of T, $T^{\mathbf{Ob}}$ is logically stronger than $(T_1{}^{\mathbf{Ob}} \wedge T_2{}^{\mathbf{Ob}})$.

Thus a theory T is unified provided it is not the union of two distinct theories T_1 and T_2 such that the observational content of T is logically equivalent to the conjunction of the observational contents of T_1 and T_2.

Here I have given an account of unity only for theories, not for propositions or classes of propositions generally. The definition *could* be applied to arbitrary classes of propositions, but there is a good reason for not doing so. As it stands it can be applied indifferently to theories construed *either* as collections of laws, *or* as conjunctions of laws. This is because whichever way we construe theories they can be uniquely factored into their component laws.[25] But with propositions generally there is no unique factorization into component propositions, and attempts to find natural factorizations, are, as we have seen in the first section, up against apparently insoluble problems.

We can now show that there are simple and intuitively acceptable examples of unified theories under this account. To this end let an attribute P be said to be *unrestricted* if there is no logical restriction on its extension: that is, for any set S of entities of the appropriate type it is logically possible that S be the extension of P. (Thus if P is an unrestricted property of individuals, all sets of individuals are possible extensions of P. If P is an unrestricted property of properties of individuals, then all sets of properties of individuals are possible extensions of P.)

We can now reformulate the example 1 the unity of which seems to be a desideratum of any account of unity.

EXAMPLE 1 (reformulated). Let P and Q be logically independent unrestricted observational attributes, and T an unrestricted theoretical attribute logically independent of the class of purely observational attributes. *A* consists of the *P-T* law and the *T-Q* law. It is the theory: {*NatPT*, *NatTQ*}.[26]

One preliminary result will prove helpful:

> (*) If *P* is an unrestricted observational attribute and *T* an unrestricted theoretical attribute logically independent of the observational attributes, then the *P-T* law and the *T-P* law are both observationally vacuous.[27]

The account of unity yields the intuitively acceptable result that *A* is a unified theory. It is not hard to see why this is so. Intuitively, the two laws of *A* mesh together in an organically fertile way. By (*) each law is observationally vacuous on its own: the observational content of each is tantamount to the tautology. However, the theory as a whole entails the purely observational proposition, *RegPQ*: hence $\{NatPT, NatTQ\}^{Ob}$ also entails *RegPQ*. Since *P* and *Q* are logically independent and unrestricted, *RegPQ* is not tautologous. Hence $\{NatPT, NatTQ\}^{Ob}$ exceeds $(NatPT)^{Ob} \wedge (NatTQ)^{Ob}$ in logical strength.

Thus the account yields an intuitively acceptable judgment of unity on examples of the first kind. But the account also ought to exclude examples of the second kind, which, as Watkins noted, are intuitively disunified.

EXAMPLE 2 (reformulated). Let *P*, *Q*, *R* and *S* be any attributes such that {*P*, *Q*} is logically independent of the class {*R*, *S*}. *B* consists of the *P-Q* law and the *R-S* law. It is thus the theory: {*NatPQ*, *NatRS*}.

Again the desired result is precisely what this account yields: that *B* is not a unified theory. The observational content of *B* is the observational content of the class of the two associated regularities $\{RegPQ, RegRS\}^{Ob}$. But that is tantamount to the conjunction of the two regularities themselves.[28]

So far we haven't spelt out what it is for one theory to be *more* unified than another. Intuitively, an observational theory (one consisting purely of observational laws) is not as unified as those theories the observational consequences of which flow from indispensable theoretical assumptions. (It is clear that observational theories could often be cooked up from the observed regularities, and so the demand for unity also militates against *ad hoc* hypotheses). A theory is thus unified if its observational consequences emerge from a theoretical framework. It is not far from this idea that a theory is the *more* unified the greater the class of observational consequences which it subsumes under its theoretical assumptions.

We can capture this idea relatively easily. Let T^{core} be the theoretical core of T: that subset which consists of all the purely theoretical laws of the theory T. And let the T^{aux} be the *auxiliary hypotheses* of T: that is, T–T^{core}. A unified theory is more unified than one that is not, so all we have to specify are the conditions under which one unified theory is more unified than another unified theory. Roughly speaking, to increase unity the theoretical core of a theory must do more work in generating observational consequences. That is to say, either the theory generates more observational content with auxiliary assumptions that are no more powerful themselves in generating observational content, or it generates the same observational content with an observationally weaker set of auxiliary assumptions. Thus:

> If T and T^* are unified, then T is more unified than T^* just in case either (i) T^{ob} exceeds T^{*ob} whilst $(T^{aux})^{ob}$ does not exceed $(T^{*aux})^{ob}$, or (ii) T^{*ob} does not exceed T^{ob} whilst $(T^{*aux})^{ob}$ does exceed $(T^{aux})^{ob}$.

EXAMPLE 3. Consider the theory C which involves in addition to P, Q, T the unrestricted theoretical property U which is logically independent of $\{P, Q, T\}$. C consists of the three laws:

$$\{NatPT, NatTU, NatUQ\}.$$

By a proof similar to that for (*) we can show that every proper subset of C has a tautologous observational content. Hence the observational content of the theory as a whole is logically stronger than the conjunctions of the observational contents of any two parts of it taken separately. Because $(A)^{aux}$ is just A itself $((A)^{aux})^{ob}$ is identical to $(A)^{ob}$. But, unlike A, theory C has a theoretical core, $\{NatTU\}$; its auxiliary hypotheses are $\{NatPT, NatUQ\}$. As already noted, the observational content of $(C)^{aux}$ is tautologous. However, the observational content of C is identical to that of A. Consequently C is more unified than A by virtue of clause (ii) of the definition.

It should not be too difficult to see how to construct theories which are more unified than C by virtue of clause (i) of the definition.

EXAMPLE 4. Let P, Q, R be unrestricted observational attributes; and S, T, U, V unrestricted theoretical attributes, all these being suitably logically independent of the others. Let D be the theory:

{*NatP*[*T&S*], *NatTU*, *NatSV*, *NatUQ*, *NatVR*}.[29]

The observational content of D includes *RegPQ* and *RegPR*. Any subset of D lacking *NatP*[*T&S*] has a tautologous observational content. Any subset of D which entails *RegPQ* includes in addition the two laws *NatTU*, *NatUQ*; and any subset of D which entails *RegPR* includes in addition the two laws *NatSV*, *NatVR*. Consequently, in any partition $\{T_1, T_2\}$ of D at least one of $T_1^{\mathbf{ob}}$ or $T_2^{\mathbf{ob}}$ is tautologous and the other fails to entail one of *RegPQ* or *RegPR*. Hence D is unified. Since its observational content exceeds that of C and its auxiliary hypotheses are observationally vacuous, it is more unified than C.

2.7. *Depth*

What is depth? From Watkins' observations it seems to involve at least two aspects; the first is theoretical involvement, the second is predictive power at the observational level. Roughly speaking, a proposition T is deeper than another T^* just in case T exceeds T^* in both respects.

We already have a rudimentary measure of theoretical involvement. Purely observational propositions have the least degree of theoretical involvement, and purely theoretical propositions have the greatest. Mixed propositions are located somewhere between the two. An objection to this last claim might be that some mixed propositions logically entail purely theoretical propositions, and as such they cannot be shallower. The point is well taken, but we can accommodate it by taking into account the other aspect of depth: logical strength. Here, then, is a preliminary account of depth:

> T is deeper than T^* just in case T exceeds T^* either in observational content or in theoretical involvement, and is not be exceeded by T^* in either.

Let us see what kind of partial ordering of propositions this account generates.

Purely observational propositions have no theoretical involvement, and so amongst these depth is gauged simply by logical strength. Observational regularities, for example, will be deeper than the singular propositions they entail. Staying within the class of theoretical propositions, depth is again measured by observational content and the same goes for the depth of mixed propositions. Comparisons between grades also seem intuitively gratifying. Mixed propositions will be deeper than their observational consequences, but are simply not comparable with observational propositions which are logically stronger or logically incomparable with them. Nor will they be comparable with the theoretical propositions they entail. Theoretical propositions are deeper than their observational consequences, but simply not comparable with mixed or observational propositions with greater or incomparable observational content.

So far so good, From the fact that theories are classes of laws we have immediately that they have greater depth than both the singular observational propositions, and the observational regularities they are invoked to explain. But it turns out that for a *theory* to be unified it must go deeper than do observational laws.

All laws are theoretical propositions: they are identical in content to their theoretical components. But it is clear that some laws have a greater degree of theoretical involvement than do others. A theory which consisted solely of observational laws, a theory without a theoretical core, would be the shallowest kind of theory possible. Let's say observational theories have theoretical involvement of degree 0. A theory which contained theoretical or mixed laws in addition to observational laws would possess a greater degree of theoretical involvement: degree 1. A theory which contained no observational laws at all would possess degree 2 theoretical involvement. Finally, a theory with a theoretical core, and no observational laws, would be the deepest of all: degree 3. We can now refine the preliminary account of depth for the special case of theories:

> Theory T is deeper than theory T^* just in case T exceeds T^* either in observational content or in theoretical depth, and is not exceeded by T^* in either.[30]

As a simple example the definition delivers the result that the theory A is deeper than the theory, E, which consists of the single observational law: *NatPQ*. Both A and E have the same observational content, *RegPQ*, but E has degree 0 theoretical involvement while A has degree 2. C is deeper than A since they share the same observational content, and the former has degree 3 theoretical involvement. D is the deepest of all, possessing degree 3 theoretical involvement and exceeding all in observational content as well.

Of greater interest than these particular judgements is the general result that *no unified theory has degree 0 theoretical involvement*, vindicating

Watkins' claim that the intuitive concepts of depth and unity are closely related.[31] We can go further: *every unified theory possesses theoretical involvement of at least degree 2.*

(*Proof.* Suppose that $T = \{L_1, \ldots L_m\}$ where L_i is $NatP_iQ_i$. Suppose that T contains an observational law, say L_1. Then T^{ob} is

$$(L_1 \wedge \ldots \wedge L_m)^{ob} = \\ = (RegP_1Q_1 \wedge RegP_2Q_2 \wedge \ldots \\ \wedge RegP_m Q_m)^{ob}.^{32}$$

If J is a purely observational proposition (i.e. $J = J^{ob}$) then

$$(J \wedge K)^{ob} = J^{ob} \wedge K^{ob}.^{33}$$

$RegP_1Q_1$ is a purely observational proposition (although $(RegP_2Q_2 \wedge RegP_mQ_m)$ need not be) and so T^{ob} is

$$(RegP_1Q_1)^{ob} \wedge \ldots \wedge (RegP_2Q_2)^{ob} \\ \wedge \ldots \wedge (RegP_mQ_m)^{ob}.$$

But then the partition

$$\{L_1\}, \{L_2, \ldots L_m\}$$

offends the organic fertility requirement.)

One further result in this vein links increase in unity with increase in depth:

> *If T and T* are unified and T is more unified than T* then T is at least as deep as T*, and is usually deeper than T*.*

(*Proof* Suppose T is more unified than T^*. There are two cases. (i) T^{ob} exceeds T^{*ob} while $(T^{aux})^{ob}$ does not exceed $(T^{*aux})^{ob}$. Since $(T^{*aux})^{ob} \neq T^{ob}$ it follows that $T^{aux} \neq T$ and so T must possess a theoretical core: its theoretical involvement is degree 3. Moreover T^{ob} exceeds T^{*ob} hence T is of greater depth than T^*. (ii)T^{*ob} does not exceed T^{ob} whilst $(T^{*aux})^{ob}$ does exceed $(T^{aux})^{ob}$. Again $(T^{aux})^{ob} \neq T^{ob}$ and so T must possess a theoretical core. Either (iia) T^{ob} does exceed T^{*ob} or (iib) they are equal in observational content. In the former case T is deeper than T^*. In the latter case T is at least as deep as T^*, having as great a degree of theoretical involvement and as much observational content. Lastly, cases (i) and (iia) would be more common in real inquiries than case (iib).

These results go a long way to vindicating Watkins' hunch that depth and unity are intimately linked. Unfortunately they don't quite justify the claim that the two concepts are 'virtually identical' (p. 218). A theory T may have a theoretical core, no observational laws, and a large observational content, and as a consequence be deeper than many unified theories, and yet T may nevertheless fail to be unified. Watkins' own example of the conjunction of Newton's laws and Bohm-Bawerk's is a case in point.

3. UNIFYING THE OPTIMUM AIM OF SCIENTIFIC INQUIRY

The depth pole of the optimum aim for scientific inquiry involves four elements (depth, unity, predictive power, and exactitude) which are quite independent of the fifth element, the connection with truth. In Watkins' scheme the connection with truth is secured by compatibility with the data: if a theory is compatible with the data then it is at least possibly true as far as we can tell. A theory may satisfy the depth pole without being compatible with the data. And it may be compatible with the data and still relatively shallow. Two questions arise. Firstly, is there some good reason why we should bother with the depth pole of the aim, or must the intrinsic value of its elements be simply asssumed? Secondly, is there some philosophical account of the aim of inquiry which would explain these distinct elements, the depth factor and the truth factor, in a unified way?

An inquiry aims at truth; not just any old truth, but the whole truth about some matter. We can idealise and say that each inquiry poses a *central* question. And each question, as we have seen, comes with a range of different possible answers: the cells of the partition which corresponds to the question. Scientific inquiry is not aimed at the truth, the whole truth and nothing but the truth about *every* aspect of the world. Rather scientific inquiry aims at the truth about the causal structure of the world, about the laws of nature.[34] The causal relation thus induces a partition of the logical space which corresponds to the central question of scientific inquiry: 'What is the causal structure of the world?'. However, it would be pretentious to claim that we will ever possess the complete, true answer to the central scientific question. At best we will have partial answers, and at best these partial answers will not be true simpliciter, but will capture something of the complete correct answer. So this aim, like Watkins', can be progressivised, provided we have an adequate account of distance between propositions. If we did have such an account an answer, T, to the central question is (objectively) better than another, T^*, just in case T is closer to the correct et C be a cell of the observational partition in $J^{ob} \wedge K^{ob}$ (and hence in both J^{ob} and ob). Since C is in K^{ob} there is some K-world in C, say u. Since $J = J^{ob}$ every world is in J; but then u is in J, and so u is in $J \wedge K$: consequently C is in $(J \wedge K)^{ob}$. iously every cell in $(J \wedge K)^{ob}$ is in $J^{ob} \wedge K^{ob}$: hence $(J \wedge K)^{ob} = J^{ob} \wedge K^{ob}$.

J. Cohen has argued that it is not verisimilitude which scientists seek but rather he calls 'legisimilitude' or closeness to the truth about natural necessity. It turns hat this can be neatly accommodated within the verisimilitude programme. See e (1982).

r the details see Oddie (1986), especially Chapter 5.

REFERENCES

trong, D.: 1983, *What is a Law of Nature?*, Cambridge University Press.

p, N.D. and Steel, T.B.: 1976, *The Logic of Questions and Answers*, Yale iversity Press.

, D.: 1988, 'Statements partly about observation' in *Philosophical Papers*, **17**, 33.

ıoto, I.: 1987, *Truthlikeness*, Reidel.

G.: 1982, 'Cohen on verisimilitude and natural necessity' in *Synthese*, **51**, –79.

G.: 1986, *Likeness to Truth*, Reidel.

G.: 1987, 'The picture theory of truthlikeness' in *What is Closer-to-the-Truth? snan Studies in the History and Philosophy of the Sciences and the Humanities: 7)*. Theo Kuipers (Ed.), Rodopi, pp 25–46.

P.: 1982, 'Foundations of partial type theory' in *Reports of Mathematical ic*, **14**, 59–72.

s, J.W.N.: 1984, *Science and scepticism*, Princeton University Press.

answer than is T^*.[35]

Assume that the collection of actual causal laws involves unobservable entities; that these laws are not isolated atoms but mesh with each other to produce the right phenomenal world; and that much of the phenomenal world is structured by such laws. Then the true theory of the world would be unified, deep, predictively powerful, and exact. And then, intuitively, depth would be a necessary (though not of course, a sufficient) condition for closeness to the truth about natural necessity. The elements of depth and truth are thus naturally united in a theory of closeness to truth. Admittedly this unification does involve a substantive, contingent assumption about the causal structure of the world. But any justification for depth must involve some such assumption: what would be the value of deep theories of the causal structure of the world unless the world itself had a deep causal structure?

Massey University

NOTES

[1] Watkins: 1984, pp. 203–225. When I mentioned some of these criticisms to John in 1986 he actively encouraged me to publish them. I hesitated because at that stage I had nothing positive to offer.

[2] References to *Science and Scepticism* in the text will be by page number.

[3] Both Ilkka Niiniluoto and I have argued that truthlikeness is the best candidate for the role of epistemic utility or value. Like any magnitude, its expected value on the evidence is a guide to its real value. Provided we have a good account of logical or inductive probability we have epistemic access to truthlikeness. See Oddie (1986), or Niiniluoto (1987).

[4] I say it is implicit: but it doesn't help if the criterion for the unity of T is weakened to the following: *some* reaxiomatisation of T satisfies the organic fertility requirement. The same kind of problems crop up.

[5] This in itself is a little strange: as far as naturalness goes A^{**} and A^* seem to be pretty much on a par. Moreover Wajsberg's requirement is highly sensitive to the range of truth functional connectives that are available in the language.

[6] The axioms C_i are logically independent, and jointly entail all four minimal disjunctions in A^{**}. Any three of the axioms must entail at least three of the minimal disjunctions. If C_1, C_2, C_3 entail more than three, then they entail all four minimal disjunctions: but then C_4 is not independent of them (contradiction). If C_1, C_2, C_3 entail exactly three minimal disjunctions then each is tantamount to the fourth. If the former, C_4 entails one of C_1, C_2, C_3 violating independence. If the latter, each axiom is equivalent to a minimal disjunction.

[7] Put slightly differently, there is only one potential falsifier of A (a conjunction of

atomic observational propositions which entails the negation of A): $P \wedge \sim Q$.

[8] They are the logical duals of Hintikka's first-order constituents. For an account of the latter see Oddie (1986).

[9] If B^*_1 entails $(P \supset Q)$ then it contains B_1. Hence B^*_2 lacks B_1 and so B^*_2 does not entail $(R \supset S)$. B^*_1 cannot entail $(R \supset S)$, for if it did it would contain all the remaining axioms as well. Hence neither of the two cells of the partition contains $(R \supset S)$ as part of its testable content. An analogous argument shows that if one of the two cells entails $(R \supset S)$ then neither entails $(P \supset Q)$.

[10] For simplicity I will ignore the temporal element.

[11] See Oddie (1986) and (1987) for an account of logical spaces generated by higher-order attributes.

[12] With Armstrong, and others, I am prepared to distinguish mere attributes from what are usually called 'universals'. To count as a universal an attribute must have something extra: it must be in some sense (which is notoriously difficult to spell out) ontologically basic. So not all properties and relations are universals. Within this framework universals can be identified with the attributes which generate the logical space. For a longer discussion of the nature of higher-order logical spaces, see Oddie (1986), Chapter 5, and Oddie (1987).

[13] We can allow that at some worlds an attribute may have no extension at all. This case must be distinguished from the case of an empty extension. If P has no extension at w when the function associated with P is undefined at w. To make all this rigorous we require a partial type theory: for the foundations of the theory in which this account is implicitly embedded, see Tichy (1982). In this paper I will make no explicit use of partiality.

[14] The identification of questions with partitions was introduced by Belnap and Steele (1976). In Oddie (1982) partitions are used to characterise what a proposition says *about* natural necessity; and in Oddie (1986) (chapter 5) this explication of aboutness is generalised. Each partition is taken to represent an *aspect* of the truth, and what P says *about* that aspect of the truth is taken to be the disjunction of all cells of that partition which overlap with P.

[15] See Armstrong (1983) for a battery of arguments against such attempts.

[16] We could call these *possible laws*, saving *law* for those possible laws that obtain in fact. But it will prove more economical to drop the modal qualifier.

[17] Here I will ignore probabilistic connections, of which *Nat* can be regarded as a special, limiting case. While probabilistic considerations would complicate the account they would not change the broad outlines.

[18] Lewis (1988) uses the partition of logical space induced by the relations of observational equivalence to characterise statements *solely about* observation. These are statements which are purely observational, in my terminology.

[19] That is $((P^{Th} \& P^{Ob}) \supset P)$. There are propositions which cannot be uniquely factored into observational and theoretical components, and which thereby have a non-trivial mixed component.

[20] If T happens to be infinite there is no problem with infinite conjunction: it is just set-theoretic intersection. Theories involving functional dependence of continuous magnitudes could thus be assimilated to the account by regarding them as non-denumerably infinite classes of laws, or conjunctions of such.

[21] We need to make special mention here of the logically necessary proposition, the

tautology. It is identical to both its observational content a
It doesn't matter much whether we say that the tautol
observational, or whether we stipulate that such pro
tautologous.

[22] More generally, the observational content of a class
content of the class of associated regularities.

[23] Suppose u and v are observationally equivalent worl
is embedded in the observational partition P and Q hav
they do in v. Hence the extension of P in u is a subset of
case the extension of P in v is a subset of the extension
in both or neither: that is *RegPQ* does not breach the
tional partition.

[24] See Section 2.2 for the definition of logical indepen

[25] In order to guarantee this we need here an ext
assumption to the effect that every law is logically
example, while *Nat* may be contingently transitive it
Humean terminology, every law is a *distinct exist*
exceptions to this strict and total independence the n
theory into a set of laws will be small.

[26] Note that P, Q and T are here attributes, not syntac

[27] For $(NatPT)^{Ob}$ is just $(RegPT)^{Ob}$ (that is, $((\forall x)(P$
the observational partition, and suppose that in world
C^* be that cell of the T-partition in which the extens
must be one such since T is unrestricted.) Since
observational attributes the T-partition shares some
cells of the observational-partition. So let U be a
Then in world U the proposition *RegPT* is true,
disjuncts generating $(RegPT)^{Ob}$: that is, $(RegPT)^{Ob}$
observational partition. An analogous proof show
vacuous.

[28] $\{NatPQ, RegRS\}^{Ob}$ is $\{NatPQ \wedge RegRS\}^{Ob}$: but
tional propositions then $(J \wedge K)^{Ob}$ is $(J^{Ob} \wedge K^{Ob})$, w

[29] Here & is the obvious property-forming ope
property which an object has just in case it ha
extension of $P\&Q$ in a possible world w is the in
Q in w.

[30] This is not *quite* Watkins' proposal rewritten
largely to the spirit of it. One half of the pro
content) is just a rewrite; the other half, to do w
capture what he is after.

[31] For suppose that T is unified and consis
observational content of T is the conjunctio
Because all of these are observational regular
tional predicates. Consider any partition of T i
the conjunction of the regularities correspondin
so $T_1^{ob} \wedge T_2^{ob}$ is tantamount to $(T_1 \wedge T_2)^{ob}$.

[32] See note 22.

APPENDIX:
BIBLIOGRAPHY OF THE PUBLISHED WORK OF J.W.N. WATKINS

(1949) 'Laski on Conscience and Counter-Revolution' (Century Essay Prize), *The Nineteenth Century and After*, ***146***, March 1949, pp. 173–179.

(1951) Letter on 'Belief', *The Times Literary Supplement*, February 16, 1951.

(1952a) 'Organisation in Science and Society' (review article of M. Polanyi, *The Logic of Liberty*), *Ethics*, **62**, April 1952, pp. 201–204.

(1952b) 'Ideal Types and Historical Explanation', *British Journal for the Philosophy of Science*, **3**, May 1952, pp. 22–34. [Reprinted in (1953f), (1973a), (1973e); German translation in (1972c); Italian translations in (1973f) and 1979h).]

(1952c) Review of John Bowle, *Hobbes and his Critics*, *The Cambridge Journal*, **5**, June 1952, pp. 569–571.

(1952d) 'The Principle of Methodological Individualism', *British Journal for the Philosophy of Science*, **3**, August 1952, pp. 186–196. [Reprinted in (1953f).]

(1952e) 'Political Tradition and Political Theory: an Examination of Professor Oakeshott's Political Philosophy', *Philosophical Quarterly*, **2**, October 1952, pp. 323–337. [Partially reprinted in (1969i).]

(1952f) 'Social Science' (review of K.G. Collier, *The Science of Humanity* and Barbara Wooton, *Testament for Social Science*), *British Journal for the Philosophy of Science*, **3**, November 1952, pp. 261–265.

(1953a) 'Walter Eucken, Philosopher-Economist', *Ethics*, **63**, January 1953, pp. 131–135.

(1953b) 'The Seventeenth Century: the age of individuality', *History Today*, **3**, March 1953, pp. 200–206.

(1953c) Review of Paul Hazard, *The European Mind*, *History Today*, **3**, June 1953.

Fred D'Agostino & I. C. Jarvie, Freedom and Rationality: Essays in Honor of John Watkins, 369–382

(1953d) Letter to *The Listener*, August 6, 1953.

(1953e) 'Scientism and Society' (review article of F.A. Hayek, *The Counter-Revolution of Science*), *Ethics*, **64**, October 1953, pp. 56–59.

(1953f) 'Ideal Types and Historical Explanation' in H. Feigl and M. Brodbeck (Eds.), *Readings in the Philosophy of Science*, Appleton-Century-Crofts, pp. 723–743. [A conflation of (1952b) and (1952d). Reprinted in (1969b), (1973a), (1973e); German translation in (1972c); Italian translations in (1973f) and (1979h); Polish translation in (1985g).]

(1954a) Review of Bertrand Russell, *The Impact of Science on Society, British Journal for the Philosophy of Science*, **4**, February 1954, pp. 352–353.

(1954b) 'The Strange Face of Evil', *The Listener*, September 10, 1954.

(1954c) 'Liberty Above All Things', *The Listener*, December 9, 1954.

(1955a) 'Methodological Individualism: A Reply', *Philosophy of Science*, **22,** January 1955, pp. 58–62. [Reprinted in (1973c).]

(1955b) 'Of Human Freedom', a discussion with Maurice Cranston, *Encounter*, **17**, February 1955, pp. 49–60.

(1955c) Review of Morris R.A. Cohen, *Reason and Nature, British Journal for the Philosophy of Science* **5**, February 1955, pp. 348–350.

(1955d) Review of P. Zagorin, *A History of Political Thought in the English Revolution, Political Studies*, **3**, February 1955, pp. 348–350.

(1955e) 'Philosophy and Politics in Hobbes', *Philosophical Quarterly*, **5**, April 1955, pp. 125–146. [Reprinted in (1965d) and (1969h).]

(1955f) 'Decisions and Uncertainty' (review article of G.L.S. Shackle, *Expectation in Economics* and C.F. Carter *et al.* (Eds.), *Uncertainty and Business Decisions*), *British Journal for the Philosophy of Science*, **6**, August 1955, pp. 66–78. [Reprinted in 1957l).]

(1955g) Review of H.B. Acton, *The Illusion of an Epoch, British Journal for the Philosophy of Science*, **6**, August 1955, pp. 169–171.

(1956a) 'Historical Truthfulness' (essay-review of P. Geyl, *Debates with Historians*), *Spectator*, March 9, 1956.

(1956b) Review of W.D. Lamont, *The Value Judgement*, *Philosophical Quarterly*, April 1956, p. 135.

(1956c) 'What Has Become of Political Philosophy?' (essay-review of P. Laslett (Ed.), *Philosophy, Politics and Society*), *Spectator*, June 29, 1956.

(1956d) 'Massification' (essay-review of Karl Mannheim, *Essays on the Sociology of Culture*), *Spectator*, August 24, 1956.

(1956e) 'Liberalism and Equality', *Spectator*, December 28, 1956.

(1957a) Review of G.L.S. Shackle, *Uncertainty in Economics*, *British Journal for the Philosophy of Science*, **7**, February 1957, pp. 362–363.

(1957b) 'Between Analytic and Empirical', *Philosophy*, **32**, April 1957, pp. 112–131.

(1957c) 'The Posthumus Career of Thomas Hobbes', *Review of Politics*, **19**, July 1957, pp. 351–360.

(1957d) 'Historical Explanation in the Social Sciences', *British Journal for the Philosophy of Science*, **8**, August 1957, pp. 104–117. [Reprinted in (1959e), (1968b), (1969c), (1969e), (1973a); Portugese translations in (1969f) and (1975c); Polish translation in (1985f).]

(1957e) Review of Maurice Cranston, *John Locke: a Biography*, *History Today*, August 1957.

(1957f) Review of Raymond Aron, *The Opium of the Intellectuals*, *Crossbow*, Autumn 1957.

(1957g) 'The Haunted Universe', *The Listener*, November 21 and 28, 1957. [Reprinted in (1961d.]

(1957h) 'Epistemology and Politics', *Proceedings of the Aristotelian Society*, 1957–8, pp. 79–102. [Reprinted in (1987b); Italian translation in (1969g); German translation in (1972d).]

(1957i/j) 'Farewell to the Paradigm Case Argument' and 'A Reply to Professor Flew's Comment', *Analysis*, **18**, December 1957, pp. 25–33 and 41–42.

(1957k) 'A Modern Samson in the Temple of Philosophy' (review article of Morton H. White, *Toward Re-Union in Philosophy*), *Ratio*, **1**, December 1957, pp. 68–84.

(1957l) 'Decisions and Uncertainty', in *Uncertainty and Business Decisions*, C.F. Carter, G.P. Meredith, G.L.S. Shackle (Eds.),

2nd edition, 1957, pp. 107–121. [Reprint of (1955f).]

(1958a) 'After the Revolution in Philosophy' (essay-review of *The Revolution in Philosophy*, introduced by Gilbert Ryle), *The Rationalist Annual*, 1958, pp. 58–66.

(1958b) 'The Alleged Inadequacy of Methodological Individualism', *The Journal of Philosophy*, **55**, 24 April, 1958, pp. 390–395. [Reprinted in (1969d); Italian translation in (1979i).]

(1958c) 'Is Political Philosophy Dead?', *Encounter*, **10**, June 1958, pp. 57–67.

(1958d) 'Confirmable and Influential Metaphysics', *Mind*, **67**, July 1958, pp. 344–365. [Italian translation in (1983a); partial Chinese translation in (1986d).]

(1958e) 'A Rejoinder to Professor Hempel's Reply', *Philosophy*, **33**, December 1958, pp. 349–355.

(1958f) Review of Morris Ginsberg, *Essays in Sociology and Social Philosophy*, vols. I and II, *Mind*, **67**, October 1958, pp. 568–569.

(1958g) 'Philosophy of History', in Raymond Klibansky (Ed.), *Philosophy in the Mid-Century: a survey*, vol. 3, Firenze, 1958, pp. 158–176.

(1959a) 'Revolutionaries and Their Principles: John Milton', *The Listener*, January 22.

(1959b) 'The Two Theses of Methodological Individualism', *British Journal for the Philosophy of Science*, **9**, February 1959, pp. 319–320.

(1959c) Review of A.R. Radcliffe-Brown, *A Natural Science of Society, British Journal for the Philosophy of Science*, **10**, August 1959, pp. 160–162.

(1959d) 'Thomas Hobbes's "Leviathan"', *The Listener*, November 12, 1959. [Reprinted in (1964e).]

(1959e) 'Historical Explanation in the Social Sciences' in Patrick Gardiner (Ed.), *Theories of History*, Free Press, 1959, pp. 503–514. [Reprint of (1957d).]

(1959f) 'Third Reply to Mr Goldstein', *British Journal for the Philosophy of Science*, **10**, November 1959, pp. 242–244.

(1959g) Review of Michael Argyle, *The Scientific Study of Social Behaviour, British Journal for the Philosophy of Science*, **10**,

November 1959, pp. 250–251.

(1959h) 'Political Prisoner in Hungary' (review article of Paul Ignotus, *Political Prisoner*), *The Humanist*, December 1959, pp. 21–22.

(1959i) 'Mr Stove's Blunders', *Australasian Journal of Philosophy*, **37**, December 1959, pp. 240–241.

(1960a) 'When Are Statements Empirical?' *British Journal for the Philosophy of Science*, **10**, February 1960, pp. 287–308.

(1960b) 'Confirmation Without Background Knowledge', *British Journal for the Philosophy of Science*, **10**, February 1960, pp. 318–320.

(1960c) 'Reply to Mr Stove's reply', *Australasian Journal of Philosophy*, **38**, May 1960, pp. 54–58.

(1960d) Review Article of Ernest Gellner, *Words and Things, Ratio*, **3**, 1960, pp. 106–110.

(1960e) 'Reply to Professor Körner', *Mind*, **69**, July 1960, pp. 406–407.

(1961a) 'Professor Scheffler's Note', *Philosophical Studies*, **12**, January–February 1961, pp. 16–19.

(1961b) Review of Howard Warrender, *The Political Philosophy of Hobbes, Philosophy*, **36**, April–July 1961, pp. 238–241.

(1961c) Chapter II ('Philosophy') in Arthur Seldon (Ed.), *Agenda for a Free Society*, IEA-Hutchinson, 1961, pp. 31–49.

(1961d) 'The Haunted Universe', *Nous*, **7**, June 1961, pp. 1–13.

(1962a) 'Reply to Professor Puccetti', *British Journal for the Philosophy of Science*, **13**, May 1962, pp. 53–54.

(1962b) 'Milton's Vision of a Reformed England' in A.F. Scott (Ed.), *Speaking of the Famous*, Macmillan, 1962, pp. 22–30. [Reprint of (1959b).]

(1963a) 'On Explaining Disaster', *The Listener*, January 10, 1963.

(1963b) 'Confession is Good for Ideas', *The Listener*, April 18, 1963. [Reprinted in (1964a) and (1964b).]

(1963c) 'Negative Utilitarianism', *Proceedings of the Aristotelian Society*, Supplementary Volume, **XXXVII**, 1963, pp. 95–114.

(1964a) 'Confession is Good for Ideas' in David Edge (Ed.), *Experiment*, BBC, 1964, pp. 64–70.

(1964b) 'Confession is Good for Ideas' in B.W.M. Young and P.D.R. Gardiner (Eds.), *Intelligent Reading*, Longmans, 1964.

(1964c) 'Galileo's Theory of Knowledge', *The Listener*, July 2, 1964.

(1964d) 'Confirmation, the Paradoxes, and Positivism' in Mario Bunge (Ed.), *The Critical Approach to Science and Philosophy: essays in honour of Karl Popper*, The Free Press, 1964, pp. 92–115. [Partially reprinted in (1970d).]

(1964e) 'Hobbes' in Maurice Cranston (Ed.), *Western Political Philosophers*, Bodley Head, 1964, pp. 45–52. [Reprint of (1959d).]

(1965a) *Hobbes's System of Ideas*, London, Hutchinson, 1965. [Spanish translation in (1974a); American edition in (1968e); second edition (1973d).]

(1965b) Review of W.B. Gallie, *Philosophy and the Historical Understanding*, *Mind*, **74**, July 1965, pp. 442–444.

(1965c) Review of W.H. Walsh, *Metaphysics*, *Philosophy*, **40**, July 1965, pp. 260–261.

(1965d) 'Philosophy and Politics in Hobbes' in Keith Brown (Ed.), *Hobbes Studies*, Blackwell, 1965, pp. 237–262. [Reprint, with corrections, of (1955e).]

(1965e) 'Mandelbaum's Critical Realism' (review article of Maurice Mandelbaum, *Philosophy, Science and Sense-Perception*), *British Journal for the Philosophy of Science*, **16**, November 1965, pp. 249–252.

(1965f) 'Word Magic and the Trivialization of Philosophy' (review article of Morris Lazerowitz, *Studies in Metaphilosophy*), *Ratio*, December 7, 1965, pp. 206–218.

(1966a) 'The U.G.C. and that Questionable Questionnaire', *The Times Review of Industry and Technology*, March 1966, pp. 74–75.

(1966b) 'John Stuart Mill and the Liberty of the Individual' in David Thomson (Ed.), *Political Ideas*, C.A. Watts, 1966, pp. 161–175.

(1966c) Review of Robert G. Colodny (Ed.), *Beyond the Edge of Certainty*, *Philosophy*, 41, October 1966, pp. 359–362.

(1967) 'Decisions and Belief' in Robin Huges (Ed.), *Decision Making*, BBC, 1967, pp. 9–26. [German translation in (1971b).]

(1968a) 'Anthropomorphism in Social Science' in Imre Lakatos and Alan Musgrave (Eds.), *Problems in the Philosophy of Science* (Proceedings of the International Colloquium in Philosophy of Science, London, 1965, vol. 3), North-Holland, 1968, pp. 423–426.

(1968b) 'Methodological Individualism and Social Tendencies' in May Brodbeck (Ed.), *Readings in the Philosophy of the Social Sciences*, 1968, pp. 269–280. [Reprint of (1957d).]

(1968c/d) 'Non-inductive corroboration' and 'Hume, Carnap and Popper' in Imre Lakatos (Ed.), *The Problem of Inductive Logic* (Proceedings of the International Colloquium in Philosophy of Science, London, 1965, vol. 2), North-Holland, 1968, pp. 61–66 and 272–282.

(1968e) *Hobbes's System of Ideas*, New York, Barnes and Noble, 1968. [American paperback edition of (1965a).]

(1969a) 'Comprehensively Critical Rationalism', *Philosophy*, **44**, January 1969, pp. 57–61.

(1969b/c/d) 'Methodological Individualism and Non-Hempelian Ideal Types', 'Social Phenomena Result from the Activities of Individual Agents', 'Methodological Individualism: A Reply to Criticism' in Leonard I. Krimerman (Ed.), *The Nature and Scope of Social Science: A critical Anthology*, Appleton-Century-Crofts, 1969, pp. 457–472, 603–611, 621–624. [Reprints of (1953f), (1957d), (1958b).]

(1969e) 'Historical Explanation in the Social Sciences', The Bobbs-Merrill Reprint Series in Philosophy, 1969. [Reprint of (1957d).]

(1969f) 'A Explicacao Historica nas Ciecas' in Patrick Gardiner (Ed.), *Teorias da Historia*, Lisbon, Fundacao Calouste Gulbenkian, 1969, pp. 622–636. [Portugese translation of (1957d).]

(1969g) 'Epistemologia e Politica' in Riccardo Campa (Ed.), *Le Implicazioni Politiche Della Scienze*, Edizioni Della Nuouva Antologia, Rome, 1969, pp. 205–234. [Italian translation of (1957h).]

(1969h) 'Philosophy and Politics in Hobbes' in Bernard H. Baumrin (Ed.), *Hobbes's Leviathan*, Wadsworth Publishing Co, 1969, pp. 83–106. [Reprint of (1955e).]

(1969i) 'A Critique of Oakeshott's Distinction' in Richard H. Cox

(Ed.), *Ideology, Politics and Political Theory*, Wadsworth Publishing Co., 1969, pp. 259–263. [Reprint of a section of (1952e).]

(1970a) 'Against "Normal Science"' in Imre Lakatos and Alan Musgrave (Eds.), *Criticism and the Growth of Knowledge*, Cambridge University Press, 1970, pp. 25–37. [German translation in (1974g).]

(1970b/c) 'Imperfect Rationality' and 'Reply' to Comment by Alan Donagan in Robert Borger and Frank Cioffi (Eds.), *Explanation in the Behavioural Sciences*, 1970, pp. 167–217, 228–230. [Reprint in (1975a); German translation in (1978c); Italian translation in (1981).]

(1970d) 'The "Paradoxes of Confirmation"' in Baruch A. Brody (Ed.), *Readings in the Philosophy of Science*, Prentice-Hall Inc., 1970, pp. 433–438. [Reprint of a section of (1964d).]

(1971a) 'CCR: A Refutation', *Philosophy*, **46**, January 1971, pp. 56–61.

(1971b) 'Entscheidung und Uberzeugung' in Hans Albert and Ernst Topitsch (Eds.), *Werturteilsstreit*, Wissenschaftliche Buchgesellschaft, 1971, pp. 313–333. [German translation of (1967).]

(1971c) 'Freedom and Predictability: an amendment to Mackay', *British Journal for the Philosophy of Science*, **22**, August 1971, pp. 263–275.

(1971d) Review of Stephan Körner, *Categorial Frameworks*, *British Journal for the Philosophy of Science*, **22**, November 1971, pp. 373–377.

(1972a) Review of Edwin Erwin, *The Concept of Meaninglessness*, *British Journal for the Philosophy of Science*, **23**, February 1972, pp. 83–84.

(1972b) 'Liberty' in Maurice Cranston and Richard S. Peters (Eds.), *Hobbes and Rousseau: a collection of critical essays*, Anchor Books, Doubleday, 1972, pp. 213–232. [Reprint of (1965a), ch. 7.]

(1972c/d) 'Idealtypen und historische Erklarung' and 'Erkenntnistheorie und Politik' in Hans Albert (Ed.), *Theorie und Realität*, J.C.B. Mohr, 1972, pp. 331–356, 393–413. [German translations of

(1953f) and (1957h).]

(1972e) 'K.R. Popper: Die Einheit seines Denkens' in Josef Speck (Ed.), *Grundprobleme, der grossen philosophen*, Vandenhoeck & Ruprecht, Göttingen, 1972, pp. 151–214. [German translation of (1974b).]

(1972f) 'Social Knowledge and the Public Interest' in William Robson (Ed.), *Man and the Social Sciences*, Allen & Unwin, 1972, pp. 173–196. [American reprint in 1974k).]

(1973a/b/c) 'Ideal Types and Historical Explanation', 'Historical Explanation in the Social Sciences', 'Methodological Individualism: A Reply' in John O'Neill (Ed.), *Modes of Individualism and Collectivism*, Heinemann, 1973, pp. 143–184. [Reprints of (1953f), (1957d), (1955a).]

(1973d) *Hobbes's System of Ideas*, second, revised edition, London, Hutchinson University Library, 1973.

(1973e) 'Ideal Types and Historical Explanation' in Alan Ryan (Ed.), *The Philosophy of Social Explanation*, Oxford University Press, 1973, pp. 82–104. [Reprint of (1953f).]

(1973f) 'Tipi ideali e spiegazinni storiche', *Controcorrente*, V, April 1973, pp. 51–74. [Italian translation of (1953f).]

(1974a) *Que ha dicho verdaderamente Hobbes*, Doncil, Madrid, 1975. [Spanish translation of (1965a).]

(1974b) 'The Unity of Popper's Thought' in Paul A. Schilpp (Ed.), *The Philosophy of Karl Popper*, The Library of Living Philosophers, Open Court, 1974, vol. i, pp. 371–412. [German translations in (1972e) and (1978c); Italian translation in (1981); French translation in (1985c); Spanish translation in (1987i).]

(1974c/d/e) 'Self-Interest and Morality', 'What does Chisholm require of us?', 'Comment' on Roy Edgley on Reason and Violence in Stephan Körner (Ed.), *Practical Reason*, Blackwell, 1974, pp. 67–77, 36–40, 156–158. [(1974c) partially incorporated in (1985f).]

(1974f) 'Professor Imre Lakatos', obituary notice, unsigned, *The Times*, February 2, 1974.

(1974g/h) 'Gegen die "Normalwissenschaft"' and 'Professor Imre Lakatos' in Imre Lakatos and Alan Musgrave (Eds.), *Kritik und Erkenntnisfortschritt*, trans. P.K. Feyerabend and A. Szabo,

Vieweg, Braunschweig, 1974, pp. 25–38, 323–324. [German translation of (1970a) and reprint of (1974f).]

(1974i) 'Otto Neurath' (review article of Otto Neurath, *Empiricism and Sociology*), *British Journal for the Philosophy of Science*, **25**, December 1974, pp. 343–352.

(1974j) Commentary on Eileen Barker, 'Facts and Values, and Social Science' in *Science and Absolute Values*, The International Cultural Foundation, Inc., 1974, vol. ii, pp. 776–778.

(1974k) 'Social Knowledge and the Public Interest' in William Robson (Ed.), *Man and the Social Sciences*, Sage, 1974, pp. 193–196. [American edition of (1972f).]

(1975a) 'Imperfect Rationality' in Robert Borger and Frank Cioffi (Eds.), *Explanation in the Behavioural Sciences*, Cambridge University Press, 1975. [Reprint of (1970b).]

(1975b) 'Metaphysics and the Advancement of Science', *British Journal for the Philosophy of Science*, **26**, June 1975, pp. 91–121. [Italian translation in (1983a); Chinese translation in (1983d).]

(1975c) 'A Explicacao Historicas nas Ciencas Sociais' in Patrick Gardiner (Ed.), *Teorias da Historia*, Fundacao Calouste Gulbenkian, Lisbon, 1975. [Reprint of (1969d).]

(1975d) 'Kolme Nakemysta Inhimillisesta Vapandesta', *Kriittisen Rationalismin Julkaisusarja*, **1/75**, 1975, pp. 6–44. [Finnish translation of (1975e).]

(1975e) 'Three Views Concerning Human Freedom' in R.S. Peters (Ed.), *Nature and Conduct*, Macmillan, 1975, pp. 200–228. [Finnish translation in 1975d)]; Italian translation in (1978b) and (1981); German translation in (1978c).]

(1976a) 'The Human Condition: Two Criticisms of Hobbes' in R.S. Cohen, P.K. Feyerabend, M.W. Wartofsky (Eds.), *Essays in Memory of Imre Lakatos*, Reidel, 1976, pp. 691–716. [German translation in (1978c); Italian translation in (1981).]

(1976b) Review of F.A. Hayek, *Law, Legislation and Liberty*, *Philosophy of the Social Sciences*, **6**, December 1976, pp. 369–372.

(1977a) 'Towards a Socialist Science' (review of Hilary and Steven Rose (Eds.), *The Political Economy of Science* and *The*

Radicalisation of Science), *The Times Literary Supplement*, February 4, 1977, p. 127.

(1977b) Chapter in Joan Abse (Ed.), *My LSE*, Robson Books, 1977, pp. 65–82.

(1977c) 'Towards a Unified Decision Theory: a non-Bayesian Approach' in R.E. Butts and J. Hintikka (Eds.), *Foundational Problems in the Special Sciences*, Reidel, 1977, pp. 345–379. [German translation in (1978c); Italian translation in (1981).]

(1977d) 'Moritz Schlick and the Mind-Body Problem' (review article of Moritz Schlick, *General Theory of Knowledge*), *British Journal for the Philosophy of Science*, **28**, December 1977, pp. 369–382.

(1978a) 'Minimal Presuppositions and Maximal Metaphysics', *Mind*, **87**, 1978, pp. 195–209. [Italian translation in (1983a).]

(1978b) 'Tre concezioni sulla liberta dell'uomo' in *Natura a Liberta*, Armando Armando, Rome, 1978, pp. 95–128. [Italian translation of (1975e).]

(1978c) *Freiheit und Entscheidung*, Tübingen, J.C.B. Mohr (Paul Siebeck), 1978. [German translations of (1976a), (1970b), (1977c), (1974b), (1975e).]

(1978d) 'Metoda falsyfikacji wobec rozwoju nanki', *Zagadriena naukoznowstwas*, **14**, 1978, pp. 226–244. [Polish translation of (1978f).]

(1978e) 'A Basic Difficulty in the Mind-Brain Identity-Hypothesis' in *Proceedings of the Sixth International Conference on the Unity of the Sciences*, 1978, pp. 943–953. [Reprinted in (1982c).]

(1978f/g) 'The Popperian Approach to Scientific Knowledge' and 'Corroboration and the Problem of Content-Comparison' in G. Radnitzky and G. Andersson (Eds.), *Progress and Rationality in Science*, Reidel, 1978, pp. 23–43, 339–378. [Polish translation of (1978f) in (1978d); German translation in (1979d/e).]

(1979a) 'Simplicity and the Problem of Induction', *Philosophical Inquiry*, **1**, Spring 1979, pp. 183–194.

(1979b) 'Does Critical Rationalism Provide a Moral-Political Philosophy?' in H. Berghel, A. Mülsner, E. Köhler (Eds.), *Wittgenstein, The Vienna Circle, and Critical Rationalism* (Proceedings of the Third Wittgenstein Symposium), 1979, pp. 381–385.

(1979c) 'Moritz Schlick and the Repudiation of Scientific Realism' in *Proceedings of the Seventh International Conference on the Unity of the Sciences*, 1979, pp. 963–972.

(1979d/e) 'Die Poppersche Analyse der wissenschaftlichen Erkenntnis' and 'Die Bewährung und das Problem des Gehaltvergleichs' in G. Radnitzky and G. Andersson (Eds.), *Fortschritt und Rationalität der Wissenschaft*, J.C.B. Mohr (Siebeck), 1979, pp. 27–49, 393–437. [German translation of (1978f/g).]

(1979f) 'Lakatos, Imre' in *International Encyclopaedia of the Social Sciences*, David L. Sills (Ed.), The Free Press, 1979, Biographical Supplement, Volume 18, pp. 399–402.

(1979g) 'Historicism and the Theory of Science' (review of Larry Laudan, *Progress and its Problems* and Thomas S. Kuhn, *The Essential Tension), Minerva*, **17**, Summer 1979, pp. 316–321.

(1979h/i) 'Tipi ideali e spiegazione storica' and 'La pretesa insufficienza dell'individualismo metodologico' in M.V.P. Magrini (Ed.), *Filosofia Analitica E Conoscenza Storica*, Firenze, La Nuova Italia, 1979, pp. 331–360 and 393–399. [Italian translations of (1953f) and (1958b).]

(1980) 'Humanism *versus* Social Science?' in *Proceedings of the Eighth International Conference on the Unity of the Sciences*, 1980, pp. 93–101.

(1981) *Liberta' E Decisione*, Rome, Armando Armando, 1981. [Italian translations of (1976a), (1970b), (1977c), (1974b) and (1975e).]

(1982) 'A Basic Difficulty in the Mind-Brain Identity-Hypothesis' in Sir John Eccles (Ed.), *Mind and Brain*, Washington, Paragon House, 1982, pp. 221–223. [Reprint of (1978e).]

(1983a) *Tre Saggi Su 'Sciencza e Metafisica'*, Rome, Borla, 1983. [Italian translation, with a new Preface, of (1958d), (1975b) and (1978a).]

(1983b) 'Lakatos, Imre' in Alan Bullock and R.B. Woodings (Eds.), *The Fontana Biographical Companion to Modern Thought*, 1983, p. 420.

(1983c) 'What Has Become of CCR?' *Proceedings of the Eleventh*

International Conference on the Unity of the Sciences, 1983, pp. 1087–1100. [Reprinted in 1987e).]

(1983d) Chinese translation of (1975b) in *Ziran Kexue Zhexue Wenti Congkan*, 3, pp. 79–90.

(1984) *Science and Scepticism*, Princeton, Princeton University Press, 1984. [Italian translation in (1986b).]

(1985a) 'Una Nuova concezione della razionalita scientifica' in Marcello Pera and Joseph Pitt (Eds.), *I Modi del Progresso*, Il Saggiatore, 1985, pp. 47–67. [Italian translation of (1987d).]

(1985b) 'Indeterminisme et Interactionisme: Deux Thèses de Popper' in J.-P. Chretien-Goni et C. Lazzeri (Eds.), *Cahiers S.T.S.*, 8, CNRS, Paris, 1985, pp. 28–73. [French translation of (1974b).]

(1985c) 'Second Thoughts on Lande's Blade', *Journal of the Indian Council of Philosophical Research*, **2**, Spring 1985, pp. 13–19.

(1985d) 'On Stove's Book, by a Fifth "Irrationalist"', *Australasian Journal of Philosophy*, **63**, 1985, pp. 259–268.

(1985e) 'Second Thoughts on Self-Interest and Morality' in Richmond Campbell and Lanning Sowden (Eds.), *Paradoxes of Rationality and Cooperation*, Vancouver, University of British Columbia Press, 1985, pp. 59–74. [Incorporates part of (1974c).]

(1985f) 'Wyjasnianie historyczne w naukach spolecznych', *Zagadnienia naukoznawstwa*, 1985, pp. 272–282. [Polish translation of (1957d).]

(1985g) 'Typy idealne a wyjasnianie historyczne', *Zagadnienia naukoznawstwa*, 1985, pp. 627–645. [Polish translation of (1953f).]

(1986a) 'A Che Serve un Grosso Cervello?', *Prometeo*, **4**, March 1986, pp. 64–69.

(1986b) *Certezza e Verita*, Roma-Bari, Laterza, 1986. [Italian translation of abridgement of (1984).]

(1986c) Chinese translation of part of Chapter 8 of (1984) in *Philosophical Problems in the Natural Sciences* (Beijing), **3**, pp. 50–56.

(1986d) Chinese translation of part of (1958d) in *Philosophical Problems in the Natural Sciences* (Beijing), **4**, pp. 34–38.

(1987a) 'A Note on Incongruent Counterparts and Verisimilitude', *Erkenntnis*, **26**, 1987, pp. 295–300.

(1987b) 'Epistemology and Politics' in J. Agassi and I.C. Jarvie (Eds.), *Rationality: The Critical View*, Dordrecht, Martinus Nijhoff, 1987, pp. 151–67. [Reprint of 1957h).]

(1987c) 'The LSE "School" of Philosophy: a Polish Interrogation', *LSE Quarterly* **1**, June 1987, pp. 213–226.

(1987d) 'A New View of Scientific Rationality' in J. Pitt and M. Pera (Eds.), *Rational Changes in Science*, Dordrecht, Reidel, 1987, pp. 63–82.

(1987e) 'Comprehensively Critical Rationalism: a Retrospect' in G. Radnitzky and W.W. Bartley, III (Eds.), *Evolutionary Epistemology, Theory of Rationality and the Sociology of Knowledge*, LaSalle, Open Court, 1987, pp. 151–167. [Reprint, with changes, of (1983c).]

(1987f) 'Nauka moze byc racjonalna', *Nowe Ksiazki*, **4**, 1987, pp. 4–10. [Polish translation of (1987c).]

(1987g) 'A New View of Scientific Rationality', *Revue de Synthese*, **108**, juillet-decembre 1987, pp. 455–473.

(1987h) 'Lawson on the Raven Paradox and Background Knowledge', *British Journal for the Philosophy of Science*, **38**, December 1987, pp. 567–571.

(1987i) 'La Unidad del Pensamiento de Popper', *Lecciones y Ensayos*, **48**, 1987, pp. 13–72. [Spanish translation of (1974b).]

(1988a) 'The Pragmatic Problem of Induction', *Analysis*, **48**, January 1988, pp. 18–20.

(1988b) Japanese translation by H. Tanaka and K. Takano of (1973d), *Hobbes's System of Ideas*, Tokyo: Miraisha.

INDEX